T0092961

Intermittent Convex Integration for the 3D Euler Equations

Tristan Buckmaster
Nader Masmoudi
Matthew Novack
Vlad Vicol

PRINCETON UNIVERSITY PRESS
PRINCETON AND OXFORD
2023

Requests for permission to reproduce material from this work should be sent to permissions@press.princeton.edu

Published by Princeton University Press,
41 William Street, Princeton, New Jersey 08540
99 Banbury Road, Oxford OX2 6JX

press.princeton.edu

ISBN: 9780691249551
ISBN (pbk): 9780691249544
ISBN (e-book): 9780691249568

British Library Cataloging-in-Publication Data is available

Editorial: Diana Gillooly and Kiran Pandey
Production Editorial: Nathan Carr
Jacket/Cover Design: Heather Hansen
Production: Lauren Reese
Publicity: William Pagdatoon
Copyeditor: Bhisham Bherwani

The publisher would like to acknowledge the authors of this volume for acting as the compositor for this book.

This book has been composed in LaTeX.

10 9 8 7 6 5 4 3 2 1

Annals of Mathematics Studies
Number 217

Contents

Intermittent Convex Integration
for the 3D Euler Equations

Chapter One

Introduction

We consider the homogeneous incompressible Euler equations

$$\partial_t v + \operatorname{div}(v \otimes v) + \nabla p = 0 \tag{1.1a}$$
$$\operatorname{div} v = 0 \tag{1.1b}$$

for the unknown velocity vector field v and scalar pressure field p, posed on the three-dimensional box $\mathbb{T}^3 = [-\pi, \pi]^3$ with periodic boundary conditions. We consider weak solutions of (1.1), which may be defined in the usual way for $v \in L_t^2 L_x^2$.

We show that within the class of weak solutions of regularity $C_t^0 H_x^{1/2-}$, the 3D Euler system (1.1) is *flexible*.[1] An example of this flexibility is provided by:

Theorem 1.1 (Main result). *Fix $\beta \in (0, 1/2)$. For any divergence-free vector fields $v_{\mathrm{start}}, v_{\mathrm{end}} \in L^2(\mathbb{T}^3)$ which have the same mean, any $T > 0$, and any $\epsilon > 0$, there exists a weak solution $v \in C([0,T]; H^\beta(\mathbb{T}^3))$ to the 3D Euler equations (1.1) such that $\|v(\cdot, 0) - v_{\mathrm{start}}\|_{L^2(\mathbb{T}^3)} \leq \epsilon$ and $\|v(\cdot, T) - v_{\mathrm{end}}\|_{L^2(\mathbb{T}^3)} \leq \epsilon$.*

Since the vector field v_{end} may be chosen to have a much higher (or much lower) kinetic energy than the vector field v_{start}, the above result shows the existence of infinitely many *non-conservative* weak solutions of 3D Euler in the regularity class $C_t^0 H_x^{1/2-}$. Theorem 1.1 further shows that the set of so-called *wild initial data* is dense in the space of L^2 periodic functions of given mean. The novelty of this result is that these weak solutions have *more than* $1/3$ *regularity*, when measured on a L_x^2-based Banach scale.

Remark 1.2. We have chosen to state the flexibility of the 3D Euler equations as in Theorem 1.1 because it is a simple way to exhibit weak solutions which are non-conservative, leaving the entire emphasis of the proof on the *regularity class* in which the weak solutions lie. Using by now standard approaches encountered in convex integration constructions for the Euler equations, we may alternatively establish the following variants of flexibility for (1.1) within the class of $C_t^0 H_x^{1/2-}$ weak solutions:

---

[1] Loosely speaking, we consider a system of partial differential equations of physical origin to be *flexible* in a certain regularity class if at this regularity level the PDEs are not anymore predictive: there exist infinitely many solutions, which behave in a non-physical way, in stark contrast to the behavior of the PDE in the smooth category. We refer the interested reader to the discussion in the surveys of De Lellis and Székelyhidi Jr. [30, 32], which draw the analogy with the flexibility in Gromov's *h*-principle [40].

1. The proof of Theorem 1.1 also shows that *given any $\beta < 1/2$, $T > 0$, and $E > 0$, there exists a weak solution $v \in C(\mathbb{R}, H^\beta(T^3))$ of the 3D Euler equations such that* $\operatorname{supp}_t v \subset [-T, T]$, *and* $\|v(\cdot, 0)\|_{L^2} \geq E$. Such weak solutions are nontrivial and have compact support in time, thereby implying the *non-uniqueness* of weak solutions to (1.1) in the regularity class $C_t^0 H_x^{1/2-}$. The argument is sketched in Remark 3.7 below.

2. The proof of Theorem 1.1 may be modified to show that *given any $\beta \in (0, 1/2)$, and any C^∞ smooth function $e\colon [0, T] \to (0, \infty)$, there exists a weak solution $v \in C^0([0, T]; H^\beta(T^3))$ of the 3D Euler equations, such that $v(\cdot, t)$ has kinetic energy $e(t)$, for all $t \in [0, T]$.* In particular, the flexibility of 3D Euler in $C_t^0 H_x^{1/2-}$ may be shown to also hold within the class of *dissipative* weak solutions, by choosing e to be a non-increasing function of time. This is further discussed in Remark 3.8 below.

1.1   CONTEXT AND MOTIVATION

Classical solutions of the Cauchy problem for the 3D Euler equations (1.1) are known to exist, locally in time, for initial velocities which lie in $C^{1,\alpha}$ for some $\alpha > 0$ (see, e.g., Lichtenstein [48]). These solutions are unique, and they conserve (in time) the kinetic energy $\mathcal{E}(t) = \frac{1}{2} \int_{T^3} |v(x, t)|^2 dx$, giving two manifestations of *rigidity* of the Euler equations within the class of smooth solutions.

Motivated by hydrodynamic turbulence, it is natural to consider a much broader class of solutions to the 3D Euler system; these are the *distributional* or *weak* solutions of (1.1), which may be defined in the natural way as soon as $v \in L_t^2 L_x^2$, since (1.1) is in divergence form. Indeed, one of the fundamental assumptions of Kolmogorov's '41 theory of turbulence [46] is that in the infinite Reynolds number limit, turbulent solutions of the 3D Navier-Stokes equations exhibit anomalous dissipation of kinetic energy; by now, this is considered to be an experimental fact; see, e.g., the book of Frisch [39] for a detailed account. In particular, this anomalous dissipation of energy necessitates that the family of Navier-Stokes solutions does not remain uniformly bounded in the topology of $L_t^3 B_{3,\infty,x}^\alpha$ for any $\alpha > 1/3$, as the Reynolds number diverges, as was alluded to in the work of Onsager [58].[2] Thus, in the infinite Reynolds number limit for turbulent solutions of 3D Navier-Stokes, one expects the convergence to *weak* solutions of 3D Euler, not classical ones.

It turns out that even in the context of weak solutions, the 3D Euler equa-

---

[2]Onsager did not use the Besov norm

$$\|v\|_{B_{p,\infty}^\alpha} = \|v\|_{L^p} + \sup_{|z| > 0} |z|^{-\alpha} \|v(\cdot + z) - v(\cdot)\|_{L^p};$$

here we use this modern notation and the sharp version of this conclusion, cf. Constantin, E, and Titi [22], Duchon and Robert [35], and Drivas and Eyink [34].

tions enjoy some conditional variants of rigidity. An example is the classical *weak-strong* uniqueness property.[3] Another example is the question of whether weak solutions of the 3D Euler equation conserve kinetic energy. This is the subject of the Onsager conjecture [58], one of the most celebrated connections between phenomenological theories in turbulence and the rigorous mathematical analysis of the PDEs of fluid dynamics. For a detailed account we refer the reader to the reviews [37, 21, 61, 30, 64, 32, 33, 12, 14] and mention here only a few of the results in the Onsager program for 3D Euler.

Constantin, E, and Titi [22] established the rigid side of the Onsager conjecture, which states that if a weak solution v of (1.1) lies in $L_t^3 B_{3,\infty,x}^\beta$ for some $\beta > 1/3$, then v conserves its kinetic energy. The endpoint case $\beta = 1/3$ was addressed by Cheskidov, Constantin, Friedlander, and Shvydkoy [16], who established a criterion which is known to be sharp in the context of 1D Burgers. By using the Bernstein inequality to transfer information from L_x^2 into L_x^3, the authors of [16] also prove energy-rigidity for weak solutions based on a regularity condition for an L_x^2-based scale: if $v \in L_t^3 H_x^\beta$ with $\beta > 5/6$, then v conserves kinetic energy (see also the work of Sulem and Frisch [63]). We emphasize the discrepancy between the energy-rigidity threshold exponents $5/6$ for the L^2-based Sobolev scale, and $1/3$ for L^p-based regularity scales with $p \geq 3$.

The first flexibility results were obtained by Scheffer [59], who constructed nontrivial weak solutions of the 2D Euler system, which lie in $L_t^2 L_x^2$ and have compact support in space and time. The existence of infinitely many dissipative weak solutions to the Euler equations was first proven by Shnirelman in [60], in the regularity class $L_t^\infty L_x^2$. Inspired by the work [53] of Müller and Šverak for Lipschitz differential inclusions, in [29] De Lellis and Székelyhidi Jr. have constructed infinitely many dissipative weak solutions of (1.1) in the regularity class $L_t^\infty L_x^\infty$ and have developed a systematic program towards the resolution of the flexible part of the Onsager conjecture, using the technique of *convex integration*. Inspired by Nash's paradoxical constructions for the isometric embedding problem [54], the first proof of flexibility of the 3D Euler system in a Hölder space was given by De Lellis and Székelyhidi Jr. in the work [31]. This breakthrough or crossing of the L_x^∞ to C_x^0 barrier in convex integration for 3D Euler [31] has in turn spurred a number of results [8, 6, 9, 27] which have used finer properties of the Euler equations to increase the regularity of the wild weak solutions being constructed. The flexible part of the Onsager conjecture was finally resolved by Isett [43, 42] in the context of weak solutions with compact support in time (see also the subsequent work by the first and last authors with De Lellis and Székelyhidi Jr. [11] for dissipative weak solutions), by showing that for any regularity parameter $\beta < 1/3$, the 3D Euler system (1.1) is flexible in the class of $C_{t,x}^\beta$ weak solutions. We refer the reader to the review

---

[3]If v is a strong solution of the Cauchy problem for (1.1) with initial datum $v_0 \in L^2$, and $w \in L_t^\infty L_x^2$ is merely a weak solution of the Cauchy problem for (1.1), which has the additional property that its kinetic energy $\mathcal{E}(t)$ is less than the kinetic energy of v_0, for a.e. $t > 0$, then in fact $v \equiv w$. See, e.g., the review [66] for a detailed account.

papers [30, 64, 32, 33, 12, 14] for more details concerning convex integration constructions in fluid dynamics, and for open problems in this area. We note that the situation in two dimensions appears considerably more difficult, as the full flexible side of the Onsager conjecture remains open in this setting [56]. Successfully extending either the homogeneous $C^{1/3-}$ constructions, or the present construction, to the 2D Euler equations appears to require new ideas.

Since the aforementioned convex integration constructions are spatially homogenous, they yield weak solutions whose Hölder regularity index cannot be taken to be larger than $1/3$ (recall that weak solutions in $L_t^3 C_x^\beta$ with $\beta > 1/3$ must conserve kinetic energy). However, *the exponent $1/3$ is not expected to be a sharp threshold for energy rigidity/flexibility if the weak solutions' regularity is measured on an L_x^p-based Banach scale with $p < 3$.* This expectation stems from the measured intermittent nature of turbulent flows; see, e.g., Frisch [39, Figure 8.8, page 132]. In broad terms, intermittency is characterized as a deviation from the Kolmogorov '41 scaling laws, which were derived under the assumptions of homogeneity and isotropy (for a rigorous way to measure this deviation, see Cheskidov and Shvydkoy [20]). A common signature of intermittency is that for $p \neq 3$, the pth order structure function[4] exponents ζ_p deviate from the Kolmogorov-predicted values of $p/3$. We note that the regularity statement $v \in C_t^0 B_{p,\infty}^s$ corresponds to a structure function exponent $\zeta_p = sp$; that is, Kolmogorov '41 predicts that $s = 1/3$ for all p. The exponent $p = 2$ plays a special role, as it allows one to measure the intermittent nature of turbulent flows on the Fourier side as a power-law decay of the energy spectrum. Throughout the last five decades, the experimentally measured values of ζ_2 (in the inertial range, for viscous flows at very high Reynolds numbers) have been consistently observed to *exceed* the Kolmogorov-predicted value of $2/3$ [1, 50, 62, 45, 15, 44, 55], thus showing a steeper decay rate in the inertial range power spectrum than the one predicted by the Kolmogorov-Obhukov 5/3 law. Moreover, in the mathematical literature, Constantin and Fefferman [23] and Constantin, Nie, and Tanveer [24] have used the 3D Navier-Stokes equations to show that the Kolmogorov '41 prediction $\zeta_2 = 2/3$ is only consistent with a lower bound for ζ_2, instead of an exact equality.

Prior to this work, it was not known whether the 3D Euler equation can sustain weak solutions which have kinetic energy that is uniformly bounded in time but not conserved, and which have spatial regularity equal to or exceeding $H_x^{1/3}$, corresponding to $\zeta_2 \geq 2/3$; see [12, Open Problem 5] and [14, Conjecture 2.6]. Theorem 1.1 gives the first such existence result. The solutions in Theo-

---

[4]In analogy with L^p-based Besov spaces, absolute p^{th} order structure functions are typically defined as $S_p(\ell) = \int_0^T \int_{\mathbb{T}^3} \int_{\mathbb{S}^2} |v(x + \ell z, t) - v(x,t)|^p dz dx dt$. The structure function exponents in Kolmogorov's '41 theory are then given by $\zeta_p = \lim \sup_{\ell \to 0+} \frac{\log S_p(\ell)}{\log(\epsilon \ell)}$, where $\epsilon > 0$ is the postulated anomalous dissipation rate in the infinite Reynolds number limit. Of course, for any non-conservative weak solution we may define a positive number $\epsilon = \int_0^T |\frac{d}{dt} \mathcal{E}(t)| dt$ as a substitute for Kolmogorov's ϵ, which allows one to define ζ_p accordingly.

rem 1.1 may be constructed to have second-order structure function exponent ζ_2 an arbitrary number in $(0, 1)$, showing that (1.1) exhibits weak solutions which severely deviate from the Kolmogorov-Obhukov 5/3 power spectrum.

We note that in a recent work [18], Cheskidov and Luo established the sharpness of the $L_t^2 L_x^\infty$ endpoint of the Prodi-Serrin criteria for the 3D Navier-Stokes equations, by constructing non-unique weak (mild) solutions of these equations in $L_t^p L_x^\infty$, for any $p < 2$.[5] As noted in [18, Theorem 1.10], their approach also applies to the 3D Euler equations, yielding weak solutions that lie in $L_t^1 C_x^\beta$ for any $\beta < 1$, and thus these weak solutions also have more than 1/3 regularity. The drawback is that the solutions constructed in [18] do not have bounded (in time) kinetic energy, in contrast to Theorem 1.1, which yields weak solutions with kinetic energy that is continuous in time.

Theorem 1.1 is proven by using an intermittent convex integration scheme, which is necessary in order to reach beyond the 1/3 regularity exponent, uniformly in time. Intermittent convex integration schemes have been introduced by the first and last authors in [13] in order to prove the non-uniqueness of weak (mild) solutions of the 3D Navier-Stokes equations with bounded kinetic energy, and then refined in collaboration with Colombo [7] to construct solutions which have partial regularity in time. Recently, intermittent convex integration techniques have been used successfully to construct non-unique weak solutions for the transport equation (cf. Modena and Székelyhidi Jr. [52, 51], Brué, Colombo, and De Lellis [5], and Cheskidov and Luo [17]), the 2D Euler equations with vorticity in a Lorentz space (cf. [4]), the stationary 4D Navier-Stokes equations (cf. Luo [49]), the α-Euler equations (cf. [3]), and the MHD equations and related variants (cf. Dai [26], the first and last authors with Beekie [2]), and the effect of temporal intermittency has recently been studied by Cheskidov and Luo [18]. We refer the reader to the reviews [12, 14] for further references, and for a comparison between intermittent and homogenous convex integration.

When applied to three-dimensional nonlinear problems, intermittent convex integration has insofar only been successful at producing weak solutions with negligible spatial regularity indices, uniformly in time. As we explain in Section 1.2, there is a fundamental obstruction to achieving high regularity: in physical space, intermittency causes concentrations that result in the formation of intermittent peaks, and to handle these peaks the existing techniques have used an extremely large separation between the frequencies in consecutive steps of the convex integration scheme.[6] This book is the first to successfully implement a high-regularity (in L^2), spatially intermittent, temporally homogenous, convex integration scheme in three space dimensions, and shows that for the 3D Euler system any regularity exponent $\beta < 1/2$ may be achieved.[7] In fact, the

---

[5]See also [19] for a proof that the space $C_t^0 L_x^p$ is critical for uniqueness at $p = 2$, in two space dimensions.

[6]This becomes less of an issue when one considers the equations of fluid dynamics in very high space dimensions; cf. Tao [65].

[7]It was known within the community (see Section 2.4.1 for a detailed explanation) that

techniques developed in this book are the backbone for the recent work [57] of the last two authors, which gives an alternative, intermittent, proof of the Onsager conjecture. In general, we expect the framework developed in the present work to inspire future iterations requiring a combination of intermittency and sharp regularity estimates.

1.2   IDEAS AND DIFFICULTIES

As alluded to in the previous paragraph, the main difficulty in reaching a high regularity exponent for weak solutions of (1.1) is that the existing intermittent convex integration schemes do not allow for consecutive frequency parameters λ_q and λ_{q+1} to be close to each other. In essence, this is because intermittency smears out the set of active frequencies in the approximate solutions to the Euler system (instead of concentric spheres, they are more akin to thick concentric annuli), and several of the key estimates in the scheme require frequency separation to achieve L^p-decoupling (see Section 2.4.1). Indeed, high regularity exponents necessitate an almost geometric growth of frequencies ($\lambda_q = \lambda_0^q$), or at least a barely super-exponential growth rate $\lambda_{q+1} = \lambda_q^b$ with $0 < b - 1 \ll 1$ (in comparison, the schemes in [13, 7] require $b \approx 10^3$). Essentially every new idea in this manuscript is aimed either directly or indirectly at rectifying this issue: how does one take advantage of intermittency, and at the same time keep the frequency separation nearly geometric?

The building blocks used in the convex integration scheme are intermittent pipe flows,[8] which we describe in Section 2.3. Due to their spatial concentration and their periodization rate, quadratic interactions of these building blocks produce both the helpful low frequency term which is used to cancel the previous Reynolds stress $\mathring{R}_q$, and a number of other errors which live at intermediate frequencies. These errors are spread throughout the frequency annulus with inner radius λ_q and outer radius λ_{q+1}, and may have size only slightly less than that of $\mathring{R}_q$. If left untreated, these errors only allow for a very small regularity parameter β. In order to increase the regularity index of our weak solutions, we need to take full advantage of the frequency separation between the slow frequency λ_q and the fast frequency λ_{q+1}. As such, the intermediate-frequency errors need to be further corrected via velocity increments designed to push these residual stresses towards the frequency sphere of radius λ_{q+1}. The quadratic interactions among these higher order velocity corrections themselves, and in principle also

---

there is a key obstruction to reaching a regularity index in L^2 for a solution to the Euler equations larger than $1/2$ via convex integration.

[8]The moniker used in [27] and the rest of the literature for these stationary solutions has been "Mikado flows." However, we rely rather heavily on the geometric properties of these solutions, such as orientation and concentration around axes, and so to emphasize the tube-like nature of these objects, we will frequently use the term "pipe flows."

with the old velocity increments, in turn create *higher order Reynolds stresses*, which live again at intermediate frequencies (slightly higher than before), but whose amplitude is slightly smaller than before. This process of adding *higher order velocity perturbations* designed to cancel intermediate-frequency higher order stresses has to be repeated many times until all the resulting errors are either small or live at frequency $\approx \lambda_{q+1}$, and thus are also small upon inverting the divergence. See Sections 2.4 and 2.6 for a more thorough account of this iteration.

Throughout the process described in the above paragraph, we need to keep adding velocity increments, while at the same time keeping the high-high-high frequency interactions under control. The fundamental obstacle here is that when composing the intermittent pipe flows with the Lagrangian flow of the slow velocity field, the resulting deformations are not spatiotemporally homogenous. In essence, the intermittent nature of the approximate velocity fields implies that a sharp global control on their Lipschitz norm is unavailable, thus precluding us from implementing a gluing technique as in [42, 11]. Additionally, we are faced with the issue that pipe flows which were added at different stages of the higher order correction process have different periodization rates and different spatial concentration rates, and may a priori overlap. Our main idea here is to implement a *placement technique* which uses the *relative intermittency* of pipe flows from previous or same generations, in conjunction with a sharp bound on their local Lagrangian deformation rate, to determine suitable spatial shifts for the placement of new pipe flows so that they dodge all other bent pipes which live in a restricted space-time region. This geometric placement technique is discussed in Section 2.5.2.

A rigorous mathematical implementation of the heuristic ideas described in the previous two paragraphs, which crucially allows us to slow down the frequency growth to be almost geometric, requires extremely sharp information on all higher order errors and their associated velocity increments. For instance, in order to take advantage of the transport nature of the linearized Euler system while mitigating the loss of derivatives issue which is characteristic of convex integration schemes, we need to keep track of essentially *infinitely many sharp material derivative estimates* for all velocity increments and stresses. Such estimates are naturally only attainable on a *local inverse Lipschitz timescale*, which in turn necessitates keeping track of the precise location in space of the peaks in the densities of the pipe flows, and performing a frequency localization with respect to both the Eulerian and the Lagrangian coordinates. In order to achieve this, we introduce carefully designed *cutoff functions*, which are defined recursively for the velocity increments (in order to keep track of overlapping pipe flows from different stages of the iteration), and iteratively for the Reynolds stresses (in order to keep track of the correct amplitude of the perturbation which needs to be added to correct these stresses); see Section 2.5. The cutoff functions we construct effectively play the role of a joint Eulerian-and-Lagrangian Littlewood-Paley frequency decomposition, which in addition keeps track of both the position in space and the amplitude of various objects (akin to a wavelet decomposition). The analysis of these cutoff functions requires estimating very high order commutators between Lagrangian and Eulerian derivatives (see Chapter 6 and Appendix A). Lastly, we mention an additional technical complication: since the sharp control of the Lipschitz norm of the approximate velocities in our scheme is local in space and time, we need to work with an inverse divergence operator (e.g., for computing higher order stresses) which, up to much lower order error terms, maintains the spatial support of the vector fields that it is applied to. Additionally, we need to be able to estimate an essentially infinite number of material derivatives applied to the output of this inverse divergence operator. This issue is addressed in Section A.8.

1.3   ORGANIZATION OF THE BOOK

The goal of this book is to prove Theorem 1.1 through an explicit construction of satisfactory weak solutions of the 3D Euler equations. Many aspects of this construction are in fact predicated on several recent advancements in the field of convex integration, particularly for the Euler and Navier-Stokes equations. Readers wishing to familiarize themselves with the important concepts can consult the survey paper [12], which provides an excellent overview of the relevant literature, along with essentially complete proofs of some fundamental results. We also refer the reader to the foundational papers [31, 8, 43, 11, 13], in which much of the aforementioned theory for the Euler and Navier-Stokes equations was developed.

As the complete proof of Theorem 1.1 is quite intricate, we have provided in Chapter 2 a broad overview of the main ideas, and how they tie together in order to prove the end result. Any path through this book, whether a short sojourn or a deep dive, should begin here. Specifically, Chapter 2 contains an outline of the convex integration scheme, in which we replace some of the actual (and more complicated) estimates and definitions appearing in the proof with heuristic ones in order to highlight the new ideas at an intuitive level. Readers familiar with the aforementioned literature may read only this chapter and still encounter the inspiration behind every new idea in the proof.

For those readers wishing to move past heuristics, the proof of Theorem 1.1 is given in Chapter 3, assuming that a number of estimates hold true inductively for the solutions of the Euler-Reynolds system at every step of the convex integration iteration. The remainder of the book is dedicated to showing that the inductive bounds stated in Section 3.2 may indeed be propagated from step q to step $q+1$. Chapter 4 contains the construction of the intermittent pipe flows used in this book and describes the careful placement required to show that these pipe flows do not overlap on a suitable space-time set. The mollification step of the proof is performed in Chapter 5. Chapter 6 contains the definitions of the cutoff functions used in the proof and establishes their properties. Readers may skip the proofs in Chapters 5 and 6, simply take the results for granted, and read the rest of the book successfully. Chapter 7 breaks down the main inductive bounds from Section 3.2 into components which take into account the higher order stresses and perturbations. Chapter 8 then proves the constituent parts of the inductive bounds outlined in Chapter 7. Chapter 9 carefully defines the many parameters in the proof, states the precise order in which they are chosen, and lists a few consequences of their definitions. Finally, Appendix A contains the analytical toolshed to which we appeal throughout the book. Readers may also wish to read the proofs in the appendix sparingly, as the statements are generally sufficient for understanding most of the arguments.

1.4 ACKNOWLEDGMENTS

T. B. was supported by the NSF grant DMS-1900149 and a Simons Foundation Mathematical and Physical Sciences Collaborative Grant.

N. M. was supported by the NSF grant DMS-1716466 and by Tamkeen under the NYU Abu Dhabi Research Institute grant of the center SITE.

V. V. was supported by the NSF grant CAREER DMS-1911413.

Chapter Two

Outline of the convex integration scheme

In Section 2.1, we list the set of parameters used throughout this chapter. The primary inductive assumptions are detailed in Section 2.2. The principal building blocks of the convex integration scheme, namely, *intermittent pipe flows*, are described in Section 2.3. We also elaborate on the intricacies of implementing these building blocks within the context of a convex integration scheme. In particular, we will describe the degree of *intermittency* used in their construction, the placement of the pipes, and the control of their Lagrangian deformation. Section 2.4, describes the new concept of *higher order stresses* that play a key role in the convex integration scheme. An additional crucial technical ingredient to the construction, specialized cutoff functions, will be described in Section 2.5. Such cutoff functions allow precise localization of scales of both the velocity and Reynolds stress. In addition, the cutoffs play a key role in the placement of pipes. Section 2.6 details the construction of the inductive perturbation to the velocity designed to correct the Reynolds stress error of the previous iteration. Finally, heuristic estimates of the new Reynolds stress error resulting from the perturbation are given in Section 2.7.

2.1 A GUIDE TO THE PARAMETERS

In order to make sharp estimates throughout the scheme, we will require numerous parameters. For the reader's convenience, we have collected in this section the *heuristic definitions* of all the parameters introduced in the following sections of the outline. The parameters are listed in Section 2.1.1 in the order corresponding to their first appearance in the outline. We give as well brief descriptions of the significance of each parameter.

2.1.1 Definitions

Definition 2.1 (Parameters introduced in Chapter 1).

1. β: *The regularity exponent corresponding to a solution* $v \in C_t H^\beta(\mathbb{T}^3)$.

Definition 2.2 (Parameters introduced in Section 2.2).

1. q: *The integer which represents the primary stages of the iterative convex integration scheme.*

2. $\lambda_q = a^{(b^q)}$: *The primary parameter used to quantify frequencies. a and b will be chosen later, with $a \in \mathbb{R}_+$ being a sufficiently large positive number and $b \in \mathbb{R}$ a real number slightly larger than 1.*

3. $\delta_q = \lambda_q^{-2\beta}$: *The primary parameter used to quantify amplitudes of stresses and perturbations.*

4. $\tau_q = (\delta_q^{1/2}\lambda_q)^{-1}$: *The primary parameter used to quantify the cost of a material derivative $\partial_t + v_q \cdot \nabla$.*[1]

Definition 2.3 (Parameters introduced in Section 2.3).

1. *n: The primary parameter which will be used to divide up the frequencies between λ_q and λ_{q+1} and which will take non-negative integer values. The divisions will be used for both the frequencies of the higher order stresses in Section 2.4 as well as the thickness of the intermittent pipe flows used to correct the higher order stresses.*

2. $n_{\max}$: *A large integer which is fixed independently of q and which sets the largest allowable value of n.*

3. $r_{q+1,n} = \left(\lambda_q\lambda_{q+1}^{-1}\right)^{\left(\frac{4}{5}\right)^{n+1}}$: *The parameter quantifying intermittency, or the thickness of a tube periodized at unit scale for values of n such that $0 \leq n \leq n_{\max}$.*[2]

4. $\lambda_{q,n} = \lambda_{q+1}r_{q+1,n} = \lambda_q^{\left(\frac{4}{5}\right)^{n+1}}\lambda_{q+1}^{1-\left(\frac{4}{5}\right)^{n+1}}$: *The minimum frequency present in an intermittent pipe flow $\mathbb{W}_{q+1,n}$. Equivalently, $(\lambda_{q+1}r_{q+1,n})^{-1}$ is the scale to which $\mathbb{W}_{q+1,n}$ is periodized. See Figure 2.1.*

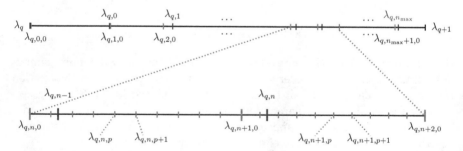

Figure 2.1: Schematic of the frequency parameters appearing in Definitions 2.2 and 2.4.

---

[1]For technical reasons, τ_q^{-1} will be chosen to be slightly shorter than $\delta_q^{\frac{1}{2}}\lambda_q$. For the heuristic calculations, one may ignore this modification and simply use $\tau_q^{-1} = \delta_q^{\frac{1}{2}}\lambda_q$.

[2]In particular, this choice gives $r_{q+1,n+1} = r_{q+1,n}^{\frac{4}{5}}$. In our proof, the inequality $r_{q+1,n}^3 \ll r_{q+1,n+1}^4$ plays a crucial role. In order to absorb q independent constants, as well as to ensure that there is a sufficient gap between these parameters to ensure decoupling, we have chosen to work with the $\frac{4}{5}$ instead of the $\frac{3}{4}$ geometric scale.

Definition 2.4 (Parameters introduced in Section 2.4).

1. *For $2 \leq n \leq n_{\max}$, $\lambda_{q,n,0} = \lambda_q^{\left(\frac{4}{5}\right)^{n-1} \cdot \frac{5}{6}} \lambda_{q+1}^{1-\left(\frac{4}{5}\right)^{n-1} \cdot \frac{5}{6}}$ is the minimum frequency present in the higher order stress $\mathring{R}_{q,n}$. Conversely, $\lambda_{q,n+1,0}$ is the maximum frequency present in $\mathring{R}_{q,n}$. When $n = 0$, we set $\lambda_{q,0,0} = \lambda_q$ to be the maximum frequency present in $\mathring{R}_{q,0} = \mathring{R}_q$, and when $n = 1$, $\lambda_{q,1,0} = \lambda_{q,0}$ is the minimum frequency present in $\mathring{R}_{q,1}$, while $\lambda_{q,2,0}$ is the maximum frequency.*

2. *p: A secondary parameter which takes positive integer values and which will be used to divide up the frequencies in between $\lambda_{q,n,0}$ and $\lambda_{q,n+1,0}$, as well as the higher order stresses.*

3. *$p_{\max}$: A large integer, fixed independently of q, which is the largest allowable value of p.*

4. *$\lambda_{q,n,p} = \lambda_{q,n,0}^{1-\frac{p}{p_{\max}}} \lambda_{q,n+1,0}^{\frac{p}{p_{\max}}}$: The maximum frequency present in the higher order stress $\mathring{R}_{q,n,p}$ for $1 \leq n \leq n_{\max}$ and $1 \leq p \leq p_{\max}$. Conversely, $\lambda_{q,n,p-1}$ is the minimum frequency in $\mathring{R}_{q,n,p}$. When $n = 0$ and p takes any value, we adopt the convention that $\lambda_{q,0,p} = \lambda_q$. See Figure 2.1.*

5. *$f_{q,n} = \lambda_{q,n+1,0}^{\frac{1}{p_{\max}}} \lambda_{q,n,0}^{-\frac{1}{p_{\max}}}$: The increment between frequencies $\lambda_{q,n,p-1}$ and $\lambda_{q,n,p}$ for $n \geq 1$. We have the equalities*

$$\lambda_{q,n,p} = \lambda_{q,n,0} f_{q,n}^p, \qquad \lambda_{q,n+1,0} = \lambda_{q,n,0} f_{q,n}^{p_{\max}}.$$

For ease of notation, when $n = 0$ we set $f_{q,n} = 1$.

6. *For $n = 0$ and $p = 1$, $\delta_{q+1,0,1} := \delta_{q+1}$ is the amplitude of $\mathring{R}_q := \mathring{R}_{q,0}$. For $n = 0$ and $p \geq 2$, $\delta_{q+1,0,p} = 0$, since there are no higher order stresses at $n = 0$. For $n \geq 1$ and any value of p, the amplitude of $\mathring{R}_{q,n,p}$ is given by*

$$\delta_{q+1,n,p} := \frac{\delta_{q+1}\lambda_q}{\lambda_{q,n,p-1}} \cdot \prod_{n'<n} f_{q,n'}.$$

One should view the product of $f_{q,n'}$ terms as a negligible error, which is justified by calculating

$$\prod_{0 \leq n' \leq n_{\max}} f_{q,n'} = \left(\frac{\lambda_{q,n_{\max}+1,0}}{\lambda_{q,1,0}}\right)^{\frac{1}{p_{\max}}} \leq \left(\frac{\lambda_{q+1}}{\lambda_q}\right)^{\frac{1}{p_{\max}}} \tag{2.1}$$

and assuming that $p_{\max}$ is large.

Definition 2.5 (Parameters introduced in Section 2.5).

1. *ε_Γ: A very small positive number.*

2. *$\Gamma_{q+1} = \left(\lambda_{q+1}\lambda_q^{-1}\right)^{\varepsilon_\Gamma}$: A parameter which will be used to quantify deviations in amplitude. In particular, Γ_q will be used to quantify amplitudes*

of both velocity fields and (higher order) stresses.

2.2   INDUCTIVE ASSUMPTIONS

For every non-negative integer q we will construct a solution $(v_q, p_q, \mathring{R}_q)$ to the Euler-Reynolds system

$$\partial_t v_q + \operatorname{div}(v_q \otimes v_q) + \nabla p_q = \operatorname{div} \mathring{R}_q \qquad (2.2a)$$

$$\operatorname{div} v_q = 0. \qquad (2.2b)$$

Here $\mathring{R}_q$ is assumed to be a trace-free symmetric matrix. The relative size of the approximate solution v_q and the Reynolds stress error $\mathring{R}_q$ will be measured in terms of the frequency parameter λ_q and the amplitude parameter δ_q, which are defined in Definition 2.2. We will propagate the following basic inductive estimates on $(v_q, \mathring{R}_q)$:[3]

$$\|v_q\|_{H^1} \leq \delta_q^{\frac{1}{2}} \lambda_q \qquad (2.3)$$

$$\|\mathring{R}_q\|_{L^1} \leq \delta_{q+1}. \qquad (2.4)$$

We shall see later that in order to build solutions belonging to $\dot{H}^\beta$ for β approaching $\frac{1}{2}$, we must propagate additional estimates on higher order material and spatial derivatives of both v_q and $\mathring{R}_q$ in L^2 and L^1, respectively. Roughly speaking, every spatial derivative on either v_q or $\mathring{R}_q$ costs a factor of λ_q. Additional material derivatives are more delicate and will be discussed further in Section 2.5, but for the time being, one may imagine that each material derivative $D_{t,q} := \partial_t + v_q \cdot \nabla$ on v_q or $\mathring{R}_q$ costs a factor of τ_q^{-1}.

2.3   INTERMITTENT PIPE FLOWS

Pipe flows, both homogeneous and intermittent, have proven to be one of the most useful components of many convex integration schemes. Homogeneous pipe flows were introduced first by Daneri and Székelyhidi Jr. [27]. The prototypical pipe flow in the $\vec{e}_3$ direction is constructed using a smooth function $\rho : \mathbb{R}^2 \to \mathbb{R}$ which is compactly supported, for example in a ball of radius 1 centered at the origin, and has zero mean. Letting $\varrho : \mathbb{T}^2 \to \mathbb{R}$ be the $\mathbb{T}^2$-periodized version of

---

[3]By $\|v_q\|_{H^1}$, we actually mean $\|v_q\|_{C_t^0 H_x^1}$. Similarly, $\|\mathring{R}_q\|_{L^1}$ stands for $\|\mathring{R}_q\|_{C_t^0 L_x^1}$. Unless stated explicitly otherwise, all the norms used in this book are uniform in time and will be abbreviated in the same way as in this example.

ρ, the $\mathbb{T}^3$-periodic pipe flow $\mathbb{W} : \mathbb{T}^3 \to \mathbb{R}^3$ is defined as

$$\mathbb{W}(x_1, x_2, x_3) = \varrho(x_1, x_3)e_2 \,. \tag{2.5}$$

It is immediate that $\mathbb{W}$ is divergence-free and a stationary solution to the Euler equations. Pipe flows such as $\mathbb{W}$ have been used in convex integration schemes which produce solutions in L^∞-based spaces [27, 43, 11]. At the qth stage of the iteration, the $\frac{\mathbb{T}^3}{\lambda_{q+1}}$-periodized pipe flow $\mathbb{W}(\lambda_{q+1}\cdot)$ is used to construct the perturbation.

By contrast, intermittent pipe flows are *not* spatially homogeneous. Intermittency in the context of convex integration schemes was introduced by the first and last authors in [13] via *intermittent Beltrami flows*, which are defined via their Fourier support and may be likened to modified and renormalized Dirichlet kernels. Intermittent pipe flows were introduced by Modena and Székelyhidi Jr. in the context of the transport and transport-diffusion equation [52] and have also been utilized for the higher dimensional (at least four dimensional[4]) Navier-Stokes equations [49, 65]. The precise objects we use are defined in (4.10) in Proposition 4.4, but let us briefly describe some of their important attributes. The intermittency is quantified by the parameter $r_{q+1,n} \ll 1$. Let $\rho_{r_{q+1,n}} : \mathbb{R}^2 \to \mathbb{R}$ be defined by $\rho_{r_{q+1,n}}(\cdot) = \rho\left(\frac{\cdot}{r_{q+1,n}}\right)$, and let $\varrho_{r_{q+1,n}}$ be the $\mathbb{T}^2$-periodized version of $\rho_{r_{q+1,n}}$. Thus one can see that $r_{q+1,n}$ describes the *thickness of the pipes at unit scale*. In order to make the intermittent pipe flows of unit size in $L^2(\mathbb{T}^3)$, one must multiply by a factor of $r_{q+1,n}^{-1}$, meaning that the Lebesgue norms of the resulting object $\mathbb{W}_{r_{q+1,n}}$ scale as

$$\left\| \mathbb{W}_{r_{q+1,n}} \right\|_{L^p(\mathbb{T}^3)} \sim r_{q+1,n}^{\frac{2}{p}-1}. \tag{2.6}$$

Let $\mathbb{W}_{q+1,n}$ be the $\frac{\mathbb{T}^3}{(r_{q+1,n}\lambda_{q+1})}$-periodic version of $\mathbb{W}_{r_{q+1,n}}$. Notice that this implies that the thickness of the pipes comprising $\mathbb{W}_{q+1,n}$ is of order λ_{q+1}^{-1} for all n, and that the Lebesgue norms of the periodized object $\mathbb{W}_{q+1,n}$ depend only on $r_{q+1,n}$. Per Definition 2.3, the thickness of the pipes used in the perturbation at stage $q + 1$ will be quantified by

$$r_{q+1,n} = \left(\frac{\lambda_q}{\lambda_{q+1}}\right)^{\left(\frac{4}{5}\right)^{n+1}} .$$

This choice will be jusified upon calculation of the heuristic bounds.

---

[4]In three dimensions, intermittent pipe flows are not sufficiently sparse to handle the error term arising from the Laplacian. This issue was addressed by Colombo and the first and last authors in [7] through the usage of *intermittent jets*, and similar objects have been used in subsequent papers as well (see work of Brue, Colombo, and De Lellis [5], Cheskidov and Luo [17, 18]).

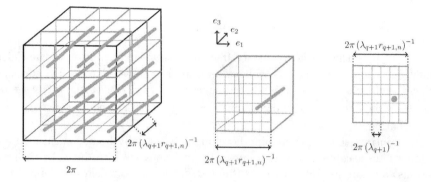

Figure 2.2: A pipe flow $\mathbb{W}_{q+1,n}$ which is periodized to scale $(\lambda_{q+1}r_{q+1,n})^{-1} = \lambda_{q,n}^{-1}$ is placed in a direction parallel to the e_2 axis. Upon taking into account periodic shifts, we note that there are $r_{q+1,n}^{-2}$ many options to place this pipe. This degree of freedom will be used later; see, e.g., Figure 2.7.

2.3.1 Lagrangian coordinates, intermittency, and placements

In order to achieve the optimum regularity β, we will define the pipe flows which comprise the perturbation at stage $q+1$ in Lagrangian coordinates corresponding to the velocity field v_q. Due to the inherent instability of Lagrangian coordinates over timescales longer than that dictated by the Lipschitz norm of the velocity field, there will be many sets of coordinates used in different time intervals which are then patched together using a partition of unity. This technique has been used frequently in recent convex integration schemes, beginning with work of Isett [41], the first author, De Lellis, and Székelyhidi Jr. [10], and Isett, the first author, De Lellis, and Székelyhidi Jr. [8], but perhaps most notably in the proof of the Onsager conjecture by Isett [43] and the subsequent strengthening to dissipative solutions by the first and last authors, De Lellis, and Székelyhidi Jr. [11].

The proof of Onsager's conjecture employs the gluing technique to prevent pipe flows defined using different Lagrangian coordinate systems from overlapping. The intermittent quality of our building blocks, and thus the approximate solution v_q, appears to obstruct the successful implementation of the gluing technique, since gluing requires a sharp control on the global Lipschitz norm of the velocity field, which will be unavailable. Thus, we cannot use the gluing technique and must control in a different fashion the possible interactions between two intermittent pipe flows defined using different Lagrangian coordinate systems.

To control these interactions, we have introduced a *placement technique* (cf. Proposition 4.8) which is used to completely prevent all such interactions. This placement technique is predicated on a simple observation about intermittent pipe flows, which to our knowledge has not yet been used in any convex integration schemes to date. When the diameter of the pipe at unit scale is of size $r_{q+1,n}$, there are $(r_{q+1,n})^{-2}$ disjoint choices for the support of pipe. These choices simply correspond to shifting the intersection of the axis of the pipe in the plane which is perpendicular to the axis; cf. Proposition 4.3. This degree of freedom is unaffected by periodization and is depicted in Figure 2.2 for a $\frac{\mathbb{T}^3}{\lambda_{q+1} r_{q+1,n}}$-periodic intermittent pipe flow $\mathbb{W}_{q+1,n}$. We will exploit this degree of freedom to choose placements for each set of pipes which *entirely avoid* other sets of pipes on small discretized regions of space-time. The space-time discretization is made possible through the usage of cutoff functions which will be discussed in more detail later in Section 2.5. We remark that De Lellis and Kwon [28] have introduced a placement technique in the context of C^α, globally dissipative solutions to the 3D Euler equations which is predicated on restricting the timescale of the Lagrangian coordinate systems to be significantly shorter than the Lipschitz timescale. This restriction significantly limits the regularity of the final solution and is thus not suited for an intermittent scheme aimed at $H^{\frac{1}{2}-}$ regularity.

2.4 HIGHER ORDER STRESSES

2.4.1 Regularity beyond $1/3$

The resolution of the flexible side of the Onsager conjecture in [43] and [11] mentioned previously shows that given some prescribed regularity index $\beta \in (0, \frac{1}{3})$, one can construct dissipative weak solutions u in C^β. Conversely, following on partial work by Eyink [36], Constantin, E, and Titi [22] have proven that conservation of energy in the Euler equations requires only that $u \in L_t^3\left(B_{3,\infty}^\alpha\right)$ for $\alpha > 1/3$. This leaves open the possibility of building dissipative weak solutions with more than $\frac{1}{3}$-many derivatives in $L^p\left(\mathbb{T}^3\right)$ (uniformly in time in our case) for $p < 3$.

Let us present a heuristic estimate which indicates a regularity limit of $H^{\frac{1}{2}}$ for solutions produced via convex integration schemes. For this purpose, let us focus on one of the principal parts of the stress in an intermittent convex integration scheme (for the familiar reader, this is part of the oscillation error). The perturbations include a coefficient function a which depends on $\mathring{R}_q$ and thus for which derivatives cost λ_q and which has amplitude $\delta_{q+1}^{1/2}$ (the square root of the amplitude of the stress). These coefficient functions are multiplied by intermittent pipe flows $\mathbb{W}_{q+1,0}$ for which derivatives cost λ_{q+1} and which have unit size in L^2, but are only periodized to scale $(\lambda_{q+1} r_{q+1,0})^{-1}$. When the divergence lands on the square of the coefficient function a^2 in the nonlinear term, the resulting error term satisfies the estimate

$$\left\| \text{div}^{-1}\left(\nabla(a^2)\mathbb{P}_{\neq 0}(\mathbb{W}_{q+1,0} \otimes \mathbb{W}_{q+1,0}) \right) \right\|_{L^1} \leq \frac{\delta_{q+1}\lambda_q}{\lambda_{q+1}r_{q+1,0}}. \tag{2.7}$$

The numerator is the size of $\nabla(a^2)$ in L^1, while the denominator is the gain induced by inverting the divergence at $\lambda_{q+1}r_{q+1,0}$, which is the minimum frequency of $\mathbb{P}_{\neq 0}(\mathbb{W}_{q+1,0} \otimes \mathbb{W}_{q+1,0}) = \mathbb{W}_{q+1,0} \otimes \mathbb{W}_{q+1,0} - \fint_{\mathbb{T}^3} \mathbb{W}_{q+1,0} \otimes \mathbb{W}_{q+1,0}$. Note that we have used implicitly that $\mathbb{W}_{q+1,0}$ has unit L^2 norm, and that by periodicity $\mathbb{P}_{\neq 0}(\mathbb{W}_{q+1,0} \otimes \mathbb{W}_{q+1,0})$ decouples from $\nabla(a^2)$. This error would be minimized when $r_{q+1,0} = 1$, in which case

$$\frac{\delta_{q+1}\lambda_q}{\lambda_{q+1}} < \delta_{q+2} \iff \lambda_{q+1}^{-2\beta+\frac{1}{b}} < \lambda_{q+1}^{-2\beta b+1}$$

$$\iff 2\beta b^2 - 2\beta b < b - 1$$

$$\iff 2\beta b(b-1) < b - 1$$

$$\iff \beta < \frac{1}{2b}. \tag{2.8}$$

Any intermittency parameter $r_{q+1,0} \ll 1$ would *weaken* this estimate since the gain induced from inverting the divergence will only be $\lambda_{q+1}r_{q+1,0} \ll \lambda_{q+1}$. On the other hand, we will see that a small choice of $r_{q+1,0}$ *strengthens all other error terms*, and because of this, in our construction we will choose $r_{q+1,0}$ as in Definition 2.3, item (3). One may refer to the blog post of Tao [65] for a slightly different argument which reaches the same apparent regularity limit. This apparent regularity limit is independent of dimension, and we believe that the method in this book cannot be modified to yield weak solutions with regularity $L_t^\infty W_x^{s,p}$ with $s > 1/2$, for any $p \in [1,2]$.

The higher order stresses mentioned in Section 1.2 will compensate for the losses incurred in this nonlinear error term when $r_{q+1,0} \ll 1$. As we shall describe in the next section, we use the phrase "higher order stresses" to describe errors which are higher in frequency and smaller in amplitude than $\mathring{R}_q$, but not sufficiently small enough or at high enough frequency to belong to $\mathring{R}_{q+1}$. Similarly, "higher order perturbations" are used to correct the higher order stresses and thus increase the extent to which an approximate solution solves the Euler equations.

2.4.2 Specifics of the higher order stresses

In convex integration schemes which measure regularity in L^∞ (i.e., using Hölder spaces C^α), pipe flows interact through the nonlinearity to produce low ($\approx \lambda_q$) and high ($\approx \lambda_{q+1}$) frequencies. We denote by $w_{q+1,0}$ the perturbation designed to correct $\mathring{R}_q$. In the absence of intermittency, the low frequencies from the self-interaction of $w_{q+1,0}$ cancel the Reynolds stress error $\mathring{R}_q$, and the high frequencies are absorbed by the pressure up to an error small enough to be placed in $\mathring{R}_{q+1}$. In an intermittent scheme, the self-interaction of the intermittent pipe flows comprising $w_{q+1,0}$ produces low, intermediate, and high frequencies. The low and high frequencies play a similar role as before. However, the intermediate frequencies cannot be written as a gradient, nor are small enough to be absorbed in $\mathring{R}_{q+1}$. This issue has limited the available regularity on the final solution in many previous intermittent convex integration schemes. In order to reach the threshold $H^{\frac{1}{2}}$, we address this issue using higher order Reynolds stress errors $\mathring{R}_{q,n}$ for $n = 1, 2, \ldots, n_{\max}$; cf. Figure 2.3.

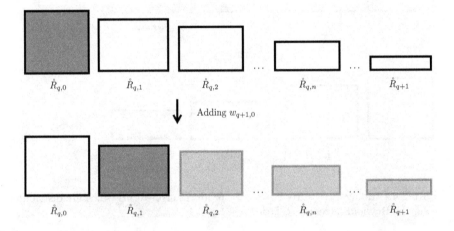

Figure 2.3: Adding the increment $w_{q+1,0}$ corrects the stress $\mathring{R}_{q,0} = \mathring{R}_q$, but produces error terms which live at frequencies that are intermediate between λ_q and λ_{q+1}, due to the intermittency of $w_{q+1,0}$. These new errors are sorted into higher order stresses $\mathring{R}_{q,n}$ for $1 \le n \le n_{\max}$, as depicted above. The heights of the boxes corresponds to the amplitude of the errors that will fall into them, while the frequency support of each box increases from λ_q for $\mathring{R}_{q,0} = \mathring{R}_q$, to λ_{q+1} for $\mathring{R}_{q+1}$.

After the addition of $w_{q+1,0}$ to correct $\mathring{R}_q$, which is labeled in Figure 2.4 as $\mathring{R}_{q,0}$, low frequency error terms are produced, which we divide into higher order stresses. To correct the error term of this type at the *lowest* frequency, which is labeled $\mathring{R}_{q,1}$ in Figure 2.4, we add a sub-perturbation $w_{q+1,1}$. The subsequent bins are lighter in color to emphasize that they are not yet full; that is, there are more error terms which have yet to be constructed but will be sorted into such bins. The emptying of the bins then proceeds inductively on n, as we add higher order perturbations $w_{q+1,n}$, which are designed to correct $\mathring{R}_{q,n}$. For

$1 \leq n \leq n_{\max}$, the frequency support of $\mathring{R}_{q,n}$ is[5]

$$\left\{ k \in \mathbb{Z}^3 : \lambda_{q,n,0} \leq |k| < \lambda_{q,n+1,0} \right\}. \tag{2.9}$$

This division will be justified upon calculation of the heuristic bounds in Section 2.7.

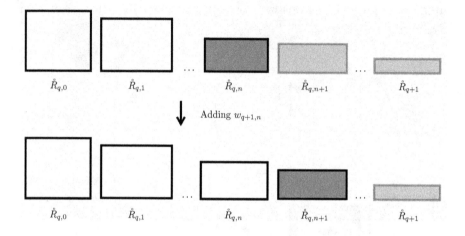

Figure 2.4: Adding $w_{q+1,n}$ to correct $\mathring{R}_{q,n}$ produces error terms which are distributed among the Reynolds stresses $\mathring{R}_{q,n'}$ for $n+1 \leq n' \leq n_{\max}$.

Let us now explain the motivation for the division of $\mathring{R}_{q,n}$ into the further subcomponents $\mathring{R}_{q,n,p}$. Suppose that we add a perturbation $w_{q+1,n}$ to correct $\mathring{R}_{q,n}$ for $n \geq 1$. The amplitude of $w_{q+1,n}$ would depend on the amplitude of $\mathring{R}_{q,n}$, which in turn depends on the gain induced by inverting the divergence to produce $\mathring{R}_{q,n}$, which depends then on the minimum frequency $\lambda_{q,n,0}$. However, derivatives on the *low* frequency coefficient function used to define $w_{q+1,n}$ would depend on the *maximum* frequency of $\mathring{R}_{q,n}$, which is $\lambda_{q,n+1,0}$. The (sharp-eyed) reader may at this point object that the first derivative on the low frequency coefficient function $\nabla(a(\mathring{R}_{q,n}))$ should be cheaper, since $\mathring{R}_{q,n}$ is obtained from inverting the divergence, and taking the gradient of the cutoff function written above should thus morally involve bounding a zero-order operator. However, constructing the low frequency coefficient function presents technical difficulties which prevent us from taking advantage of this intuition. In fact, the failure of this intuition is the sole reason for the introduction of the parameter p, as

---

[5]In reality, the higher order stresses are not compactly supported in frequency. However, they will satisfy derivative estimates to very high order which are characteristic of functions with compact frequency support.

one may see from the heuristic estimates later. In any case, increasing the regularity β of the final solution requires minimizing this gap between the gain in amplitude provided by inverting the divergence and the cost of a derivative, and so we subdivide $\mathring{R}_{q,n}$ into further components $\mathring{R}_{q,n,p}$ for $1 \leq p \leq p_{max}$.[6] Both $n_{\max}$ and $p_{\max}$ are fixed independently of q. Each component $\mathring{R}_{q,n,p}$ then will have frequency support in the set

$$\left\{k \in \mathbb{Z}^3 : \lambda_{q,n,p-1} \leq |k| < \lambda_{q,n,p}\right\} = \left\{k \in \mathbb{Z}^3 : \lambda_{q,n,0} f_{q,n}^{p-1} \leq |k| < \lambda_{q,n,0} f_{q,n}^{p}\right\}.$$
$$(2.10)$$

Notice that by the definition of $f_{q,n}$ in Definition 2.4, 2.10 defines a partition of the frequencies in between $\lambda_{q,n,0}$ and $\lambda_{q,n+1,0}$ for $1 \leq p \leq p_{\max}$. Figure 2.5 depicts this division, and we shall describe in the heuristic estimates how each subcomponent $\mathring{R}_{q,n,p}$ is corrected by $w_{q+1,n,p}$, with all resulting errors absorbed into either $\mathring{R}_{q+1}$ or $\mathring{R}_{q,n'}$ for $n' > n$.

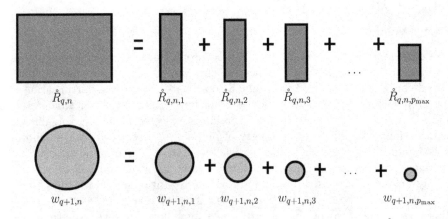

Figure 2.5: The higher order stress $\mathring{R}_{q,n}$ is decomposed into components $\mathring{R}_{q,n,p}$, which increase in frequency and decrease in amplitude as p increases. We use the bases of the boxes to indicate support in frequency, where frequency is increasing from left to right, and the heights to indicate amplitudes. Each subcomponent $\mathring{R}_{q,n,p}$ is corrected by its own corresponding sub-perturbation $w_{q+1,n,p}$, which has a commensurate frequency and amplitude.

Thus, the net effect of the higher order stresses is that one may take errors for which the inverse divergence provides a weak estimate due to the presence of relatively low frequencies and push them to higher frequencies for which the inverse divergence estimate is stronger. We will repeat this process until all errors are moved (almost) all the way to frequency λ_{q+1}, at which point

---

[6]There are certainly a multitude of ways to manage the bookkeeping for amplitudes and frequencies. Using both n and p is convenient because then n is the only index which quantifies the rate of periodization.

they are absorbed into $\mathring{R}_{q+1}$. Heuristically, this means that in constructing the perturbation w_{q+1} at stage q, we have *eliminated* all the higher order error terms which arise from self-interactions of intermittent pipe flows, thus producing a solution v_{q+1} to the Euler-Reynolds system at level $q+1$ which is *as close as possible* to a solution of the Euler equations. We point out that one side effect of the higher order perturbations is that the total perturbation w_{q+1} has spatial support which is *not* particularly sparse, since as n increases the perturbations $w_{q+1,n}$ become successively less intermittent and thus more homogeneous. At the same time, the frequency support of our solution is also not too sparse, since b is close to 1 and $r_{q+1,0} = \left(\lambda_q \lambda_{q+1}^{-1}\right)^{\frac{4}{5}}$, so that many of the frequencies between λ_q and λ_{q+1} are active.

2.5   CUTOFF FUNCTIONS

2.5.1   Velocity and stress cutoffs

The concept of a turnover time, which is proportional to the inverse of the gradient of the mean flow v_q, is crucial to the convex integration schemes mentioned earlier which utilized Lagrangian coordinates. Since the perturbation is expected to be roughly flowed by the mean flow v_q, the turnover time determines a timescale on which the perturbation is expected to undergo significant deformations. An important property of pipe flows, first noted by Daneri and Székelyhidi Jr. in [27] and utilized crucially by Isett [43] towards the proof of Onsager's conjecture, is that the length of time for which pipe flows written in Lagrangian coordinates remain approximately stationary solutions to Euler depends *only* on the Lipschitz norm of the transport velocity v_q and not on the Lipschitz norms of the original (undeformed) pipe flow. However, the timescale under which pipe flows transported by an intermittent velocity field remain coherent is space-time dependent, in contrast to previous convex integration schemes in which the timescale was uniform across $\mathbb{R} \times \mathbb{T}^3$. As such, we will need to introduce space-time cutoffs $\psi_{i,q}$ in order to determine the local turnover time. In particular, the cutoff $\psi_{i,q}$ will be defined such that

$$\|\nabla v_q\|_{L^\infty(\operatorname{supp}\psi_{i,q})} \lesssim \delta_q^{1/2} \lambda_q \Gamma_{q+1}^i := \tau_q^{-1} \Gamma_{q+1}^i. \tag{2.11}$$

With such cutoffs defined, we then define in addition a family of temporal cutoffs $\chi_{i,k,q}$ which will be used to restrict the timespan of the intermittent pipe flows in terms of the local turnover. Each cutoff function $\chi_{i,k,q}$ will have temporal support contained in an interval of length

$$\tau_q \Gamma_{q+1}^{-i}. \tag{2.12}$$

It should be noted that we will design the cutoffs so that we can deduce much more on its support than (2.11). Since the material derivative $D_{t,q} := \partial_t + v_q \cdot \nabla$ will play an important role, we will require estimates involving material derivatives $D_{t,q}^N$ of very high order.[7] We expect the cost of a material derivative to be related to the turnover time, which itself is local in nature. As such, high order material derivative estimates will be done on the support of the cutoff functions and will be of the form

$$\left\| \psi_{i,q} D_{t,q}^N \mathring{R}_{q,n,p} \right\|_{L^r}.$$

In addition to the family of cutoffs $\psi_{i,q}$ and $\chi_{i,k,q}$, we will also require stress cutoffs $\omega_{i,j,q,n,p}$ which determine the local size of the Reynolds stress errors $\mathring{R}_{q,n,p}$; in particular $\omega_{i,j,q,n,p}$ will be defined such that

$$\left\| \nabla^M \mathring{R}_{q,n,p} \right\|_{L^\infty(\text{supp}\,\omega_{i,j,q,n,p})} \le \delta_{q+1,n,p} \Gamma_{q+1}^{2j} \lambda_{q,n,p}^M. \tag{2.13}$$

Previous intermittent convex integration schemes have managed to successfully cancel intermittent stress terms with much simpler stress cutoff functions than the ones we use. However, mitigating the loss of spatial derivative in the oscillation error means that we have to propagate sharp spatial derivative estimates of arbitrarily high order on the stress in order to produce solutions with regularity approaching $\dot{H}^{\frac{1}{2}}$. Due to this requirement, we then have to estimate the *second* derivative (and higher) of the stress cutoff function

$$\left\| \nabla^2 \left(\omega^2 \left(\mathring{R}_{q,n,p} \right) \right) \right\|_{L^1},$$

which in turn necessitates bounding the local L^2 norm of $\nabla \mathring{R}_{q,n,p}$ due to the term

$$\left\| (\nabla^2(\omega^2)) \left(\mathring{R}_{q,n,p} \right) \left| \nabla \mathring{R}_{q,n,p} \right|^2 \right\|_{L^1}.$$

Given inductive estimates about the derivatives of $\mathring{R}_q$ *only in* L^1 which have not been upgraded to L^p for $p > 1$, this term will obey a fatally weak estimate, which is why we must estimate $\mathring{R}_{q,n,p}$ in L^∞ as in (2.13).

2.5.2 Checkerboard cutoffs

As mentioned in the discussion of intermittent pipe flows, we must prevent pipes originating from different Lagrangian coordinate systems from intersecting. The first step is to reduce the complexity of this problem by restricting the size of the spatial domain on which intersections must be prevented. Towards this end,

---

[7]The loss of material derivative in the transport error means that to produce solutions with regularity approaching $\dot{H}^{\frac{1}{2}}$, we have to propagate material derivative estimates of arbitrarily high order on the stress.

consider the maximum frequency of the original stress $\mathring{R}_q = \mathring{R}_{q,0}$, or any of the higher order stresses $\mathring{R}_{q,n}$ for $n \geq 1$. We may write these frequencies as $\lambda_{q+1} r_1$ for $\lambda_q \lambda_{q+1}^{-1} \leq r_1 < 1$. We then decompose $\mathring{R}_{q,n}$ using a checkerboard partition of unity comprised of bump functions which follow the flow of v_q and have support of diameter $(\lambda_{q+1} r_1)^{-1}$. These two properties ensure that we have *preserved* the derivative bounds on $\mathring{R}_{q,n}$. Thus, we fix the set Ω to be the support of an individual checkerboard cutoff function in this partition of unity at a fixed time; cf. (4.28).

Suppose furthermore that Ω is inhabited by disjoint sets of deformed intermittent pipe flows which are periodized to spatial scales no finer than $(\lambda_{q+1} r_2)^{-1}$ for $0 < r_1 < r_2 < 1$. In practice, r_2 will be $r_{q+1,n}$, where $r_{q+1,n}$ is the amount of intermittency used in the pipes which comprise the perturbation $w_{q+1,n}$ which is used to correct $\mathring{R}_{q,n}$. The pipes which already inhabit Ω may first be from previous generations of perturbations $w_{q+1,n'}$ for $n' < n$, in which case they are periodized to spatial scales much broader than $(\lambda_{q+1} r_2)^{-1}$, or from an overlapping checkerboard cutoff function used to decompose $\mathring{R}_{q,n}$ on which a placement of pipes periodized to spatial scale $(\lambda_{q+1} r_2)^{-1}$ has already been chosen. In either case, these pipes will have been deformed by the velocity field v_q on the timescale given by the inverse of the local Lipschitz norm. We represent the support of these deformed pipe flows in terms of axes $\{A_i\}_{i \in \mathcal{I}}$ around which the pipes $\{P_i\}_{i \in \mathcal{I}}$ are concentrated to thickness λ_{q+1}^{-1} (recall from Section 2.3 that all intermittent pipe flows used in our scheme have this thickness).

We will now explain that *under appropriate restrictions on r_1 and r_2*, one may choose a new set of (straight, i.e. not deformed) intermittent pipe flows $\mathbb{W}_{r_2, \lambda_{q+1}}$ periodized to scale $(\lambda_{q+1} r_2)^{-1}$ which are disjoint from each deformed pipe P_i and are *on the support of* Ω. Heuristically, this task becomes easier when r_2 is smaller, since this means both that we have more choices of placement for the new set, and that there are fewer pipes P_i inhabiting Ω. Conversely, this task becomes more difficult when r_1 is smaller, since then Ω is larger and will contain more pipes P_i. We assume throughout that the deformations of the P_is are mild enough to preserve the expected length, curvature, and spacing bounds between neighboring pipes that arise from writing pipes in Lagrangian coordinates and flowing for a length of time which is strictly less than the inverse of the Lipschitz norm of the velocity field.

First, we can estimate the cardinality of the set $\mathcal{I}$ (which indexes the axes A_i and pipes P_i) from above by $r_2^2 r_1^{-2}$. To understand this bound, first note that if we had *straight* pipes P_i periodized to scale $(\lambda_{q+1} r_2)^{-1}$ inhabiting a *cube* of side length $(\lambda_{q+1} r_1)^{-1}$, this bound would hold. Using the fact that our deformed pipes obey similar length, curvature, and spacing bounds as straight pipes and that our set Ω can be considered as a subset of a cube with side length proportional to $(\lambda_{q+1} r_1)^{-1}$, the same bound will hold up to dimensional constants. Secondly, by the intermittency of the desired set of new pipes, we have r_2^{-2} choices for the placement of the new set, as indicated in Figure 2.2.

To finish the argument, we must estimate how many of these r_2^{-2} choices would lead to non-empty intersections between the new pipes and any P_i. To calculate this bound, we will imagine the placement of the new set of straight pipes as occurring on a two-dimensional plane which is perpendicular to the axes of the pipes. After projecting each P_i onto this two-dimensional plane, our task is to choose the intersection points of the new pipes with the plane so that the new pipes do not intersect the shadows of the P_i's.

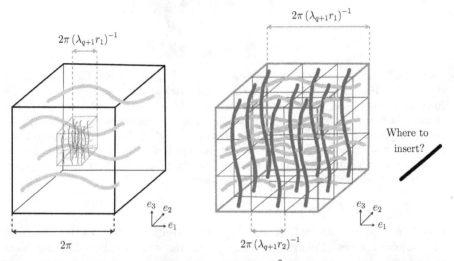

Figure 2.6: In the figure on the left we display $\mathbb{T}^3$, in which we have four large deformed pipes, representing a very sparse set of pipe flows from an old generation that were deformed by v_q, and a small subcube depicting the support of a cutoff function $\zeta_{q,i,k,n,\vec{l}}$, whose diameter is $\approx (\lambda_{q+1}r_1)^{-1}$. Due to the sparseness, very few (if any!) of the old generation pipes intersect the support of the cutoff. The figure on the right further zooms into the support of the cutoff, to emphasize its contents. On the support of $\zeta_{q,i,k,n,\vec{l}}$ we have displayed two sets of deformed pipe flows, in lighter and darker shades. These pipes flows were deformed also by v_q, from a nearby time at which they were straight and periodic at scale $(\lambda_{q+1}r_2)^{-1}$. At the current time, at which the above figure is considered, these pipe flows aren't quite periodic anymore, but they are close. The question now is: can we place a straight pipe flow, periodic at scale $(\lambda_{q+1}r_2)^{-1}$, whose axis is orthogonal to the front face of the box on the right, and which does not intersect *any of the existing pipes in this region?* To see that this is possible, in Figure 2.7 we will estimate the area of shadows on this face of the cube.

Given one of the deformed pipes P_i, since its thickness is λ_{q+1}^{-1} and its length inside Ω is proportional to the diameter of Ω, specifically $(\lambda_{q+1}r_1)^{-1}$, we may cover the shadow of P_i on the plane with $\approx r_1^{-1}$ many balls of diameter λ_{q+1}^{-1}. Covering all the P_i's thus requires $\approx r_2^2 r_1^{-2} \cdot r_1^{-1}$ balls of diameter λ_{q+1}^{-1}. Now, imagine the intersection of the new set of pipes with the plane. Each choice of placement defines this intersection as essentially a set of balls of diameter $\approx \lambda_{q+1}^{-1}$

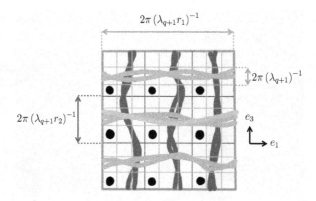

Figure 2.7: As mentioned in the caption of Figure 2.6, we consider the image on the right and *project* all of the pipes present in the box onto the front face of the cube (parallel to the $e_3 - e_1$ plane). Because these existing pipes were deformed by v_q, the shadow does not consist of straight lines, and in fact the projections can overlap. By estimating the area of this projection, we see that if $r_2^4 \ll r_1^3$ then there is enough room left to insert a *new* pipe flow with orientation axis e_2 (represented by the dark disks in the above figure), which will not intersect any of the projections of the existing pipes, and thus not intersect the existing pipes themselves.

equally spaced at distance $(\lambda_{q+1} r_2)^{-1}$. The intermittency ensures that there are r_2^{-2} disjoint choices of placement, i.e., r_2^{-2} disjoint sets of balls which represent the intersection of a particularly placed new set of pipes with the plane. As long as

$$r_2^2 r_1^{-2} \cdot r_1^{-1} \ll r_2^{-2} \quad \Longleftrightarrow \quad r_2^4 \ll r_1^3 \,,$$

there must exist at least one choice of placement which does not produce *any* intersections between $\mathbb{W}_{r_2, \lambda_{q+1}}$ and the P_is. Notice that if r_1 is too small or if r_2 is too large, this inequality will not be satisfied, thus validating our previous heuristics about r_1 and r_2.

To obey the relative intermittency inequality between r_1 and r_2 derived above for placements of new intermittent pipes on sets of a certain diameter, we will utilize cutoff functions

$$\zeta_{q,i,k,n,\vec{l}},$$

which are defined using a variety of parameters. The index q describes the stage of the convex integration scheme, while i and k refer to the velocity and temporal cutoffs defined above. The parameter n corresponds to a higher order stress $\mathring{R}_{q,n}$ and refers to its minimum frequency $\lambda_{q,n,0}$, quantifying the value of $(\lambda_{q+1} r_1)^{-1}$ and the diameter of the support as described earlier. The parameter $\vec{l} = (l, w, h)$ depends on q and n and provides an enumeration of the (three-dimensional) checkerboard covering $\mathbb{T}^3$ at scale $(\lambda_{q,n,0})^{-1}$. On the support of one of these checkerboard cutoff functions, we can inductively place pipes periodized to scale $(\lambda_{q+1} r_2)^{-1} = \lambda_{q,n}^{-1}$ which are disjoint. The checkerboard cutoff functions and the pipes themselves all follow the same velocity field, ensuring that the disjointness at a single time slice is sufficient.

2.5.3 Cumulative cutoff function

Finally, the variety of cutoffs described above will be combined into the family of cutoffs

$$\eta_{i,j,k,q,n,p,\vec{l}} := \eta_{i,j,k,q,n,p} := \chi_{i,k,q}\psi_{i,q}\omega_{i,j,q,n,p}\zeta_{q,i,k,n,\vec{l}},$$

which have timespans of $\tau_q \Gamma_{q+1}^{-i}$ and L^2 norms

$$\left\| \eta_{i,j,k,q,n,p,\vec{l}} \right\|_{L^2} \lesssim \Gamma_{q+1}^{-\frac{i}{2}} \cdot \Gamma_{q+1}^{-\frac{j}{2}}. \tag{2.14}$$

We will also require a cutoff $\eta_{i\pm,j\pm,k\pm,q,n,p,\vec{l}}$ which is defined to be 1 on the support of $\eta_{i,j,k,q,n,\vec{l}}$ and satisfies the estimate

$$\left\| \eta_{i\pm,j\pm,k\pm,q,n,\vec{l}} \right\|_{L^2} \lesssim \Gamma_{q+1}^{-\frac{i}{2}} \cdot \Gamma_{q+1}^{-\frac{j}{2}}. \tag{2.15}$$

We remark that (2.14) and (2.15) are only heuristics (see Lemma 6.41 for the precise estimate). Designing the cutoffs turned out to be for the authors perhaps the most significant technical challenge of the book. Their definition will be inductive and estimates involving them will involve several layers of induction.

2.6 THE PERTURBATION

The intermittent pipe flows of Section 2.3, the higher order stresses of Section 2.4, and the cutoff functions of Section 2.5 provide the key ingredients in the construction of the perturbation

$$w_{q+1} := \sum_{n=0}^{n_{\max}} \sum_{p=1}^{p_{\max}} w_{q+1,n,p} := \sum_{n=0}^{n_{\max}} w_{q+1,n}.$$

In the above double sum, we will adopt the convention that $w_{q+1,0,p} = 0$ unless $p = 1$ to streamline notation. Let us emphasize that w_{q+1} is constructed *inductively* on n for the following reason. Each perturbation $w_{q+1,n} = \sum_{p=1}^{p_{\max}} w_{q+1,n,p}$ will contribute error terms to all higher order stresses $\mathring{R}_{q,\tilde{n},p}$ for $\tilde{n} > n$ and $1 \le p \le p_{\max}$, and so $\mathring{R}_{q,\tilde{n}} = \sum_{p=1}^{p_{\max}} \mathring{R}_{q,\tilde{n},p}$ is not a well-defined object until each $w_{q+1,n'}$ has been constructed for all $n' < n$. For the purposes of the following heuristics, we will abbreviate the cutoff functions by $a_{n,p}$, and ignore summation over many of the indexes which parametrize the cutoff functions, as they are not necessary to understand the heuristic estimates. We will freely use the heuristic that the cutoff functions allow us to use the $L_t^\infty H_x^1$ norm of v_q to control terms (usually related to the turnover time) which previously required global Lipschitz bounds on v_q.

Let $\Phi_{q,k} : \mathbb{R} \times \mathbb{T}^3 \to \mathbb{T}^3$ be the solution to the transport equation

$$\partial_t \Phi_{q,k} + v_q \cdot \nabla \Phi_{q,k} = 0$$

with initial data given to be the identity at time $t_k = k\tau_q$. We mention that this definition is *purely* heuristic, since as mentioned previously, the Lagrangian coordinate systems will have to be indexed by another parameter which encodes the fact that ∇v_q is spatially inhomogeneous.[8] For the time being let us ignore this issue. Each map $\Phi_{q,k}$ has an effective timespan $\tau_q = (\delta_q^{\frac{1}{2}} \lambda_q)^{-1}$, at which point one resets the coordinates and defines a new transport map $\Phi_{q,k+1}$ starting from the identity. Let $\mathbb{W}_{q+1,n}$ denote the pipe flow with intermittency $r_{q+1,n}$ periodized to scale $(\lambda_{q+1} r_{q+1,n})^{-1}$. The perturbation $w_{q+1,n,p}$ is then defined heuristically by

$$w_{q+1,n,p}(x,t) = \sum_k a_{n,p}\left(\mathring{R}_{q,n,p}(x,t)\right) (\nabla\Phi_{q,k}(x,t))^{-1} (x,t) \mathbb{W}_{q+1,n}(\Phi_{q,k}(x,t)).$$

We have adopted the convention that $\mathring{R}_q = \mathring{R}_{q,0} = \mathring{R}_{q,0,1}$ and $\mathring{R}_{q,0,p} = 0$ if $p \geq 2$. Composing with $\Phi_{q,k}$ adapts the pipe flows to the Lagrangian coordinate system associated to v_q so that $(\nabla\Phi_{q,k})^{-1}\mathbb{W}_{q+1,n}(\Phi_{q,k})$ is Lie-advected and remains divergence-free to leading order. The perturbation $w_{q+1,n,p}$ has the following properties:

1. The thickness (at unit scale) of the pipes on which $w_{q+1,n,p}$ is supported depends only on q and n and is quantified by

$$r_{q+1,n} = \left(\frac{\lambda_q}{\lambda_{q+1}}\right)^{\left(\frac{4}{5}\right)^{n+1}}. \tag{2.16}$$

Thus, the perturbations become *less* intermittent as n increases, since the thickness of the pipes (periodized at unit scale) becomes larger as n increases. Notice that the maximum frequency of $\mathring{R}_{q,n,p}$ is $\lambda_{q,n,p}$ for $n \geq 1$ per (2.10), and λ_q for $n = 0$, while the minimum frequency of the intermittent pipe flow $\mathbb{W}_{q+1,n}$ used to construct $w_{q+1,n,p}$ is $\lambda_{q,n}$. Referring back to Definition 2.3 and Definition 2.4, we have that for $1 \leq n \leq n_{\max}$ and $1 \leq p \leq p_{\max}$,

$$\lambda_{q,n,p} = \lambda_{q,n,0}^{1-\frac{p}{p_{\max}}} \lambda_{q,n+1,0}^{\frac{p}{p_{\max}}} \leq \lambda_{q,n+1,0}$$

---

[8]The actual transport maps used in the proof are defined in Definition 6.26.

$$= \lambda_q^{\left(\frac{4}{5}\right)^n \cdot \frac{5}{6}} \lambda_{q+1}^{1-\left(\frac{4}{5}\right)^n \cdot \frac{5}{6}} \ll \lambda_q^{\left(\frac{4}{5}\right)^{n+1}} \lambda_{q+1}^{1-\left(\frac{4}{5}\right)^{n+1}} = \lambda_{q,n},$$

which ensures that the low frequency portion of $w_{q+1,n,p}$ decouples from the high frequency intermittent pipe flow $\mathbb{W}_{q+1,n}$. For $n = 0$, the maximum frequency of $\mathring{R}_{q,0} = \mathring{R}_q$ is λ_q, which is much less than $\lambda_{q,0}$ per Definition 2.3.

2. The L^2 size of $w_{q+1,n,p}$ is equal to the square root of the L^1 norm of $\mathring{R}_{q,n,p}$, which in turn depends on the minimum frequency of $\mathring{R}_{q,n,p}$ and will be $\delta_{q+1,n,p}$, where we define $\delta_{q+1,0,p} = \delta_{q+1}$. For $n \geq 1$ and $1 \leq p \leq p_{\max}$, we have from Definition 2.5 that

$$\delta_{q+1,n,p} = \frac{\delta_{q+1}\lambda_q}{\lambda_{q,n,p-1}} \prod_{n'<n} f_{q,n'}.$$

3. For $n \geq 1$, derivatives on the low frequency coefficient function of $w_{q+1,n,p}$ cost the maximum frequency of $\mathring{R}_{q,n,p}$, which is $\lambda_{q,n,p}$. For $n = 0$, $\mathring{R}_{q,0} = \mathring{R}_q$, so that each spatial derivative on the coefficient function of $w_{q+1,0}$ costs λ_q.

4. The transport error and Nash error created by the addition of $w_{q+1,n,p}$ are small enough to be absorbed into $\mathring{R}_{q+1}$ for every n .

5. Per Definition 2.3, the oscillation error which results from $w_{q+1,n,p}$ interacting with itself has minimum frequency

$$\lambda_{q,n} = \lambda_{q+1}r_{q+1,n} = \lambda_q^{\left(\frac{4}{5}\right)^{n+1}} \lambda_{q+1}^{1-\left(\frac{4}{5}\right)^{n+1}} .$$

2.7 THE REYNOLDS STRESS ERROR AND HEURISTIC ESTIMATES

Note that since the relation (2.2) is linear in the Reynolds stress, replacing q with $q + 1$, the right-hand side can be split into three components:

$$\text{div} \left(w_{q+1} \otimes w_{q+1} + \mathring{R}_q \right)$$
$$\partial_t w_{q+1} + v_q \cdot \nabla w_{q+1} \tag{2.17}$$
$$w_{q+1} \cdot \nabla v_q ,$$

which we call the *oscillation error*, *transport error*, and *Nash error* respectively.

2.7.1 Type 1 oscillation error

In this section, we sketch the heuristic estimates which justify the following principle: the low frequency, high amplitude errors arising from the self-interaction of an intermittent pipe flow can be transferred to higher frequencies and smaller amplitudes through the higher order stresses and perturbations. We shall show that the following estimates are self-consistent and allow for the construction of solutions approaching the regularity threshold $\dot{H}^{\frac{1}{2}}$:

$$\left\| \nabla^M \mathring{R}_q \right\|_{L^1} \leq \delta_{q+1} \lambda_q^M \tag{2.18}$$

$$\left\| \nabla^M \mathring{R}_{q,n,p} \right\|_{L^1} \leq \frac{\delta_{q+1}\lambda_q}{\lambda_{q,n,p-1}} \prod_{n'<n} f_{q,n'} \lambda_{q,n,p}^M = \delta_{q+1,n,p} \lambda_{q,n,p}^M . \tag{2.19}$$

The higher order stress $\mathring{R}_{q,n,p}$ is defined using the spatial Littlewood-Paley projection operator

$$\mathbb{P}_{[q,n,p]} := \mathbb{P}_{[\lambda_{q,n,p-1}, \lambda_{q,n,p})} = \mathbb{P}_{\geq \lambda_{q,n,p-1}} \mathbb{P}_{< \lambda_{q,n,p}},$$

which projects onto the frequencies from (2.10). We define $\mathring{R}_{q,n,p}$ as follows:

$$\mathring{R}_{q,n,p} := \sum_{n'<n} \sum_{p'=1}^{p_{\max}} \operatorname{div}^{-1} \left(\nabla \left(a_{n',p'}^2 (\mathring{R}_{q,n',p'}) \nabla \Phi_{q,k}^{-1} \otimes \nabla \Phi_{q,k}^{-T} \right) \right.$$

$$\left. : \left(\mathbb{P}_{[q,n,p]} \left(\mathbb{W}_{q+1,n'} \otimes \mathbb{W}_{q+1,n'} \right) \right) \left(\Phi_{q,k} \right) \right). \tag{2.20}$$

We pause here to point out an important consequence of this definition. Let n' be fixed, and consider the right side of the above equality. Then, due to the periodicity of $\mathbb{W}_{q+1,n'}$ at scale $(\lambda_{q+1} r_{q+1,n'})^{-1}$, we have[9]

$$\mathbb{W}_{q+1,n'} \otimes \mathbb{W}_{q+1,n'}$$
$$= \mathbb{P}_{=0} \left(\mathbb{W}_{q+1,n'} \otimes \mathbb{W}_{q+1,n'} \right) + \mathbb{P}_{\neq 0} \left(\mathbb{W}_{q+1,n'} \otimes \mathbb{W}_{q+1,n'} \right)$$
$$= \mathbb{P}_{=0} \left(\mathbb{W}_{q+1,n'} \otimes \mathbb{W}_{q+1,n'} \right) + \mathbb{P}_{\geq \lambda_{q+1} r_{q+1,n'}} \left(\mathbb{W}_{q+1,n'} \otimes \mathbb{W}_{q+1,n'} \right).$$

For $n' \geq 1$, we have that

$$\lambda_{q+1} r_{q+1,n'} = \lambda_q^{\left(\frac{4}{5}\right)^{n'+1}} \lambda_{q+1}^{1-\left(\frac{4}{5}\right)^{n'+1}} \gg \lambda_q^{\left(\frac{4}{5}\right)^{n'} \cdot \frac{5}{6}} \lambda_{q+1}^{1-\left(\frac{4}{5}\right)^{n'} \cdot \frac{5}{6}} = \lambda_{q,n'+1,0} = \lambda_{q,n',p_{\max}},$$

where $\lambda_{q,n'+1,0}$ is the minimum frequency of $\mathring{R}_{q,n'+1} = \sum_{p'=0}^{p_{\max}} \mathring{R}_{q,n'+1,p'}$, while for $n' = 0$ we have that

$$\lambda_{q+1} r_{q+1,0} = \lambda_{q,1} = \lambda_q^{\left(\frac{4}{5}\right)} \lambda_{q+1}^{1-\left(\frac{4}{5}\right)} = \lambda_{q,1,0},$$

which is the minimum frequency of $\mathring{R}_{q,1}$. Therefore, we have shown that the error terms arising from *all* non-zero modes of $\mathbb{W}_{q+1,n'} \otimes \mathbb{W}_{q+1,n'}$ are accounted for in the higher order stresses $\mathring{R}_{q,\tilde{n}}$ for $\tilde{n} > n'$. Thus, the higher order stresses

---

[9]We denote by $\mathbb{P}_{\neq 0}$ the operator which subtracts from a function its mean in space.

created by the interaction of $w_{q+1,n'}$ will be absorbed into higher order stresses with *strictly larger* values of n.

Now assuming that $\mathring{R}_{q,n',p'}$ and $w_{q+1,n',p'}$ are well-defined for all $n' < n$ and $1 \leq p' \leq p_{\max}$ and using the heuristic estimates from the previous section for $w_{q+1,n',p'}$, we can estimate the component of $\mathring{R}_{q,n,p}$ coming from $w_{q+1,n',p'}$ by recalling (2.20) and writing

$$\left\| \mathring{R}_{q,n,p} \right\|_{L^1} \leq \sum_{n' < n} \frac{\delta_{q+1,n',p'} \lambda_{q,n',p'}}{\lambda_{q,n,p-1}}$$

$$= \sum_{n' < n} \frac{\frac{\delta_{q+1} \lambda_q}{\lambda_{q,n',p'-1}} \prod_{n'' < n'} f_{q,n''} \lambda_{q,n',p'}}{\lambda_{q,n,p-1}}$$

$$\leq \sum_{n' < n} \frac{\delta_{q+1} \lambda_q}{\lambda_{q,n,p-1}} \prod_{n'' \leq n'} f_{q,n''}$$

$$\lesssim \frac{\delta_{q+1} \lambda_q}{\lambda_{q,n,p-1}} \prod_{n'' < n} f_{q,n''} = \delta_{q+1,n,p} .$$

The denominator comes from the gain induced by the combination of the inverse divergence and the Littlewood-Paley projector $\mathbb{P}_{[q,n,p]}$. The numerator is the amplitude of $\nabla |a_{n',p'} (\mathring{R}_{q,n',p'})|^2$, computed using the chain rule and the assumption (2.19) on $\nabla \mathring{R}_{q,n',p'}$. We have used the fact that the L^2 norm of $\mathbb{W}_{q+1,n'}$ is normalized to unit size. Any derivatives on $\mathring{R}_{q,n,p}$ will cost $\lambda_{q,n,p}$, which is the maximum frequency in the Littlewood-Paley projector $\mathbb{P}_{[q,n,p]}$. Thus, all terms which will land in $\mathring{R}_{q,n,p}$ will satisfy the correct estimates *given that $\mathring{R}_{q,n',p'}$ satisfies the correct estimates for $n' < n$ and $1 \leq p' \leq p_{\max}$. Since $\mathring{R}_q =: \mathring{R}_{q,0}$ satisfies the inductive assumptions, we can initiate this iteration at level $n = 0$ while satisfying (2.18).

Now that $\mathring{R}_{q,n,p}$ satisfies the appropriate estimates, we can correct it with a perturbation $w_{q+1,n,p}$ as described in the previous section. As before, since $\mathbb{W}_{q+1,n}$ has minimum frequency

$$\lambda_{q,n} = \lambda_{q+1} r_{q+1,n} = \lambda_q^{\left(\frac{4}{5}\right)^{n+1}} \lambda_{q+1}^{1-\left(\frac{4}{5}\right)^{n+1}} \gg \lambda_q^{\left(\frac{4}{5}\right)^n \cdot \frac{5}{6}} \lambda_{q+1}^{1-\left(\frac{4}{5}\right)^n \cdot \frac{5}{6}} = \lambda_{q,n+1,0} ,$$

and the minimum frequency in $\mathring{R}_{q,n+1}$ is $\lambda_{q,n+1,0}$, *every error term resulting from the self-interaction of $w_{q+1,n,p}$ will be absorbed into higher order stresses $\mathring{R}_{q,\tilde{n}}$ for $\tilde{n} > n$.* Therefore, we can induct on n to add a sequence of perturbations $w_{q+1,n} = \sum_{p=1}^{p_{\max}} w_{q+1,n,p}$ such that all nonlinear error terms are canceled by subsequent perturbations. Upon reaching $n_{\max}$ and recalling (2.1), we can estimate the final nonlinear error term by

$$\frac{\delta_{q+1} \lambda_q}{\lambda_{q+1} r_{q+1,n_{\max}}} \prod_{n' < n_{\max}} f_{q,n'} \leq \delta_{q+2}$$

$$\Longleftarrow \delta_{q+1} \left(\frac{\lambda_q}{\lambda_{q+1}}\right)^{1-\left(\frac{4}{5}\right)^{n_{\max}+1}-\frac{1}{p_{\max}}} \leq \delta_{q+2}$$

$$\Longleftrightarrow \lambda_{q+1}^{-2\beta} \lambda_{q+1}^{\left(\frac{1}{b}-1\right)\left(1-\left(\frac{4}{5}\right)^{n_{\max}+1}-\frac{1}{p_{\max}}\right)} \leq \lambda_{q+1}^{-2\beta b}$$

$$\Longleftrightarrow 2\beta b(b-1) \leq (b-1)\left(1-\left(\frac{4}{5}\right)^{n_{\max}+1}-\frac{1}{p_{\max}}\right)$$

$$\Longleftrightarrow \beta \leq \frac{1}{2b}\left(1-\left(\frac{4}{5}\right)^{n_{\max}+1}-\frac{1}{p_{\max}}\right).$$

Choosing b to be close to 1 and $n_{\max}$ and $p_{\max}$ sufficiently large shows that these error terms are commensurate with $\mathring{H}^{\frac{1}{2}-}$ regularity.

2.7.2 Type 2 oscillation error

We now consider the second type of oscillation error, which would arise as a result of two *distinct* pipes intersecting and thus serves no purpose in the cancellation of stresses. Beginning with $\mathring{R}_q = \mathring{R}_{q,0}$, we have that every derivative on $\mathring{R}_{q,0}$ costs λ_q. Therefore, we may decompose $\mathring{R}_{q,0}$ using a checkerboard partition of unity at scale λ_q^{-1}. Referring back to the discussion of the checkerboard cutoff functions, this sets the value of r_1 to be $\lambda_q \lambda_{q+1}^{-1}$. Now, suppose that on a single square of this checkerboard, we have placed a set of intermittent pipe flows $\mathbb{W}_{q+1,0}$ which are periodized to scale $(\lambda_{q+1} r_{q+1,0})^{-1}$. After flowing the pipes and the checkerboard square by v_q for a short length of time,[10] we must place a new set of pipes $\mathbb{W}'_{q+1,0}$ which are disjoint from the flowed pipes $\mathbb{W}_{q+1,0}$. Given the choice of r_1, this will be possible provided that

$$r_{q+1,0} = r_2 \ll r_1^{\frac{3}{4}}. \tag{2.21}$$

Thus, the *minimum* amount of intermittency needed to successfully place disjoint sets of intermittent pipes is $\left(\lambda_q \lambda_{q+1}^{-1}\right)^{\frac{3}{4}}$. Per Definition 2.3, our choice of $r_{q+1,0}$ is $\left(\lambda_q \lambda_{q+1}^{-1}\right)^{\frac{4}{5}}$, which is then sufficiently small.

Let us now assume that we have successfully corrected $\mathring{R}_{q,n'}$ for $n' < n$, and that we wish to correct $\mathring{R}_{q,n} = \sum_{p=1}^{p_{\max}} \mathring{R}_{q,n,p}$ with a perturbation $w_{q+1,n} = \sum_{p=1}^{p_{\max}} w_{q+1,n,p}$. First, we recall that

$$\left\|\nabla^M \mathring{R}_{q,n,p}\right\|_{L^1} \lesssim \delta_{q+1,n,p} \lambda_{q,n,p}^M.$$

Therefore, we can multiply $\mathring{R}_{q,n,p}$ by a checkerboard partition of unity at scale

---

[10]The length of time is equal to the local Lipschitz norm of v_q on the support of the cutoff $\psi_{i,q}$, given by the time cutoff hidden in $a_{n,p}$.

$\lambda_{q,n,0}^{-1} \gg \lambda_{q,n,p}^{-1}$ while preserving these bounds. We must choose values of r_1 and r_2, as in Section 2.5.2. Since for $n \geq 2$

$$\lambda_{q+1} r_1 = \lambda_{q,n,0} = \lambda_q^{\left(\frac{4}{5}\right)^{n-1} \cdot \frac{5}{6}} \lambda_{q+1}^{1 - \left(\frac{4}{5}\right)^{n-1} \cdot \frac{5}{6}} = \lambda_{q+1} \cdot \left(\frac{\lambda_q}{\lambda_{q+1}}\right)^{\left(\frac{4}{5}\right)^{n-1} \cdot \frac{5}{6}},$$

and for $n = 1$

$$\lambda_{q,1,0} = \lambda_q^{\frac{4}{5}} \lambda_{q+1}^{\frac{1}{5}} \gg \lambda_{q+1} \cdot \left(\frac{\lambda_q}{\lambda_{q+1}}\right)^{\left(\frac{4}{5}\right)^{1-1} \cdot \frac{5}{6}},$$

we have that for all $n \geq 1$

$$r_1 \geq \left(\frac{\lambda_q}{\lambda_{q+1}}\right)^{\left(\frac{4}{5}\right)^{n-1} \cdot \frac{5}{6}}.$$

Recall that $\mathring{R}_{q,n,p}$ will be corrected by $w_{q+1,n,p}$, which is constructed using intermittent pipe flows $\mathbb{W}_{q+1,n}$ with intermittency

$$r_{q+1,n} = \left(\frac{\lambda_q}{\lambda_{q+1}}\right)^{\left(\frac{4}{5}\right)^{n+1}} = r_2.$$

Thus in order to succeed in placing pipes $\mathbb{W}_{q+1,n}$ which avoid both previous generations of pipes, which are periodized to scales rougher than $\mathbb{W}_{q+1,n}$, and pipes from the same generation on overlapping cutoff functions, we must ensure that

$$r_2 \ll r_1^{\frac{3}{4}}$$

$$\iff \left(\frac{\lambda_q}{\lambda_{q+1}}\right)^{\left(\frac{4}{5}\right)^{n+1}} \ll \left(\frac{\lambda_q}{\lambda_{q+1}}\right)^{\left(\frac{4}{5}\right)^{n-1} \cdot \frac{5}{6} \cdot \frac{3}{4}}$$

$$\iff \left(\frac{4}{5}\right)^{n-1} \cdot \frac{5}{6} \cdot \frac{3}{4} < \left(\frac{4}{5}\right)^{n+1}$$

$$\iff \frac{1}{2} < \left(\frac{4}{5}\right)^3 = \frac{64}{125}.$$

So our choice of $r_{q+1,n}$ is sufficient to ensure that we can successfully place intermittent pipe flows when constructing $w_{q+1,n,p}$ which are disjoint from all other pipe flows from either previous generations ($n' < n$) or the same generation (the same value of n).

2.7.3 Nash and transport errors

The heuristic for the Nash and transport errors is that our choice of $r_{q+1,n}$ provides much more intermittency than is needed to ensure that linear errors arising from $w_{q+1,n,p}$ can be absorbed into $\mathring{R}_{q+1}$.[11] In other words, the Type 2 oscillation errors required much more intermittency than the Nash and transport errors will.

Let us start with the Nash error arising from the addition of $w_{q+1,0,1}$, which is designed to correct $\mathring{R}_q$. Using decoupling, the cost of a derivative on $\mathbb{W}_{q+1,0}$ being λ_{q+1} (so that inverting the divergence gains a factor of λ_{q+1}), the size of ∇v_q in L^2, and the L^1 size of $\mathbb{W}_{q+1,n}$ being $r_{q+1,0}$, the size of this error is

$$\frac{1}{\lambda_{q+1}}\delta_{q+1}^{1/2}\delta_q^{1/2}\lambda_q r_{q+1,0} = \frac{1}{\lambda_{q+1}}\delta_{q+1}^{1/2}\delta_q^{1/2}\lambda_q\left(\frac{\lambda_q}{\lambda_{q+1}}\right)^{\left(\frac{4}{5}\right)}.$$

This is (much) less than δ_{q+2} since

$$\frac{\delta_{q+1}^{1/2}\delta_q^{1/2}\lambda_q^{3/2}}{\lambda_{q+1}^{3/2}} \leq \delta_{q+2} \iff \lambda_{q+1}^{-\beta}\lambda_{q+1}^{-\frac{\beta}{b}}\lambda_{q+1}^{\frac{1}{b}\cdot\frac{3}{2}}\lambda_{q+1}^{-\frac{3}{2}} \leq \lambda_{q+1}^{-2\beta b}$$

$$\iff 2\beta b^2 - \beta b - \beta \leq (b-1)\cdot\frac{3}{2}$$

$$\iff \beta(2b+1)(b-1) \leq (b-1)\cdot\frac{3}{2}. \qquad (2.22)$$

Choosing b close to 1 will make this error commensurate with $\dot{H}^{\frac{1}{2}-}$ regularity.

Let us now estimate the Nash error arising from the addition of $w_{q+1,n,p}$ for $n \geq 2$, given by

$$\left\|\operatorname{div}^{-1}\left(\left(a_{n,p}\nabla\Phi_{q,k}^{-1}\mathbb{W}_{q+1,n}(\Phi_{q,k})\right)\cdot\nabla v_q\right)\right\|_{L^1}.$$

Using again decoupling, the cost of a derivative on $\mathbb{W}_{q+1,n}$ being λ_{q+1} (so that inverting the divergence gains a factor of λ_{q+1}), the size of ∇v_q in L^2, the L^1 size of $\mathbb{W}_{q+1,n}$ being $r_{q+1,n}$, and (2.1), we have that for $n \geq 2$, the size of this

---

[11]One may verify that in three dimensions, the minimum amount of intermittency needed to absorb the Nash and transport errors arising from $w_{q+1,0}$ into $\mathring{R}_{q+1}$ at regularity approaching $\dot{H}^{\frac{1}{2}}$ is $r_{q+1,0} = \lambda_q^{\frac{1}{2}}\lambda_{q+1}^{-\frac{1}{2}}$. In general, one can further verify that given errors supported at frequency $\lambda_q^{\alpha}\lambda_{q+1}^{1-\alpha}$, one could correct them using intermittent pipe flows with minimum frequency $\lambda_q^{\frac{\alpha}{2}}\lambda_{q+1}^{1-\frac{\alpha}{2}}$ while absorbing the resulting Nash and transport errors into $\mathring{R}_{q+1}$. One should compare this with (2.21), which shows that the placement technique requires more intermittency, which at level $n = 0$ corresponds to $\lambda_q^{\frac{3}{4}}\lambda_{q+1}^{-\frac{3}{4}}$.

error is

$$\frac{1}{\lambda_{q+1}} \cdot \delta_{q+1,n,p}^{\frac{1}{2}} r_{q+1,n} \cdot \delta_q^{\frac{1}{2}} \lambda_q$$

$$\leq \frac{1}{\lambda_{q+1}} \cdot \delta_{q+1,n,1}^{\frac{1}{2}} r_{q+1,n} \cdot \delta_q^{\frac{1}{2}} \lambda_q$$

$$= \frac{1}{\lambda_{q+1}} \left(\frac{\delta_{q+1} \lambda_q}{\lambda_{q,n,0}} \right)^{\frac{1}{2}} \left(\prod_{n' < n} f_{q,n'} \right)^{\frac{1}{2}} \left(\frac{\lambda_q}{\lambda_{q+1}} \right)^{\left(\frac{4}{5}\right)^{n+1}} \delta_q^{\frac{1}{2}} \lambda_q$$

$$\leq \frac{1}{\lambda_{q+1}} \left(\frac{\delta_{q+1} \lambda_q}{\lambda_q^{\left(\frac{4}{5}\right)^{n-1} \cdot \frac{5}{6}} \lambda_{q+1}^{1-\left(\frac{4}{5}\right)^{n-1} \cdot \frac{5}{6}}} \right)^{\frac{1}{2}} \left(\frac{\lambda_q}{\lambda_{q+1}} \right)^{\left(\frac{4}{5}\right)^{n+1} - \frac{1}{2p_{\max}}} \delta_q^{\frac{1}{2}} \lambda_q .$$

Since

$$\frac{1}{2p_{\max}} + \frac{1}{2} \cdot \frac{5}{6} \cdot \left(\frac{4}{5}\right)^{n-1} < \left(\frac{4}{5}\right)^{n+1}$$

independently of $n \geq 2$ if $p_{\max}$ is sufficiently large, the Nash error will be smaller than δ_{q+2} based on the preceding estimates. Furthermore, one may check that $\delta_{q+1,1,1}^{\frac{1}{2}} r_{q+1,1} < \delta_{q+1,2,1}^{\frac{1}{2}} r_{q+1,2}$, so that the Nash error arising from the addition of $w_{q+1,1,p}$ is also satisfactorily small for all p.

Now let us consider the transport error. The size of the transport error arising from the addition of $w_{q+1,n,p}$ is

$$\left\| \mathrm{div}^{-1} \left((D_{t,q} a_{n,p}) \nabla \Phi_{q,l}^{-1} \mathbb{W}_{q+1,n} \right) \right\|_{L^1} \leq \frac{1}{\lambda_{q+1}} \tau_q^{-1} \delta_{q+1,n,p}^{\frac{1}{2}} r_{q+1,n}$$

$$= \frac{1}{\lambda_{q+1}} \cdot \delta_{q+1,n,p}^{\frac{1}{2}} r_{q+1,n} \cdot \delta_q^{\frac{1}{2}} \lambda_q. \quad (2.23)$$

Thus, the transport error is the same size as the Nash error and is sufficiently small to be put into $\mathring{R}_{q+1}$.

Chapter Three

Inductive assumptions

While in Chapter 2 we have outlined in broad terms the main steps in the proof of Theorem 1.1, along with the heuristics for some of the choices we have made in our proof, starting with the current section, we work with precise estimates.

In Section 3.1 we introduce some of the notation used in the proof, such as the Euler-Reynolds system, the mollified velocity, velocity increments, material/directional derivatives, our notation for geometric upper bounds with two different bases, and our notation for $\|\cdot\|_{L^p}$.

In Section 3.2 we introduce the principal amplitude and frequency parameters used in proof (the precise definitions and the order of choosing these parameters is detailed in Section 9.1). Next, in Sections 3.2.1 and 3.2.2 we state the *primary inductive assumptions* for the velocity, velocity increments, and Reynolds stress. These primary estimates hold on the support of previous generation velocity cutoff functions, which are inductively assumed to satisfy a number of properties, listed in Section 3.2.3. Lastly, in Section 3.2.4 we list a number of bounds for the velocity increments and mollified velocities, which involve all possible combinations of space and material derivatives, up to a certain order. These bounds are technical in nature, and should be ignored at a first reading; their sole purpose is to allow us to bound commutators between D^n and $D_{t,q}^m$ for very high values of n and m.

In conclusion, in Section 3.4 we show that if we are able to propagate the previously stated inductive estimates from step q to step $q + 1$, for every $q \geq 0$, then Theorem 1.1 follows. At the end of the section we discuss the modifications to the proof which would be necessary in order to obtain other types of flexibility statements.

3.1 GENERAL NOTATIONS

As is standard in convex integration schemes for the Euler system [29], we introduce the Euler-Reynolds system for the unknowns $(v_q, \mathring{R}_q)$:

$$\partial_t v_q + \operatorname{div}(v_q \otimes v_q) + \nabla p_q = \operatorname{div} \mathring{R}_q \qquad (3.1a)$$

$$\operatorname{div} v_q = 0. \qquad (3.1b)$$

Here and throughout the book, the pressure p_q is uniquely defined by solving $\Delta p_q = \operatorname{div}\operatorname{div}(\mathring{R}_q - v_q \otimes v_q)$, with $\int_{\mathbb{T}^3} p_q dx = 0$.

In order to avoid the usual derivative loss issue in convex integration schemes, for $q \geq 0$ we use the space-time mollification operator defined in Section 9.4, equation (9.64), to smooth out the velocity field v_q as:

$$v_{\ell_q} := \mathcal{P}_{q,x,t} v_q . \tag{3.2}$$

In particular, we note that spatial mollification is performed at scale $\widetilde{\lambda}_q^{-1}$ (which is just slightly smaller than λ_q^{-1}), while temporal mollification is at scale $\widetilde{\tau}_{q-1}$ (which is a lot smaller than τ_{q-1}).

Next, for all $q \geq 1$, define

$$w_q := v_q - v_{\ell_{q-1}}, \qquad u_q := v_{\ell_q} - v_{\ell_{q-1}}. \tag{3.3}$$

For consistency of notation, define $w_0 = v_0$ and $u_0 = v_{\ell_0}$. Note that

$$u_q = \mathcal{P}_{q,x,t} w_q + (\mathcal{P}_{q,x,t} v_{\ell_{q-1}} - v_{\ell_{q-1}}) \tag{3.4}$$

so that we may morally think that $u_q = w_q + (\text{a small error term})$; the smallness of this error term will be ensured by choosing a mollifier with a large number of vanishing moments; cf. (9.62).

We use the following notation for the material derivative corresponding to the vector field v_{ℓ_q}:

$$D_{t,q} := \partial_t + v_{\ell_q} \cdot \nabla . \tag{3.5}$$

With this notation, we have that

$$D_{t,q} = D_{t,q-1} + u_q \cdot \nabla . \tag{3.6}$$

We also introduce the directional derivatives

$$D_q := u_q \cdot \nabla , \tag{3.7}$$

which allow us to transfer information between $D_{t,q-1}$ and $D_{t,q}$ via $D_{t,q} = D_{t,q-1} + D_q$.

Remark 3.1. If for a sequence of numbers $\{a_n\}_{n \geq 0}$ and for two parameters $0 < \lambda < \Lambda$ we have the bounds

$$a_n \leq \lambda^n, \quad \text{for all} \quad n \leq N_*$$
$$a_n \leq \lambda^{N_*} \Lambda^{n - N_*} \quad \text{for all} \quad n > N_*,$$

for some $N_* \geq 1$, we will abbreviate these bounds as

$$a_n \leq \mathcal{M}(n, N_*, \lambda, \Lambda) ,$$

where we define

$$\mathcal{M}(n, N_*, \lambda, \Lambda) := \lambda^{\min\{n, N_*\}} \Lambda^{\max\{n - N_*, 0\}} \tag{3.8}$$

for all $n \geq 0$. The first entry of $\mathcal{M}(\cdot, \cdot, \lambda, \Lambda)$ measures the index in the sequence (typically number of derivatives considered) and the second entry determines the index after which the base of the geometric bound changes from λ to Λ. This notation has the following consequence, which will be used throughout the book: if $1 \leq \lambda \leq \Lambda$, then

$$\mathcal{M}(a, N_*, \lambda, \Lambda) \mathcal{M}(b, N_*, \lambda, \Lambda) \leq \mathcal{M}(a + b, N_*, \lambda, \Lambda). \tag{3.9}$$

When either a or b are larger than N_* the above inequality creates a loss; for $a + b \leq N_*$, it is an equality.

Remark 3.2. Throughout this section, and the remainder of the book, in order to abbreviate notation we shall use the notation $\|f\|_{L^p}$ to denote $\|f\|_{L_t^\infty(L^p(\mathbb{T}^3))}$. That is, all L^p norms stand for L^p *norms in space, uniformly in time.* Similarly, when we wish to emphasize a set dependence of an L^p norm, we write $\|f\|_{L^p(\Omega)}$, for some space-time set $\Omega \subset \mathbb{R} \times \mathbb{T}^3$, to stand for $\|\mathbf{1}_\Omega f\|_{L_t^\infty(L^p(\mathbb{T}^3))}$.

3.2 INDUCTIVE ESTIMATES

The proof is based on propagating estimates for solutions $(v_q, \mathring{R}_q)$ of the Euler-Reynolds system (3.1), inductively for $q \geq 0$. In order to state these bounds, we first need to fix a number of parameters in terms of which these inductive estimates are stated. We start by picking a regularity exponent $\beta \in (1/3, 1/2)$, and a super-exponential rate parameter $b \in (1, 3/2)$ such that $2\beta b < 1$. In terms of this choice of β and b, a number of additional parameters $(n_{\max}, \ldots N_{\text{fin}})$ are fixed, whose precise definition is summarized for convenience in items (3)–(12) of Section 9.1. Note that at this point the parameter $a_*(\beta, b)$ from item (13) in Section 9.1 is not yet fixed. With this choice, we then introduce the fundamental q-dependent frequency and amplitude parameters from Section 9.2. We state here for convenience the main q-dependent parameters defined in (9.15), (9.17), (9.18), and (9.21):

$$\lambda_q = 2^{\lceil (b^q) \log_2 a \rceil} \approx \lambda_{q-1}^b, \qquad \delta_q = \lambda_1^{\beta(b+1)} \lambda_q^{-2\beta}, \tag{3.10a}$$

$$\tau_q^{-1} = \delta_q^{1/2} \lambda_q \Gamma_{q+1}^{c_0 + 11}, \qquad \Gamma_{q+1} = \left(\frac{\lambda_{q+1}}{\lambda_q} \right)^{\varepsilon_\Gamma} \approx \lambda_q^{(b-1)\varepsilon_\Gamma}, \tag{3.10b}$$

where the constant c_0 is defined by (9.6). The $\approx$ symbols in (3.10) mean that the left side of the $\approx$ symbol lies between two (universal) constant multiples of the right side (see e.g. (9.16)).

Remark 3.3. Throughout the subsequent sections, we will make frequent use of the symbol $\lesssim$. We emphasize that any implicit constants indicated by $\lesssim$ are allowed to depend only on the parameters defined in Section 9.1, items (1)–(12). The implicit constants in $\lesssim$ are, however, always independent of the parameters a and q, which appear in (3.10). This allows us at the end of the proof– cf. item (13) in Section 9.1– to choose $a_*(\beta, b)$ to be sufficiently large so that for all $a \geq a_*(\beta, b)$ and all $q \geq 0$, the parameter Γ_{q+1} appearing in (3.10) is larger than all the implicit constants in $\lesssim$ symbols encountered throughout the book. That is, upon choosing a_* sufficiently large, any inequality of the type $A \lesssim B$ which appears in this book may be rewritten as $A \leq \Gamma_{q+1} B$, for any $q \geq 0$.

In order to state the inductive assumptions we use four large integers, defined precisely in Section 9.1. For the moment it is important to note that these fixed parameters are independent of q and that they satisfy the ordering

$$1 \ll \mathsf{N}_{\mathrm{cut},t} \ll \mathsf{N}_{\mathrm{ind},t} \ll \mathsf{N}_{\mathrm{ind},v} \ll \mathsf{N}_{\mathrm{fin}} . \tag{3.11}$$

The precise definitions of these integers, and the meaning of the $\ll$ symbols in (3.11), are given in (9.9), (9.10), (9.11), and (9.14). Roughly speaking, the role of these parameters is as follows:

- $\mathsf{N}_{\mathrm{cut},t}$ is the number of sharp material derivatives which are built into the velocity and stress cutoff functions.
- $\mathsf{N}_{\mathrm{ind},t}$ is the number of sharp material derivatives propagated for velocities and stresses.
- $\mathsf{N}_{\mathrm{ind},v}$ is used to quantify the number of (lossy) higher order space and time derivatives for velocities and stresses.
- $\mathsf{N}_{\mathrm{fin}}$ is used to quantify the highest order derivatives appearing in the proof.

Next, we state the inductive assumptions for the velocity increments and stresses at various levels $q \geq 0$. Throughout the section we frequently refer to the notation $\mathcal{M}(n, N_*, \lambda, \Lambda)$ from (3.8).

3.2.1 Primary inductive assumption for velocity increments

We make L^2 inductive assumptions for $u_{q'} = v_{\ell_{q'}} - v_{\ell_{q'-1}}$ at levels q' strictly below q. For all $0 \leq q' \leq q - 1$ we assume that

$$\left\| \psi_{i,q'-1} D^n D_{t,q'-1}^m u_{q'} \right\|_{L^2}$$
$$\leq \delta_{q'}^{1/2} \mathcal{M}\left(n, 2\mathsf{N}_{\mathrm{ind},v}, \lambda_{q'}, \widetilde{\lambda}_{q'}\right) \mathcal{M}\left(m, \mathsf{N}_{\mathrm{ind},t}, \Gamma_{q'}^i \tau_{q'-1}^{-1}, \widetilde{\tau}_{q'-1}^{-1}\right) \tag{3.12}$$

holds for all $n + m \leq \mathsf{N}_{\mathrm{fin}}$.

At level q, we assume that the velocity increment w_q satisfies

$$\left\| \psi_{i,q-1} D^n D_{t,q-1}^m w_q \right\|_{L^2} \leq \Gamma_q^{-1} \delta_q^{1/2} \lambda_q^n \mathcal{M}\left(m, \mathsf{N}_{\mathrm{ind},t}, \Gamma_q^{i-1} \tau_{q-1}^{-1}, \Gamma_q^{-1} \widetilde{\tau}_{q-1}^{-1}\right) \tag{3.13}$$

for $n, m \leq 7\mathsf{N}_{\mathrm{ind},v}$. Moreover, recalling from (9.67) that $\mathrm{supp}_t f$ denotes the temporal support of a function f, we inductively assume that

$$\mathrm{supp}_t(\mathring{R}_{q-1}) \subset [T_1, T_2]$$
$$\Rightarrow \quad \mathrm{supp}_t(w_q) \subset \left[T_1 - (\lambda_{q-1}\delta_{q-1}^{1/2})^{-1}, T_2 + (\lambda_{q-1}\delta_{q-1}^{1/2})^{-1}\right]. \quad (3.14)$$

3.2.2 Inductive assumption for the stress

For the Reynolds stress $\mathring{R}_q$, we make L^1 inductive assumptions

$$\left\| \psi_{i,q-1} D^n D_{t,q-1}^m \mathring{R}_q \right\|_{L^1} \leq \Gamma_q^{-\mathsf{C}_{\mathsf{R}}} \delta_{q+1} \lambda_q^n \mathcal{M}\left(m, \mathsf{N}_{\mathrm{ind},t}, \Gamma_q^{i+1}\tau_{q-1}^{-1}, \Gamma_q^{-1}\widetilde{\tau}_{q-1}^{-1}\right) \tag{3.15}$$

for all $0 \leq n, m \leq 3\mathsf{N}_{\mathrm{ind},v}$.

3.2.3 Inductive assumption for the previous generation velocity cutoff functions

More assumptions are needed in relation to the previous velocity perturbations and old cutoff functions. First, we assume that the velocity cutoff functions form a partition of unity for $q' \leq q - 1$:

$$\sum_{i \geq 0} \psi_{i,q'}^2 \equiv 1, \quad \text{and} \quad \psi_{i,q'}\psi_{i',q'} = 0 \quad \text{for} \quad |i - i'| \geq 2. \tag{3.16}$$

Second, we assume that there exists an $i_{\max} = i_{\max}(q) > 0$, which is bounded uniformly in q as

$$\frac{b+1}{b-1} \leq i_{\max}(q) \leq \frac{4}{\varepsilon_\Gamma(b-1)}, \tag{3.17}$$

such that

$$\psi_{i,q'} \equiv 0 \quad \text{for all} \quad i > i_{\max}(q'), \quad \text{and} \quad \Gamma_{q'+1}^{i_{\max}(q')} \leq \lambda_{q'}^{5/3}, \tag{3.18}$$

for all $q' \leq q - 1$. For all $0 \leq q' \leq q - 1$ and $0 \leq i \leq i_{\max}$ we assume the following pointwise derivative bounds for the cutoff functions $\psi_{i,q'}$. For mixed space and material derivatives (recall the notation from (3.5)) we assume that

$$\frac{\mathbf{1}_{\mathrm{supp}\,\psi_{i,q'}}}{\psi_{i,q'}^{1-(K+M)/\mathsf{N}_{\mathrm{fin}}}} \left| \left(\prod_{l=1}^k D^{\alpha_l} D_{t,q'-1}^{\beta_l} \right) \psi_{i,q'} \right|$$
$$\lesssim \mathcal{M}\left(K, \mathsf{N}_{\mathrm{ind},v}, \Gamma_{q'}\lambda_{q'}, \Gamma_{q'}\widetilde{\lambda}_{q'}\right) \mathcal{M}\left(M, \mathsf{N}_{\mathrm{ind},t} - \mathsf{N}_{\mathrm{cut},t}, \Gamma_{q'+1}^{i+3}\tau_{q'-1}^{-1}, \Gamma_{q'+1}^{-1}\widetilde{\tau}_{q'}^{-1}\right) \tag{3.19}$$

for $K, M, k \geq 0$ with $0 \leq K + M \leq N_{fin}$, where $\alpha, \beta \in \mathbb{N}^k$ are such that $|\alpha| = K$ and $|\beta| = M$. Lastly, we consider mixtures of space, material, and directional derivatives (recall the notation from (3.7)). Then with K, M, α, β and k as above, and with $N \geq 0$, we assume that

$$
\frac{1_{\operatorname{supp} \psi_{i,q'}}}{\psi_{i,q'}^{1-(N+K+M)/N_{fin}}} \left| D^N \left(\prod_{l=1}^{k} D_{q'}^{\alpha_l} D_{t,q'-1}^{\beta_l} \right) \psi_{i,q'} \right|
$$
$$
\lesssim \mathcal{M} \left(N, N_{ind,v}, \Gamma_{q'} \lambda_{q'}, \Gamma_{q'} \widetilde{\lambda}_{q'} \right) (\Gamma_{q'+1}^{i-c_0} \tau_{q'}^{-1})^K
$$
$$
\times \mathcal{M} \left(M, N_{ind,t} - N_{cut,t}, \Gamma_{q'+1}^{i+3} \tau_{q'-1}^{-1}, \Gamma_{q'+1}^{-1} \widetilde{\tau}_{q'}^{-1} \right) \tag{3.20}
$$

as long as $0 \leq N + K + M \leq N_{fin}$.

In addition to the above pointwise estimates for the cutoff functions $\psi_{i,q'}$, we also assume that we have a good L^1 control. More precisely, we postulate that

$$
\|\psi_{i,q'}\|_{L^1} \lesssim \Gamma_{q'+1}^{-2i+C_b} \qquad \text{where} \qquad C_b = \frac{4+b}{b-1} \tag{3.21}
$$

holds for $0 \leq q' \leq q - 1$ and all $0 \leq i \leq i_{max}(q')$.

3.2.4   Secondary inductive assumptions for velocities

Next, for $0 \leq q' \leq q - 1$, $0 \leq i \leq i_{max}$, $k \geq 1$, $K, M \geq 0$, and $\alpha, \beta \in \mathbb{N}^k$ with $|\alpha| = K$ and $|\beta| = M$, we assume that the following mixed space and material derivative bounds hold:

$$
\left\| \left(\prod_{i=1}^{k} D^{\alpha_i} D_{t,q'-1}^{\beta_i} \right) u_{q'} \right\|_{L^\infty(\operatorname{supp} \psi_{i,q'})}
$$
$$
\lesssim (\Gamma_{q'+1}^{i+1} \delta_{q'}^{1/2}) \mathcal{M} \left(K, 2N_{ind,v}, \Gamma_{q'} \lambda_{q'}, \widetilde{\lambda}_{q'} \right) \mathcal{M} \left(M, N_{ind,t}, \Gamma_{q'+1}^{i+3} \tau_{q'-1}^{-1}, \Gamma_{q'+1}^{-1} \widetilde{\tau}_{q'}^{-1} \right) \tag{3.22}
$$

for $K + M \leq 3N_{fin}/2 + 1$,

$$
\left\| \left(\prod_{i=1}^{k} D^{\alpha_i} D_{t,q'}^{\beta_i} \right) Dv_{\ell_{q'}} \right\|_{L^\infty(\operatorname{supp} \psi_{i,q'})}
$$
$$
\lesssim (\Gamma_{q'+1}^{i+1} \delta_{q'}^{1/2} \widetilde{\lambda}_{q'}) \mathcal{M} \left(K, 2N_{ind,v}, \Gamma_{q'} \lambda_{q'}, \widetilde{\lambda}_{q'} \right) \mathcal{M} \left(M, N_{ind,t}, \Gamma_{q'+1}^{i-c_0} \tau_{q'}^{-1}, \Gamma_{q'+1}^{-1} \widetilde{\tau}_{q'}^{-1} \right) \tag{3.23}
$$

for $K + M \leq 3N_{fin}/2$, and

$$\left\| \left(\prod_{i=1}^{k} D^{\alpha_i} D_{t,q'}^{\beta_i} \right) v_{\ell_{q'}} \right\|_{L^\infty(\mathrm{supp}\,\psi_{i,q'})}$$
$$\lesssim (\Gamma_{q'+1}^{i+1} \delta_{q'}^{1/2} \lambda_{q'}^2) \mathcal{M} \left(K, 2N_{ind,v}, \Gamma_{q'} \lambda_{q'}, \tilde{\lambda}_{q'} \right) \mathcal{M} \left(M, N_{ind,t}, \Gamma_{q'+1}^{i-c_0} \tau_{q'}^{-1}, \Gamma_{q'+1}^{-1} \tilde{\tau}_{q'}^{-1} \right)$$
$$(3.24)$$

for $K + M \leq 3N_{fin}/2 + 1$. Additionally, for $N \geq 0$ we postulate that mixed space, material, and directional derivatives satisfy

$$\left\| D^N \left(\prod_{i=1}^{k} D_{q'}^{\alpha_i} D_{t,q'-1}^{\beta_i} \right) u_{q'} \right\|_{L^\infty(\mathrm{supp}\,\psi_{i,q'})}$$
$$\lesssim (\Gamma_{q'+1}^{i+1} \delta_{q'}^{1/2})^{K+1} \mathcal{M} \left(N + K, 2N_{ind,v}, \Gamma_{q'} \lambda_{q'}, \tilde{\lambda}_{q'} \right)$$
$$\times \mathcal{M} \left(M, N_{ind,t}, \Gamma_{q'+1}^{i+3} \tau_{q'-1}^{-1}, \Gamma_{q'+1}^{-1} \tilde{\tau}_{q'}^{-1} \right) \qquad (3.25a)$$
$$\lesssim (\Gamma_{q'+1}^{i+1} \delta_{q'}^{1/2}) \mathcal{M} \left(N, 2N_{ind,v}, \Gamma_{q'} \lambda_{q'}, \tilde{\lambda}_{q'} \right) (\Gamma_{q'+1}^{i-c_0} \tau_{q'}^{-1})^K$$
$$\times \mathcal{M} \left(M, N_{ind,t}, \Gamma_{q'+1}^{i+3} \tau_{q'-1}^{-1}, \Gamma_{q'+1}^{-1} \tilde{\tau}_{q'}^{-1} \right) \qquad (3.25b)$$

whenever $N + K + M \leq 3N_{fin}/2 + 1$.

Remark 3.4. Identity (A.39) shows that (3.25b) automatically implies the bound

$$\left\| D^N D_{t,q'}^M u_{q'} \right\|_{L^\infty(\mathrm{supp}\,\psi_{i,q'})}$$
$$\lesssim (\Gamma_{q'+1}^{i+1} \delta_{q'}^{1/2}) \mathcal{M} \left(N, 2N_{ind,v}, \Gamma_{q'} \lambda_{q'}, \tilde{\lambda}_{q'} \right) \mathcal{M} \left(M, N_{ind,t}, \Gamma_{q'+1}^{i-c_0} \tau_{q'}^{-1}, \Gamma_{q'+1}^{-1} \tilde{\tau}_{q'}^{-1} \right)$$
$$(3.26)$$

for all $N + M \leq 3N_{fin}/2 + 1$. To see this, we take $B = D_{t,q'-1}$ and $A = D_{q'}$, so that $A + B = D_{t,q'}$. The estimate (3.26) now is a consequence of identity (A.39) and the parameter inequalities $\Gamma_{q'+1}^{c_0+3} \tau_{q'-1}^{-1} \leq \tau_{q'}^{-1}$ (which follows from (9.40)) and $\Gamma_{q'+1}^{i-c_0+1} \tau_{q'}^{-1} \leq \tilde{\tau}_{q'}^{-1}$ (which is a consequence of (3.18) and (9.43)). In a similar fashion, the bound (3.20) and identity (A.39) imply that

$$\frac{\mathbf{1}_{\mathrm{supp}\,\psi_{i,q'}}}{\psi_{i,q'}^{1-(N+M)/N_{fin}}} \left| D^N D_{t,q'}^M \psi_{i,q'} \right|$$
$$\lesssim \mathcal{M} \left(N, N_{ind,v}, \Gamma_{q'} \lambda_{q'}, \Gamma_{q'} \tilde{\lambda}_{q'} \right) \mathcal{M} \left(M, N_{ind,t} - N_{cut,t}, \Gamma_{q'+1}^{i-c_0} \tau_{q'}^{-1}, \Gamma_{q'+1}^{-1} \tilde{\tau}_{q'}^{-1} \right)$$
$$(3.27)$$

for all $N + M \leq N_{fin}$. Indeed, the above estimates follow from the same parameter inequalities mentioned above, and from identity (A.39) with $A = D_{q'}$ and

$B = D_{t,q'-1}$.

Remark 3.5. The inductive assumptions for the velocities given in Sections 3.2.1 and 3.2.4, with the definition of the mollifier operator $\mathcal{P}_{q,x,t}$ in Section 9.4, imply that the new velocity field $v_q = w_q + v_{\ell_{q-1}}$ is very close to its mollification v_{ℓ_q}, uniformly in space and time. That is, we have

$$\left\| D^n D_{t,q-1}^m (v_{\ell_q} - v_q) \right\|_{L^\infty} \leq \lambda_q^{-2} \delta_q^{1/2} \mathcal{M} \left(n, 2\mathsf{N}_{\text{ind},v}, \lambda_q, \widetilde{\lambda}_q \right)$$
$$\times \mathcal{M} \left(m, \mathsf{N}_{\text{ind},t}, \tau_{q-1}^{-1} \Gamma_q^{i-1}, \widetilde{\tau}_{q-1}^{-1} \Gamma_q^{-1} \right) \quad (3.28)$$

for all $n, m \leq 3\mathsf{N}_{\text{ind},v}$. The proof of the above bound is given in Lemma 5.1; cf. estimate (5.4).

3.3   MAIN INDUCTIVE PROPOSITION

The main inductive proposition, which propagates the inductive estimates in Section 3.2 from step q to step $q + 1$, is as follows.

Proposition 3.6. *Fix $\beta \in [1/3, 1/2)$ and choose $b \in (1, 1/2\beta)$. Solely in terms of β and b, define the parameters $n_{\max}$, C_b, C_R, c_0, ε_Γ, α_R, $\mathsf{N}_{\text{cut},t}$, $\mathsf{N}_{\text{cut},x}$, $\mathsf{N}_{\text{ind},t}$, $\mathsf{N}_{\text{ind},v}$, N_{dec}, d, and N_{fin} by the definitions in Section 9.1, items (1)–(12). Then, there exists a sufficiently large $a_* = a_*(\beta, b) \geq 1$, such that for any $a \geq a_*$, the following statement holds for any $q \geq 0$. Given a velocity field v_q which solves the Euler-Reynolds system with stress $\mathring{R}_q$, define v_{ℓ_q}, w_q, and u_q via (3.2)–(3.3). Assume that $\{u_{q'}\}_{q'=0}^{q-1}$ satisfies (3.12), w_q obeys (3.13)–(3.14), $\mathring{R}_q$ satisfies (3.15), and that for every $q' \leq q-1$ there exists a partition of unity $\{\psi_{i,q'}\}_{i \geq 0}$ such that properties (3.16)–(3.18) and estimates (3.19)–(3.25) hold. Then, there exists a velocity field v_{q+1}, a stress $\mathring{R}_{q+1}$, and a partition of unity $\{\psi_{i,q}\}_{q \geq 0}$, such that v_{q+1} solves the Euler-Reynolds system with stress $\mathring{R}_{q+1}$, u_q satisfies (3.12) for $q' \mapsto q$, w_{q+1} obeys (3.13)–(3.14) for $q \mapsto q+1$, $\mathring{R}_{q+1}$ satisfies (3.15) for $q \mapsto q+1$, and the $\psi_{i,q}$ are such that (3.16)–(3.25) hold when $q' \mapsto q$.*

3.4   PROOF OF THEOREM 1.1

Choose the parameters $\beta, b, \ldots, a_*$, as described in Section 9.1, and assume that with these parameter choices, and for *any* $a \geq a_*$, we are able to propagate the inductive bounds claimed in Sections 3.2.1–3.2.4 from step q to step $q + 1$, for all $q \geq 0$; this is achieved in Sections 6–8. We next show that if $a \geq a_*$ is chosen sufficiently large, depending additionally on the $v_{\text{start}}, v_{\text{end}}, T > 0$, and $\epsilon > 0$ from the statement of Theorem 1.1, then the inductive assumptions imply Theorem 1.1.

Without loss of generality, assume that $\int_{\mathbb{T}^3} v_{\text{start}}(x)dx = \int_{\mathbb{T}^3} v_{\text{end}}(x)dx = 0$. Since these functions lie in $L^2(\mathbb{T}^3)$, there exists $R > 0$ such that upon defining

$$v_0^{(1)} := \mathbb{P}_{\leq R} v_{\text{start}}, \qquad \text{and} \qquad v_0^{(2)} := \mathbb{P}_{\leq R} v_{\text{end}},$$

where $\mathbb{P}_{\leq R}$ denotes the Fourier truncation operator to frequencies $|\xi| \leq R$, we have that

$$\|v_0^{(1)} - v_{\text{start}}\|_{L^2(\mathbb{T}^3)} + \|v_0^{(2)} - v_{\text{end}}\|_{L^2(\mathbb{T}^3)} \leq \frac{\epsilon}{2}. \tag{3.29}$$

Note that $v_0^{(1)}, v_0^{(2)} \in C^\infty(\mathbb{T}^3)$, and thus by the classical local well-posedness theory plus propagation of regularity (see Foias, Frisch, and Temam [38]), there exists $T_0 > 0$ and unique strong solutions $v^{(1)} \in C^\infty((-T_0, T_0) \times \mathbb{T}^3)$ and $v^{(2)} \in C^\infty((T - T_0, T + T_0) \times \mathbb{T}^3)$ of the 3D Euler system (1.1), such that $v^{(1)}(x, 0) = v_0^{(1)}(x)$ and $v^{(2)}(x, T) = v_0^{(2)}(x)$. Without loss of generality, we may take $T_0 \leq T/2$.

Next, let $\varphi \colon [0, T] \to [0, 1]$ be a non-increasing C^∞ smooth function such that $\varphi \equiv 1$ on $[0, T_0/2]$ and $\varphi \equiv 0$ on $[T_0, T]$. Define the C^∞-smooth function

$$v_0(x, t) := \varphi(t)v^{(1)}(x, t) + \varphi(T - t)v^{(2)}(x, t). \tag{3.30}$$

On $[0, T]$, v_0 solves the Euler-Reynolds system (3.1) for a suitable zero mean scalar pressure p_0, with the C^∞-smooth stress $\mathring{R}_0$ defined by

$$\begin{aligned} \mathring{R}_0(x, t) := {} & (\partial_t\varphi)(t)\mathcal{R}v^{(1)}(x, t) - (\partial_t\varphi)(T - t)\mathcal{R}v^{(2)}(x, t) \\ & + \varphi(t)(\varphi(t) - 1)(v^{(1)} \mathring{\otimes} v^{(1)})(x, t) \\ & + \varphi(T - t)(\varphi(T - t) - 1)(v^{(2)} \mathring{\otimes} v^{(2)})(x, t), \end{aligned} \tag{3.31}$$

where $\mathcal{R}$ is the classical nonlocal inverse divergence operator (see (A.100) for the definition). From the above definition and the fact that $\varphi \equiv 1$ on $[0, T_0/2]$, we deduce that

$$\text{supp}_t(\mathring{R}_0) \subset [T_0/2, T - T_0/2]. \tag{3.32}$$

This fact will be needed towards the end of the proof.

For consistency of notation, we also define $v_{-1} = v_{\ell_{-1}} = u_{-1} = 0$, so that $v_0 = w_0$ holds by (3.3). For the velocity cutoffs, we let $\psi_{0,-1} = 1$ and $\psi_{i,-1} = 0$ for all $i \geq 1$. It is then immediate to check that the $\{\psi_{i,-1}\}_{i \geq 0}$ satisfy the inductive assumptions (3.16)–(3.21), for $q' = -1$, with the derivative bounds (3.19) and (3.20) being empty statements respectively for when $K + M \geq 1$ and $N + K + M \geq 1$. Moreover, the bounds (3.12) and (3.22)–(3.25b) hold for $q' = -1$ since the left side of these inequalities vanishes identically. Lastly, the assumption (3.14) is empty since there is no $\mathring{R}_{-1}$ stress to speak of.

It thus remains to verify that the pair $(v_0, \mathring{R}_0)$ defined in (3.30)–(3.31) satisfies the estimates (3.13) and (3.15), where by the above choices we have $D_{t,-1} = \partial_t$. Note that the parameter $\mathsf{N}_{\mathrm{ind},v}$ was already chosen; thus, we have that

$$
C_{\mathrm{datum}} := \max_{0 \leq n, m \leq 7\mathsf{N}_{\mathrm{ind},v}} \left\| D^n \partial_t^m v_0 \right\|_{L^\infty(0,T;L^2(\mathbb{T}^3))}
$$

$$
+ \max_{0 \leq n, m \leq 3\mathsf{N}_{\mathrm{ind},v}} \left\| D^n \partial_t^m \mathring{R}_0 \right\|_{L^\infty(0,T;L^1(\mathbb{T}^3))} < \infty. \tag{3.33}
$$

Note that C_{datum} only depends on v_{start}, v_{end}, the cutoff frequency $R > 0$, the choice of the cutoff function φ, $T > 0$, and the parameter $\mathsf{N}_{\mathrm{ind},v}$. In particular, C_{datum} does not depend on the parameter a, which is the base of the exponential defining λ_q in (3.10). Defining $\tau_{-1} = \Gamma_0^{-1} = \lambda_0^{-\varepsilon_\Gamma}$ and $\tilde{\tau}_{-1} = \Gamma_0^{-3} = \lambda_0^{-3\varepsilon_\Gamma}$ (these parameters are never used again) and using that $\lambda_0 \geq a \geq a_* \geq 1$, we thus have that (3.13) and (3.15) hold if we ensure that

$$
C_{\mathrm{datum}} \leq \Gamma_0^{-1} \delta_0^{1/2} \quad \text{and} \quad C_{\mathrm{datum}} \leq \Gamma_0^{-C_R} \delta_1. \tag{3.34}
$$

Using the fact that ε_Γ is sufficiently small with respect to β and b, we have that $\Gamma_0^{-1} \delta_0^{1/2} = \lambda_0^{-\varepsilon_\Gamma} \lambda_1^{(b+1)\beta/2} \lambda_0^{-\beta} \geq (\lambda_1 \lambda_0^{-1})^{(b+1)\beta/2} \geq (a^{b-1}/2)^\beta$. Also, by using the fact that ε_Γ is chosen to be sufficiently small with respect to β and b, we have that $\Gamma_0^{-C_R} \delta_1 = \lambda_0^{(4b+1)\varepsilon_\Gamma} \lambda_1^{(b-1)\beta} \geq (\lambda_1 \lambda_0^{-1})^{(b-1)\beta} \geq (a^{b-1}/2)^{(b-1)\beta}$. Thus, if in addition to $a \geq a_*$, as specified by item (13) in Section 9.1, we choose $a \geq a_*$ to be sufficiently large in terms of β, b and the constant C_{datum} from (3.33) in order to ensure that

$$
a^{(b-1)^2\beta} \geq 4 C_{\mathrm{datum}},
$$

then the condition (3.34) is satisfied. We make this choice of a, and thus all the estimates claimed in Sections 3.2.1–3.2.4 hold true for the base step in the induction, the case $q = 0$.

Proceeding inductively, these estimates thus hold true for all $q \geq 0$. This allows us to define a function $v \in C^0(0,T;H^{\beta'}(\mathbb{T}^3))$ for any $\beta' < \beta$ via the absolutely convergent series[1]

$$
v = \lim_{q \to \infty} v_q = v_0 + \sum_{q \geq 0} (v_{q+1} - v_q) = v_0 + \sum_{q \geq 0} \left(w_{q+1} + (v_{\ell_q} - v_q) \right), \tag{3.35}
$$

where we recall the notation (3.2) and (3.3). Indeed, by (3.13), (3.16), and interpolation, we have that $\|w_q\|_{H^{\beta'}} \leq 2\Gamma_q^{-1} \delta_q^{1/2} \lambda_q^{\beta'} = 2\Gamma_q^{-1} \lambda_1^{(b+1)\beta/2} \lambda_q^{-(\beta-\beta')}$, which is summable for $q \geq 0$ whenever $\beta' < \beta$. By appealing to the bound (3.28), we furthermore obtain that $\|v_{\ell_q} - v_q\|_{H^{\beta'}} \lesssim \lambda_q^{-2} \delta_q^{1/2} \lambda_q^{\beta'} \lesssim \lambda_1^{(b+1)\beta/2} \lambda_q^{-2-(\beta-\beta')}$, which is again summable over $q \geq 0$. This justifies the definition of v in (3.35),

---

[1] We may equivalently define $v = \lim_{q \to \infty} v_q = \lim_{q \to \infty} w_q + \sum_{q'=0}^{q-1} u_{q'} = \sum_{q' \geq 0} u_{q'}$. We choose to work with (3.35) because it highlights the dependence on v_0.

and the fact that $v \in C^0(0, T; H^{\beta'}(\mathbb{T}^3))$ for any $\beta' < \beta$. Finally, we note that by additionally appealing to (3.15), which yields $\|\mathring{R}_q\|_{L^1} \lesssim \Gamma_q^{-C_R} \delta_{q+1} \to 0$ as $q \to \infty$, in view of (3.1) the function v defined in (3.35) is a weak solution of the Euler equations on $[0, T]$.

In order to complete the proof, we return to (3.35) and note that due to (3.14) (with $q = 1$), the property (3.32) of $\mathring{R}_0$, and the fact that $\lambda_0 \delta_0^{1/2} = \lambda_0^{1-\beta} \lambda_1^{(b+1)\beta/2} \geq 4/T_0$ (which holds upon choosing a sufficiently large with respect to T_0, β, b), we have that $w_1 \equiv 0$ on the set $[0, T_0/4] \times \mathbb{T}^3 \cup [T - T_0/4, T] \times \mathbb{T}^3$. Thus, from (3.35) and the previously established bounds for w_q (via (3.13), (3.16)) and $v_{\ell_q} - v_q$ (via (3.28)), we have that

$$
\begin{aligned}
&\|v - v_0\|_{L^\infty([0,T_0/4]\cup[T-T_0/4,T];L^2(\mathbb{T}^3))} \\
&\leq \sum_{q\geq 2} \|w_q\|_{L^\infty([0,T];L^2(\mathbb{T}^3))} + \sum_{q\geq 0} \|v_{\ell_q} - v_q\|_{L^\infty([0,T];L^2(\mathbb{T}^3))} \\
&\leq 2\lambda_1^{(b+1)\beta/2} \sum_{q\geq 2} \Gamma_q^{-1} \lambda_q^{-\beta} + \lambda_1^{(b+1)\beta} \sum_{q\geq 0} \lambda_q^{-2-\beta} \\
&\leq 4\lambda_1^{(b+1)\beta/2} \Gamma_2^{-1} \lambda_2^{-\beta} + 2\lambda_1^{(b+1)\beta} \lambda_0^{-2-\beta} \\
&\leq 8\Gamma_2^{-1} \lambda_1^{(b+1)\beta/2} \lambda_1^{-\beta b} + 4\lambda_0^{(b+1)\beta} \lambda_0^{-2-\beta} \\
&\leq \lambda_1^{-(b-1)\beta/2} + 4\lambda_0^{-1/2} \\
&\leq \frac{\epsilon}{2}
\end{aligned}
\tag{3.36}
$$

once a (and thus λ_0 and λ_1) is taken to be sufficiently large with respect to b, β, and ϵ. Here, in the second to last inequality we have used that $\beta(b^2 + b - 1) \leq 3/2$, which holds since $\beta < 1/2$ and $b < 3/2$. Combining (3.36) with the definition of the functions $v^{(1)}$, $v^{(2)}$, and v_0, and the bound (3.29), we deduce that $\|v(\cdot, 0) - v_{\text{start}}\|_{L^2(\mathbb{T}^3)} \leq \epsilon$ and $\|v(\cdot, T) - v_{\text{end}}\|_{L^2(\mathbb{T}^3)} \leq \epsilon$. This concludes the proof of Theorem 1.1, with β being replaced by an arbitrary $\beta' \in (0, \beta)$.

Remark 3.7. The proof outlined above may be easily modified to show the existence of infinitely many weak solutions in $C_t^0 H_x^{1/2-}$ which are nontrivial and have compact support in time, as mentioned in Remark 1.2. The argument is as follows. Let $\varphi(t)$ be a C^∞ smooth cutoff function, with $\varphi \equiv 1$ on $-[T/4, T/4]$ and $\varphi \equiv 0$ on $\mathbb{R} \setminus [-T/2, T/2]$. Then, instead of (3.30), we define define $v_0(x, t) = E\varphi(t)(\sin(x_3), 0, 0)$. Note that the kinetic energy of v_0 at time $t = 0$ is larger than $E(2\pi)^{3/2}/2 \geq 2E$, and that v_0 has time support in $[-T/2, T/2]$. Since $(\sin(x_3), 0, 0)$ is a shear flow, the zero order stress $\mathring{R}_0$ is given by $E\varphi'(t)$ multiplied by a matrix whose entries are zero, except for the $(1,3)$ and $(3,1)$ entries, which equal $-\cos(x_3)$ (see [12, Section 5.2] for details). The point is that $\mathring{R}_0$ is smooth, and its time support lies in the interval $T/4 \leq |t| \leq T/2$, which plays the role of (3.32). Using the same argument used in the proof of Theorem 1.1, we may show that for a sufficiently large, the above defined pair $(v_0, \mathring{R}_0)$ satisfies the inductive assumptions at level $q = 0$, and that these inductive assumptions may be propagated to all $q \geq 0$. As in (3.36), we deduce that the limiting weak solution solution v has kinetic energy at time $t = 0$ which is strictly larger than E. The fact that $\operatorname{supp}_t v_0, \operatorname{supp}_t \mathring{R}_0 \subset [-T/2, T/2]$, combined with the inductive assumption (3.14) and the fact that the mollification procedure in Lemma 5.1 expands time supports by at most a factor of $\widetilde{\tau}_{q-1} \ll (\lambda_{q-1} \delta_{q-1}^{1/2})^{-1}$, implies that the weak solution v has time support in the set $|t| \leq T/2 + 4 \sum_{q \geq 0} (\lambda_q \delta_q^{1/2})^{-1} \leq T/2 + 8\lambda_0^{\beta-1}$. Choosing a sufficiently large shows that $\operatorname{supp}_t v \subset [-T, T]$.

Remark 3.8. The intermittent convex integration scheme described in this book may be modified to show that within the regularity class $C_t^0 H_x^{1/2-}$, weak solutions of 3D Euler may be constructed to attain any given smooth energy profile, as mentioned in Remark 1.2. The main modifications required to prove this fact are as follows. As in previous schemes (see, e.g., De Lellis and Székelyhidi Jr. [31], equations (7) and (9), or [13], equations (2.5) and (2.6)) we need to measure the distance between the energy resolved at step q in the iteration and the desired energy profile $e(t)$. The energy pumping produced in steps $q \mapsto q+1$ by the additions of pipe flows which comprise the velocity increments w_{q+1}, and the error due to mollification, was already understood in detail in Daneri and Székelyhidi Jr. [27] and in [11]. An additional difficulty in this book is due to the presence of the higher order stresses: the energy profile would have to be inductively adjusted also throughout the steps $n \mapsto n + 1$ and $p \mapsto p + 1$. The other difficulty is the presence of the cutoff functions. This issue was, however, already addressed in [13]– cf. Sections 4.5, 4.5, 6– albeit for a simpler version of the cutoff functions, which only included the stress cutoffs. With some effort, the argument in [13] may be indeed modified to deal with the cutoff functions present in this work.

Chapter Four

Building blocks

4.1 A CAREFUL CONSTRUCTION OF INTERMITTENT PIPE FLOWS

We recall from [54, Lemma 1] or [27, Lemma 2.4] a version of the following geometric decomposition:

Proposition 4.1 (Choosing vectors for the axes). *Let $B_{1/2}(\mathrm{Id})$ denote the ball of symmetric 3×3 matrices, centered at Id, of radius $1/2$. Then, there exists a finite subset $\Xi \subset \mathbb{S}^2 \cap \mathbb{Q}^3$, such that for every $\xi \in \Xi$ there exists a smooth positive function $\gamma_\xi \colon C^\infty\left(B_{1/2}(\mathrm{Id})\right) \to \mathbb{R}$, such that for each $R \in B_{1/2}(\mathrm{Id})$ we have the identity*

$$R = \sum_{\xi \in \Xi} \left(\gamma_\xi(R)\right)^2 \xi \otimes \xi. \tag{4.1}$$

Additionally, for every ξ in Ξ, there exist vectors $\xi^{(2)}, \xi^{(3)} \in \mathbb{S}^2 \cap \mathbb{Q}^3$ such that $\{\xi, \xi^{(2)}, \xi^{(3)}\}$ is an orthonormal basis of $\mathbb{R}^3$, and there exists a least positive integer n_* such that $n_*\xi, n_*\xi^{(2)}, n_*\xi^{(3)} \in \mathbb{Z}^3$, for every $\xi \in \Xi$.

In order to adapt the proof of Proposition 4.8 to pipe flows oriented around axes which are not parallel to the standard basis vectors e_1, e_2, or e_3, it is helpful to consider functions which are periodic not only with respect to $\mathbb{T}^3$, but also with respect to a torus for which one face is perpendicular to the axis of the pipe (i.e., one edge of the torus is parallel to the axis).

Definition 4.2 ($\mathbb{T}^3_\xi$-periodicity). *Let $\{\xi, \xi^{(2)}, \xi^{(3)}\} \subset \mathbb{S}^2 \cap \mathbb{Q}^3$ be an orthonormal basis for $\mathbb{R}^3$, and let $f : \mathbb{R}^3 \to \mathbb{R}^n$. We say that f is $\mathbb{T}^3_\xi$-periodic if for all $(k_1, k_2, k_3) \in \mathbb{Z}^3$ and $(x_1, x_2, x_3) \in \mathbb{R}^3$*

$$f\left((x_1, x_2, x_3) + 2\pi\left(k_1\xi + k_2\xi^{(2)} + k_3\xi^{(3)}\right)\right) = f(x_1, x_2, x_3), \tag{4.2}$$

and we write $f : \mathbb{T}^3_\xi \to \mathbb{R}^n$. If $\{\xi, \xi^{(2)}, \xi^{(3)}\} = \{e_1, e_2, e_3\}$, i.e., the standard basis for $\mathbb{R}^3$, we drop the subscript ξ and write $\mathbb{T}^3$. For sets $\mathcal{S} \subset \mathbb{R}^3$, we say that $\mathcal{S}$ is $\mathbb{T}^3_\xi$-periodic if the indicator function of $\mathcal{S}$ is $\mathbb{T}^3_\xi$-periodic. Additionally, if L is a positive number, we say that f is $\left(\frac{\mathbb{T}^3_\xi}{L}\right)$-periodic if

$$f\left((x_1, x_2, x_3) + \frac{2\pi}{L}\left(k_1\xi + k_2\xi^{(2)} + k_3\xi^{(3)}\right)\right) = f(x_1, x_2, x_3)$$

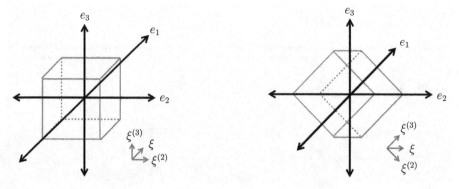

Figure 4.1: The torus on the left, $\mathbb{T}^3$, has axes parallel to the usual coordinate axes, while the torus on the right, denoted by $\mathbb{T}^3_\xi$, has been rotated and has axes parallel to a new set of vectors ξ, $\xi^{(2)}$, and $\xi^{(3)}$.

for all $(k_1, k_2, k_3) \in \mathbb{Z}^3$ and $(x_1, x_2, x_3) \in \mathbb{R}^3$. Note that if L is a positive integer, $\frac{\mathbb{T}^3_\xi}{L}$-periodicity implies $\mathbb{T}^3_\xi$-periodicity. See Figure 4.1.

We can now construct shifted intermittent pipe flows concentrated around axes with a prescribed vector direction ξ while imposing that each flow be supported in a single member of a large collection of disjoint sets. For the sake of clarity, we split the construction into two steps. First, in Proposition 4.3 we construct the shifts and then periodize and rotate the scalar-valued flow profiles and potentials associated to the pipe flows $\mathbb{W}_{\xi, \lambda, r}$. The support and placement properties are ensured at the level of the flow profile and potential. Next, we use the flow profiles to construct the pipe flows themselves in Proposition 4.4.

Proposition 4.3 (Rotating, shifting, and periodizing). *Fix $\xi \in \Xi$, where Ξ is as in Proposition 4.1. Let $r^{-1}, \lambda \in \mathbb{N}$ be given such that $\lambda r \in \mathbb{N}$. Let $\varkappa : \mathbb{R}^2 \to \mathbb{R}$ be a smooth function with support contained inside a ball of radius $\frac{1}{4}$. Then for $k \in \{0, ..., r^{-1} - 1\}^2$, there exist functions $\varkappa^k_{\lambda, r, \xi} : \mathbb{R}^3 \to \mathbb{R}$ defined in terms of $\varkappa$, satisfying the following additional properties:*

1. *We have that $\varkappa^k_{\lambda, r, \xi}$ is simultaneously $\left(\frac{\mathbb{T}^3}{\lambda r} \right)$-periodic and $\left(\frac{\mathbb{T}^3_\xi}{\lambda r n_*} \right)$-periodic.*

2. *Let F_ξ be one of the two faces of the cube $\frac{\mathbb{T}^3_\xi}{\lambda r n_*}$ which is perpendicular to ξ. Let $\mathbb{G}_{\lambda, r} \subset F_\xi \cap 2\pi \mathbb{Q}^3$ be the grid consisting of r^{-2}-many points spaced evenly at distance $2\pi(\lambda n_*)^{-1}$ on F_ξ and containing the origin. Then each grid point g_k for $k \in \{0, ..., r^{-1} - 1\}^2$ satisfies*

$$\left(\operatorname{supp} \varkappa^k_{\lambda, r, \xi} \cap F_\xi \right) \subset \left\{ x : |x - g_k| \leq 2\pi \left(4\lambda n_* \right)^{-1} \right\}. \qquad (4.3)$$

3. *The support of $\varkappa^k_{\lambda, r, \xi}$ consists of a pipe (cylinder) centered around a $\left(\frac{\mathbb{T}^3}{\lambda r} \right)$-*

periodic and $\left(\frac{\mathbb{T}_\xi^3}{\lambda r n_*}\right)$-periodic line parallel to ξ, which passes through the point g_k. The radius of the cylinder's cross-section is as given in (4.3).

4. *For* $k \neq k'$, supp $\varkappa^k_{\lambda,r,\xi} \cap \mathrm{supp}\, \varkappa^{k'}_{\lambda,r,\xi} = \emptyset$. *See Figure* 4.2.

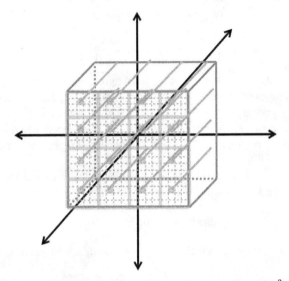

Figure 4.2: We have pictured above a grid on the front face of $\mathbb{T}^3$, in which there are $4^2 = (\lambda r)^2$ many periodic cells, each with $4^2 = r^{-2}$ many subcells of diameter $16^{-1} = \lambda^{-1}$. The periodized axes of the pipes are the line segments extending from the front face of the torus.

Proof of Proposition 4.3. For $r^{-1} \in \mathbb{N}$, which quantifies the rescaling, and for $k = (k_1, k_2) \in \{0, ..., r^{-1} - 1\}^2$, which quantifies the shifts, define $\varkappa^k_r$ to be the rescaled and shifted function

$$\varkappa^k_r (x_1, x_2) := \frac{1}{2\pi r} \varkappa \left(\frac{x_1}{2\pi r} - k_1, \frac{x_2}{2\pi r} - k_2 \right). \tag{4.4}$$

Then $(x_1, x_2) \in \mathrm{supp}\, \varkappa^k_r$ if and only if

$$\left| \frac{x_1}{2\pi r} - k_1 \right|^2 + \left| \frac{x_2}{2\pi r} - k_2 \right|^2 \leq \frac{1}{16}. \tag{4.5}$$

This implies that

$$k_1 - \frac{1}{4} \leq \frac{x_1}{2\pi r} \leq k_1 + \frac{1}{4}, \qquad k_2 - \frac{1}{4} \leq \frac{x_2}{2\pi r} \leq k_2 + \frac{1}{4}. \tag{4.6}$$

Since these inequalities cannot be satisfied by a single pair (x, y) for both $k =$

(k_1, k_2) and $k' = (k'_1, k'_2)$ simultaneously when $k \neq k'$, it follows that

$$\text{supp } \varkappa_r^k \cap \text{supp } \varkappa_r^{k'} = \emptyset \tag{4.7}$$

for all $k \neq k'$. Also, notice that plugging $k_1 = 0$ and $k_1 = r^{-1} - 1$ into (4.6) shows that the set of x_1 for which there exists (k_1, k_2) such that $\varkappa_r^k(x) \neq 0$ is contained in

$$\left\{ -\frac{r\pi}{2} \leq x_1 \leq 2\pi - \frac{3r\pi}{2} \right\},$$

which is a set with diameter strictly less than 2π. Therefore, periodizing in x_1 will not cause overlap in the supports of the periodized objects. Arguing similarly for x_2 and enumerating the pairs (k_1, k_2) with $k \in \{0, ..., r^{-1} - 1\}^2$, we overload notation and denote by $\varkappa_r^k$ the $\mathbb{T}^2$-periodized version of $\varkappa_r^k$. Thus we have produced r^{-2}-many functions which are $\mathbb{T}^2$-periodic and which have disjoint supports.

Now define $\mathbb{G}_r \subset \mathbb{T}^2$ to be the grid containing r^{-2}-many points evenly spaced at distance $2\pi r$ and containing the origin. Then

$$\mathbb{G}_r = \left\{ g_k^0 := 2\pi rk : k \in \{0, ..., r^{-1} - 1\}^2 \right\} \subset 2\pi \mathbb{Q}^2 \,.$$

Thus the support of each function $\varkappa_r^k$ contains g_k^0 as its center, but no other grid points.

Let $\xi \in \Xi$ be fixed, with the associated orthonormal basis $\{\xi, \xi^{(2)}, \xi^{(3)}\}$. For $x = (x_1, x_2, x_3) \in \mathbb{R}^3$ and $\lambda r \in \mathbb{N}$, define

$$\varkappa_{\lambda, r, \xi}^k(x) := \varkappa_r^k \left(n_* \lambda r x \cdot \xi^{(2)}, n_* \lambda r x \cdot \xi^{(3)} \right). \tag{4.8}$$

Then for $(k_1, k_2, k_3) \in \mathbb{Z}^3$,

$$\varkappa_{\lambda, r, \xi}^k \left(x + \frac{2\pi}{\lambda r}(k_1, k_2, k_3) \right)$$
$$= \varkappa_r^k \left(n_* \lambda r \left(x + \frac{2\pi}{\lambda r}(k_1, k_2, k_3) \right) \cdot \xi^{(2)}, n_* \lambda r \left(x + \frac{2\pi}{\lambda r}(k_1, k_2, k_3) \right) \cdot \xi^{(3)} \right)$$
$$= \varkappa_r^k \left(n_* \lambda r x \cdot \xi^{(2)}, n_* \lambda r x \cdot \xi^{(3)} \right)$$
$$= \varkappa_{\lambda, r, \xi}^k(x)$$

since $n_* \xi^{(2)}, n_* \xi^{(3)} \in \mathbb{Z}^3$ and $\varkappa_r^k$ is $\mathbb{T}^2$-periodic, and thus $\varkappa_{\lambda, r, \xi}^k$ is $\frac{\mathbb{T}^3}{\lambda r}$-periodic. Similarly,

$$\varkappa_{\lambda, r, \xi}^k \left(x + \frac{2\pi}{\lambda r n_*}(k_1 \xi + k_2 \xi^{(2)} + k_3 \xi^{(3)}) \right)$$
$$= \varkappa_r^k \left(n_* \lambda r \left(x + \frac{2\pi}{\lambda r n_*}(k_1 \xi + k_2 \xi^{(2)} + k_3 \xi^{(3)}) \right) \cdot \xi^{(2)},$$

$$n_*\lambda r\left(x + \frac{2\pi}{\lambda r n_*}(k_1\xi + k_2\xi^{(2)} + k_3\xi^{(3)})\right)\cdot\xi^{(3)}\Bigg)$$

$$= \varkappa_r^k\left(n_*\lambda r x\cdot\xi^{(2)}, n_*\lambda r x\cdot\xi^{(3)}\right)$$

$$= \varkappa_{\lambda,r,\xi}^k(x)$$

since

$$2\pi(k_1\xi + k_2\xi^{(2)} + k_3\xi^{(3)})\cdot\xi^{(2)} = 2\pi k_2, \quad 2\pi(k_1\xi + k_2\xi^{(2)} + k_3\xi^{(3)})\cdot\xi^{(3)} = 2\pi k_3,$$

and $\varkappa_r^k$ is $\mathbb{T}^2$-periodic. Thus $\varkappa_{\lambda,r,\xi}^k$ is $\frac{\mathbb{T}_\xi^3}{\lambda r n_*}$-periodic, and as a consequence $\frac{\mathbb{T}_\xi^3}{\lambda r}$-periodic as well. Therefore, we have proved point 1.

To prove point 2, define

$$\mathbb{G}_{\lambda,r} = \left\{g_k := 2\pi\left(k_1\left(\lambda n_*\right)^{-1}\xi^{(2)} + k_2\left(\lambda n_*\right)^{-1}\xi^{(3)}\right) : k_1, k_2 \in \{0, ..., {}^1\!/\!r - 1\}\right\}. \tag{4.9}$$

We claim that $\varkappa_{\lambda,r,\xi}^k|_{F_\xi}$ is supported in a $2\pi(4\lambda n_*)^{-1}$-neighborhood of g_k. To prove the claim, let $x \in F_\xi$ be such that $\varkappa_{\lambda,r,\xi}^k(x) \neq 0$. Then since

$$\varkappa_{\lambda,r,\xi}^k(x) = \varkappa_r^k\left(n_*\lambda r x\cdot\xi^{(2)}, n_*\lambda r x\cdot\xi^{(3)}\right),$$

we can use (4.5) to assert that $x \in \operatorname{supp}\varkappa_{\lambda,r,\xi}^k$ if and only if $x = (x_1, x_2, x_3)$ satisfies

$$\left|\frac{n_*\lambda r x\cdot\xi^{(2)}}{2\pi r} - k_1\right|^2 + \left|\frac{n_*\lambda r x\cdot\xi^{(3)}}{2\pi r} - k_2\right|^2 \leq \frac{1}{16}$$

$$\Longleftrightarrow \left|(x_1, x_2, x_3) - \left(\frac{2\pi}{n_*\lambda}k_1\xi^{(2)} + \frac{2\pi}{n_*\lambda}k_2\xi^{(3)}\right)\right|^2 \leq \left(\frac{2\pi}{4n_*\lambda}\right)^2,$$

which proves the claim.

Items 3 and 4 follow immediately after noting that $\varkappa_{\lambda,r,\xi}^k$ is constant on every plane parallel to F_ξ, and that the grid points $g_k \in \mathbb{G}_{\lambda,r}$ around which the supports of $\varkappa_{\lambda,r,\xi}^k$ are centered are spaced at a distance which is twice the diameters of the supports. $\qquad\square$

Proposition 4.4 (Construction and properties of shifted intermittent pipe flows). *Fix a vector ξ belonging to the set of rational vectors $\Xi \subset \mathbb{Q}^3$ from Proposition 4.3, $r^{-1}, \lambda \in \mathbb{N}$ with $\lambda r \in \mathbb{N}$, and large integers $2\mathsf{N}_{\mathrm{fin}}$ and d. There exist vector fields $\mathsf{W}_{\xi,\lambda,r}^k : \mathbb{T}^3 \to \mathbb{R}^3$ for $k \in \{0, ..., r^{-1} - 1\}^2$ and implicit constants depending on $\mathsf{N}_{\mathrm{fin}}$ and d but not on λ or r such that:*

1. *There exists $\varrho : \mathbb{R}^2 \to \mathbb{R}$ given by the iterated Laplacian $\Delta^{\mathsf{d}}\vartheta =: \varrho$ of a potential $\vartheta : \mathbb{R}^2 \to \mathbb{R}$ with compact support in a ball of radius $\frac{1}{4}$ such that the following holds. Let $\varrho_{\xi,\lambda,r}^k$ and $\vartheta_{\xi,\lambda,r}^k$ be defined as in Proposition 4.3.*

Then there exist $U^k_{\xi,\lambda,r} : \mathbb{T}^3 \to \mathbb{R}^3$ *such that*

$$\operatorname{curl} U^k_{\xi,\lambda,r} = \xi\lambda^{-2d}\Delta^d \left(\vartheta^k_{\xi,\lambda,r}\right) = \xi\varrho^k_{\xi,\lambda,r} =: W^k_{\xi,\lambda,r}. \tag{4.10}$$

2. *Each of the sets of functions* $\{U^k_{\xi,\lambda,r}\}_k$, $\{\varrho^k_{\xi,\lambda,r}\}_k$, $\{\vartheta^k_{\xi,\lambda,r}\}_k$, *and* $\{W^k_{\xi,\lambda,r}\}_k$ *satisfy items 1–4. In particular, when* $k \neq k'$, *we have that the intersection of the supports of* $W^{\xi,\lambda,r}_k$ *and* $W^{k'}_{\xi,\lambda,r}$ *is empty, and similarly for the other sets of functions.*

3. $W^k_{\xi,\lambda,r}$ *is a stationary, pressureless solution to the Euler equations, i.e.,*

$$\operatorname{div} W^k_{\xi,\lambda,r} = 0, \qquad \operatorname{div}\left(W^k_{\xi,\lambda,r} \otimes W^k_{\xi,\lambda,r}\right) = 0.$$

4. $\dfrac{1}{|\mathbb{T}^3|}\displaystyle\int_{\mathbb{T}^3} W^k_{\xi,\lambda,r} \otimes W^k_{\xi,\lambda,r} = \xi \otimes \xi$

5. *For all* $n \leq 2N_{\mathrm{fin}}$,

$$\left\|\nabla^n\vartheta^k_{\xi,\lambda,r}\right\|_{L^p(\mathbb{T}^3)} \lesssim \lambda^n r^{\left(\frac{2}{p}-1\right)}, \qquad \left\|\nabla^n\varrho^k_{\xi,\lambda,r}\right\|_{L^p(\mathbb{T}^3)} \lesssim \lambda^n r^{\left(\frac{2}{p}-1\right)} \tag{4.11}$$

and

$$\left\|\nabla^n U^k_{\xi,\lambda,r}\right\|_{L^p(\mathbb{T}^3)} \lesssim \lambda^{n-1} r^{\left(\frac{2}{p}-1\right)}, \qquad \left\|\nabla^n W^k_{\xi,\lambda,r}\right\|_{L^p(\mathbb{T}^3)} \lesssim \lambda^n r^{\left(\frac{2}{p}-1\right)}. \tag{4.12}$$

6. *Let* $\Phi : \mathbb{T}^3 \times [0,T] \to \mathbb{T}^3$ *be the periodic solution to the transport equation*

$$\partial_t\Phi + v \cdot \nabla\Phi = 0, \tag{4.13a}$$
$$\Phi_{t=t_0} = x, \tag{4.13b}$$

with a smooth, divergence-free, periodic velocity field v. *Then*

$$\nabla\Phi^{-1} \cdot \left(W^k_{\xi,\lambda,r} \circ \Phi\right) = \operatorname{curl}\left(\nabla\Phi^T \cdot \left(U^k_{\xi,\lambda,r} \circ \Phi\right)\right). \tag{4.14}$$

7. *For* $\mathbb{P}_{[\lambda_1,\lambda_2]}$ *a Littlewood-Paley projector,* Φ *as in* (4.13)*, and* $A = (\nabla\Phi)^{-1}$,

$$\left[\nabla \cdot \left(A\,\mathbb{P}_{[\lambda_1,\lambda_2]}\left(W_{\xi,\lambda,r} \otimes W_{\xi,\lambda,r}\right)(\Phi)A^T\right)\right]_i$$
$$= A^j_k\,\mathbb{P}_{[\lambda_1,\lambda_2]}\left(W^k_{\xi,\lambda,r}W^l_{\xi,\lambda,r}\right)(\Phi)\partial_j A^i_l$$
$$= A^j_k\xi^k\xi^l\partial_j A^i_l\,\mathbb{P}_{[\lambda_1,\lambda_2]}\left(\left(\varrho^k_{\xi,\lambda,r}\right)^2\right) \tag{4.15}$$

for $i = 1, 2, 3$.

Remark 4.5. The identity (4.15) is one of the main advantages of pipe flows over Beltrami flows. The utility of this identity is that when checking whether a pipe flow $W_{\xi,\lambda,r}$ which has been deformed by Φ is still an approximately stationary solution of the pressureless Euler equations, one does not need to estimate any derivatives of $\mathbb{W}_{\xi,\lambda,r}$—only derivatives on the flow map Φ, which will cost much less than λ.

Remark 4.6. The formulation of (4.15) is useful for our inversion of the divergence operator, which is presented in Proposition A.17 and the subsequent remark. We refer to the statement of that proposition and the subsequent remark for further properties related to (4.15).

Proof of Proposition 4.4. With the definition $\mathbb{W}_{\xi,\lambda,r}^k := \xi \varrho_{\xi,\lambda,r}^k$, the equality $\lambda^{-2\mathsf{d}} \Delta^{\mathsf{d}}(\vartheta_{\xi,\lambda,r}^k) = \varrho_{\xi,\lambda,r}^k$ follows from the proof of Proposition 4.3, specifically equations (4.4), (4.4), and (4.8). The equality $\mathrm{curl}\, \mathbb{U}_{\xi,\lambda,r}^k = \mathbb{W}_{\xi,\lambda,r}$ follows as well using the standard vector calculus identity $\mathrm{curl} \circ \mathrm{curl} = \nabla \circ \mathrm{div} - \Delta$. Secondly, properties (1), (2), and (4) from Proposition 4.3 for $\vartheta_{\xi,\lambda,r}^k$ follow from Proposition 4.3 applied to $\varkappa = \vartheta$. The same properties for $\varrho_{\xi,\lambda,r}^k$, $\mathbb{U}_{\xi,\lambda,r}^k$, and $\mathbb{W}_{\xi,\lambda,r}^k$ follow from differentiating. Next, it is clear that $\mathbb{W}_{\xi,\lambda,r}^k$ solves the pressureless Euler equations since $\xi \cdot \nabla \varrho_{\xi,\lambda,r}^k = 0$. The normalization in (4) follows from imposing that

$$\frac{1}{(2\pi)^2} \int_{\mathbb{R}^2} (\Delta^{\mathsf{d}} \vartheta(x_1, x_2))^2 \, dx_1 \, dx_2 = 1,$$

recalling that orthogonal transformations, shifts, and scaling do not alter the L^p norms of $\mathbb{T}^3$-periodic functions, and using (4.4). The estimates in (5) follow similarly, using (4.4). The proof of (4.14) in (6) can be found in the paper of Daneri and Székelyhidi Jr. [27].

The proof of (4.15) from (7) is simple and similar in spirit to (6) but perhaps not standard, and so we will check it explicitly here. We first set $\mathcal{P}$ to be the $\mathbb{T}^3$-periodic convolution kernel associated with the projector $\mathbb{P}_{[\lambda_1,\lambda_2]}$ and write

$$\nabla \cdot \left((\nabla \Phi)^{-1} \mathbb{P}_{[\lambda_1,\lambda_2]} \left(\mathbb{W}_{\xi,\lambda,r} \otimes \mathbb{W}_{\xi,\lambda,r} \right) (\Phi)(\nabla \Phi)^{-T} \right)(x)$$

$$= \nabla_x \cdot \left((\nabla \Phi)^{-1}(x) \left(\int_{\mathbb{T}^3} \mathcal{P}(y)(\mathbb{W}_{\xi,\lambda,r} \otimes \mathbb{W}_{\xi,\lambda,r})(\Phi(x-y)) \, dy \right) (\nabla \Phi)^{-T}(x) \right)$$

$$= \nabla_x \cdot \left(\int_{\mathbb{T}^3} (\nabla \Phi)^{-1}(x) \mathcal{P}(y)(\mathbb{W}_{\xi,\lambda,r} \otimes \mathbb{W}_{\xi,\lambda,r})(\Phi(x-y))(\nabla \Phi)^{-T}(x) \, dy \right)$$

$$= \nabla_x \cdot \left(\int_{\mathbb{T}^3} \mathcal{P}(y) \left((\nabla \Phi)^{-1}(x) \mathbb{W}_{\xi,\lambda,r}(\Phi(x-y)) \right) \right.$$

$$\left. \otimes \left((\nabla \Phi)^{-1}(x) \mathbb{W}_{\xi,\lambda,r}(\Phi(x-y)) \right) \, dy \right). \tag{4.16}$$

Then applying (4.14), we obtain that (4.16) is equal to

$$\int_{\mathbb{T}^3} \mathcal{P}(y) \left((\nabla\Phi)^{-1}(x) \mathbb{W}_{\xi,\lambda,r}(\Phi(x-y)) \right) \cdot \nabla_x \left((\nabla\Phi)^{-1}(x) \mathbb{W}_{\xi,\lambda,r}(\Phi(x-y)) \right) \, dy.$$

Writing out the ith component of this vector and using the notation $A = (\nabla\Phi)^{-1}$, we obtain

$$\left[\int_{\mathbb{T}^3} \mathcal{P}(y) \left(A(x) \mathbb{W}_{\xi,\lambda,r}(\Phi(x-y)) \right) \cdot \nabla_x \left(A(x) \mathbb{W}_{\xi,\lambda,r}(\Phi(x-y)) \right) \, dy \right]_i$$

$$= \int_{\mathbb{T}^3} \mathcal{P}(y) A_k^j(x) \mathbb{W}_{\xi,\lambda,r}^k(\Phi(x-y)) A_l^i(x) \partial_n \mathbb{W}_{\xi,\lambda,r}^l(\Phi(x-y)) \partial_j \Phi_n(x) \, dy$$

$$+ \int_{\mathbb{T}^3} \mathcal{P}(y) A_k^j(x) \mathbb{W}_{\xi,\lambda,r}^k(\Phi(x-y)) \partial_j A_l^i(x) \mathbb{W}_{\xi,\lambda,r}^l(\Phi(x-y)) \, dy. \quad (4.17)$$

Since the second term in (4.17) can be rewritten as

$$\int_{\mathbb{T}^3} \mathcal{P}(y) A_k^j(x) \mathbb{W}_{\xi,\lambda,r}^k(\Phi(x-y)) \partial_j A_l^i(x) \mathbb{W}_{\xi,\lambda,r}^l(\Phi(x-y)) \, dy$$

$$= A_k^j(x) \mathbb{P}_{[\lambda_1,\lambda_2]} \left(\mathbb{W}_{\xi,\lambda,r}^k \mathbb{W}_{\xi,\lambda,r}^l \right) (\Phi(x)) \partial_j A_l^i(x),$$

to conclude the proof, we must show that the first term in (4.17) is equal to 0. Using that

$$A_k^j \partial_j \Phi^n = \delta_{nk}$$

and

$$\mathbb{W}_{\xi,\lambda,r}^k \partial_k \mathbb{W}_{\xi,\lambda,r}^l = 0$$

for all l, we can simplify the first term as

$$\int_{\mathbb{T}^3} \mathcal{P}(y) A_k^j(x) \mathbb{W}_{\xi,\lambda,r}^k(\Phi(x-y)) A_l^i(x) \partial_n \mathbb{W}_{\xi,\lambda,r}^l(\Phi(x-y)) \partial_j \Phi_n(x) \, dy$$

$$= \int_{\mathbb{T}^3} \mathcal{P}(y) \delta_{nk} \mathbb{W}_{\xi,\lambda,r}^k(\Phi(x-y)) A_l^i(x) \partial_n \mathbb{W}_{\xi,\lambda,r}^l(\Phi(x-y)) \, dy$$

$$= \int_{\mathbb{T}^3} \mathcal{P}(y) \mathbb{W}_{\xi,\lambda,r}^k(\Phi(x-y)) A_l^i(x) \partial_k \mathbb{W}_{\xi,\lambda,r}^l(\Phi(x-y)) \, dy$$

$$= 0,$$

proving (4.15). $\square$

4.2 DEFORMED PIPE FLOWS AND CURVED AXES

Lemma 4.7 (Control on axes, support, and spacing). *Consider a convex neighborhood of space $\Omega \subset \mathbb{T}^3$. Let v be an incompressible velocity field, and* *define the flow $X(x,t)$*

$$\partial_t X(x,t) = v\left(X(x,t),t\right) \tag{4.18a}$$

$$X_{t=t_0} = x\,, \tag{4.18b}$$

and inverse $\Phi(x,t) = X^{-1}(x,t)$

$$\partial_t \Phi + v \cdot \nabla \Phi = 0 \tag{4.19a}$$

$$\Phi_{t=t_0} = x\,. \tag{4.19b}$$

Define $\Omega(t) := \{x \in \mathbb{T}^3 : \Phi(x,t) \in \Omega\} = X(\Omega,t)$. For an arbitrary $C > 0$, let $\tau > 0$ be a parameter such that

$$\tau \le \left(\delta_q^{1/2} \lambda_q \Gamma_{q+1}^{C+2}\right)^{-1}. \tag{4.20}$$

Furthermore, suppose that the vector field v satisfies the Lipschitz bound[1]

$$\sup_{t\in[t_0-\tau,t_0+\tau]} \|\nabla v(\cdot,t)\|_{L^\infty(\Omega(t))} \lesssim \delta_q^{1/2} \lambda_q \Gamma_{q+1}^C. \tag{4.21}$$

Let $\mathbb{W}_{\lambda_{q+1},r,\xi}^k : \mathbb{T}^3 \to \mathbb{R}^3$ be a set of straight pipe flows constructed as in Proposition 4.3 and Proposition 4.4, which are $\frac{\mathbb{T}^3}{\lambda_{q+1}r}$-periodic for $\frac{\lambda_q}{\lambda_{q+1}} \le r \le 1$ and concentrated around axes $\{A_i\}_{i\in\mathcal{I}}$ oriented in the vector direction ξ for $\xi \in \Xi$. Then $\mathbb{W} := \mathbb{W}_{\lambda_{q+1},r,\xi}^k(\Phi(x,t)) : \Omega(t) \times [t_0 - \tau, t_0 + \tau]$ satisfies the following conditions:

1. *We have the inequality*

$$\operatorname{diam}(\Omega(t)) \le \left(1 + \Gamma_{q+1}^{-1}\right)\operatorname{diam}(\Omega). \tag{4.22}$$

2. *If x and y with $x \ne y$ belong to a particular axis $A_i \subset \Omega$, then*

$$\frac{X(x,t) - X(y,t)}{|X(x,t) - X(y,t)|} = \frac{x - y}{|x - y|} + \delta_i(x,y,t), \tag{4.23}$$

   where $|\delta_i(x,y,t)| < \Gamma_{q+1}^{-1}$.

3. *Let x and y belong to a particular axis $A_i \subset \Omega$. Denote the length of the axis $A_i(t) := X(A_i \cap \Omega, t)$ in between $X(x,t)$ and $X(y,t)$ by $L(x,y,t)$. Then*

$$L(x,y,t) \le \left(1 + \Gamma_{q+1}^{-1}\right)|x - y|. \tag{4.24}$$

---

[1]The implicit constant in this inequality is assumed to be independent of q; cf. (6.60).

4. *The support of* $\mathbb{W}$ *is contained in a* $\left(1 + \Gamma_{q+1}^{-1}\right) \dfrac{2\pi}{4n_*\lambda_{q+1}}$ *-neighborhood of*

$$\bigcup_i A_i(t). \tag{4.25}$$

5. $\mathbb{W}$ *is "approximately periodic" in the sense that for distinct axes* A_i, A_j *with* $i \neq j$ *and* dist $(A_i \cap \Omega, A_j \cap \Omega) = d$,

$$\left(1 - \Gamma_{q+1}^{-1}\right) d \leq \text{dist} \left(A_i(t), A_j(t)\right) \leq \left(1 + \Gamma_{q+1}^{-1}\right) d. \tag{4.26}$$

Proof of Lemma 4.7. First, we have that for $x, y \in \Omega$,

$$|X(x,t) - X(y,t)| = \left| x - y + \int_{t_0}^{t} \partial_s X(x,s) - \partial_s X(y,s)\, ds \right|$$

$$\leq |x - y| + \int_{t_0}^{t} |v\left(X(x,s),s\right) - v\left(X(y,s),s\right)|\, ds.$$

Furthermore,

$$\left| v^\ell \left(X(x,s),s\right) - v^\ell \left(X(y,s),s\right) \right|$$

$$= \left| \int_0^1 \partial_j v^\ell \left(X(x+t(y-x),s),s\right) \partial_k X^j(x+t(y-x),s)(y-x)^k\, dt \right|$$

$$\leq \|\nabla v\|_{L^\infty(\Omega(s))} \|\nabla X\|_{L^\infty(\Omega(s))} |x - y|$$

$$\leq \frac{3}{2} \delta_q^{\frac{1}{2}} \lambda_q \Gamma_{q+1}^C |x - y|.$$

Integrating this bound from t_0 to t and using a factor of Γ_{q+1} to absorb the constant, we deduce that

$$\left(1 - \Gamma_{q+1}^{-1}\right) |x - y| \leq |X(x,t) - X(y,t)| \leq \left(1 + \Gamma_{q+1}^{-1}\right) |x - y|. \tag{4.27}$$

The inequality in (4.22) follows immediately.

To prove (4.23), we will show that for $x, y \in \Omega \cap A_i$ for a chosen axis A_i,

$$\left| \frac{x - y}{|x - y|} - \frac{X(x,t) - X(y,t)}{|X(x,t) - X(y,t)|} \right| < \Gamma_{q+1}^{-1}.$$

At time t_0, the above quantity vanishes. Differentiating inside the absolute value in time, we have that

$$\frac{d}{dt} \left[\frac{X(x,t) - X(y,t)}{|X(x,t) - X(y,t)|} \right]$$

$$= \frac{\partial_t X(x,t) - \partial_t X(y,t)}{|X(x,t) - X(y,t)|}$$

$$- \frac{X(x,t) - X(y,t)}{|X(x,t) - X(y,t)|^2} \frac{(\partial_t X(x,t) - \partial_t X(y,t)) \cdot (X(x,t) - X(y,t))}{|X(x,t) - X(y,t)|}$$

$$= \frac{v(X(x,t),t) - v(X(y,t),t)}{|X(x,t) - X(y,t)|}$$

$$- \frac{X(x,t) - X(y,t)}{|X(x,t) - X(y,t)|} \frac{(v(X(x,t),t) - v(X(y,t),t)) \cdot (X(x,t) - X(y,t))}{|X(x,t) - X(y,t)|^2}.$$

Utilizing the mean value theorem and the Lipschitz bound on v and (4.27), we deduce

$$\left| \frac{v(X(x,t),t) - v(X(y,t),t)}{|X(x,t) - X(y,t)|} \right.$$

$$\left. - \frac{X(x,t) - X(y,t)}{|X(x,t) - X(y,t)|} \frac{(v(X(x,t),t) - v(X(y,t),t)) \cdot (X(x,t) - X(y,t))}{|X(x,t) - X(y,t)|^2} \right|$$

$$\leq 2 \|\nabla v\|_{L^\infty}$$

$$\leq 2 \delta_q^{\frac{1}{2}} \lambda_q \Gamma_{q+1}^C.$$

Integrating in time from t_0 to t for $|t - t_0| \leq \left(\delta_q^{\frac{1}{2}} \lambda_q \Gamma_{q+1}^{C+2} \right)^{-1}$ and using the extra factors of Γ_{q+1} to again kill the constants, we obtain (4.23).

To prove (4.24), we parametrize the curve using X to obtain

$$L(x,y,t) = \int_0^1 |\nabla X(x + r(y - x), t) \cdot (x - y)| \, dr \leq \left(1 + \Gamma_{q+1}^{-1}\right) |x - y|.$$

The claims in (4.25) and (4.26) follow immediately from (4.27) and (4.3). □

4.3   PLACEMENTS VIA RELATIVE INTERMITTENCY

We now state and prove the main proposition regarding the placement of a new set of intermittent pipe flows which do not intersect with previously placed and possibly deformed pipes *within* a subset Ω of the full torus $\mathbb{T}^3$. We do not claim that intersections do not occur outside of Ω. In applications, Ω will be the support of a cutoff function.[2] We state the proposition for new pipes periodized to spatial scale $(\lambda_{q+1} r_2)^{-1}$ with axes parallel to a direction vector $\xi \in \Xi$. By "relative intermittency," we mean the inequality (4.31) satisfied by r_1 and r_2. The proof proceeds, first in the case $\xi = e_3$, by an elementary but rather tedious counting argument for the number of cells in a two-dimensional grid which may intersect a set concentrated around a smooth curve. In applications,

---

[2]Technically, Ω will be a set slightly larger than the support of a cutoff function. See (8.117), (8.120), and (8.131).

this set corresponds to a piece of a periodic pipe flow concentrated around its deformed axis and then projected onto a plane. Then using (1) and (2) from Proposition 4.3, we describe the minor adjustments needed to obtain the same result for new pipes with axes parallel to arbitrary direction vectors $\xi \in \Xi$.

Proposition 4.8 (Placing straight pipes which avoid bent pipes). *Consider a neighborhood of space $\Omega \subset \mathbb{T}^3$ such that*

$$\text{diam}(\Omega) \leq 16(\lambda_{q+1}r_1)^{-1}, \tag{4.28}$$

where $\lambda_q/\lambda_{q+1} \leq r_1 \leq 1$. Assume that there exist smooth $\mathbb{T}^3$-periodic curves $\{A_n\}_{n=1}^{N_\Omega} \subset \Omega^3$ and $\mathbb{T}^3$-periodic sets $\{S_n\}_{n=1}^{N_\Omega} \subset \Omega$ satisfying the following properties:

1. *There exists a positive constant $\mathcal{C}_A$ and a parameter r_2, with $r_1 < r_2 < 1$, such that*

$$N_\Omega \leq \mathcal{C}_A r_2^2 r_1^{-2}. \tag{4.29}$$

2. *For any $x, x' \in A_n$, let the length of the curve A_n which lies between x and x', be denoted by $L_{n,x,x'}$. Then, for every $1 \leq n \leq N_\Omega$ we have*

$$L_{n,x,x'} \leq 2|x - x'|. \tag{4.30}$$

3. *For every $1 \leq n \leq N_\Omega$, S_n is contained in a $2\pi(1 + \Gamma_{q+1}^{-1})(4n_*\lambda_{q+1})^{-1}$-neighborhood of A_n.*

Then, there exists a geometric constant $C_* \geq 1$ such that if

$$C_* \mathcal{C}_A r_2^4 \leq r_1^3, \tag{4.31}$$

then, for any $\xi \in \Xi$ (recall the set Ξ from Proposition 4.1), we can find a set of pipe flows $\mathbb{W}_{\lambda_{q+1},r_2,\xi}^{k_0}: \mathbb{T}^3 \to \mathbb{R}^3$ which are $\frac{\mathbb{T}^3}{\lambda_{q+1}r_2}$-periodic, concentrated to width $\frac{2\pi}{4\lambda_{q+1}n_*}$ around axes with vector direction ξ, that satisfy the properties listed in Proposition 4.4, and for all $n \in \{1, ..., N_\Omega\}$,

$$\text{supp}\, \mathbb{W}_{\lambda_{q+1},r_2,\xi}^{k_0} \cap S_n = \emptyset. \tag{4.32}$$

Remark 4.9. As mentioned previously, the sets S_n will be supports of previously placed pipes oriented around *deformed* axes A_n. The properties of S_n and A_n will follow from Lemma 4.7.

Proof of Proposition 4.8. For simplicity, we first give the proof for $\xi = e_3$, and explain how to treat the case of general $\xi \in \Xi$ at the end of the proof.

---

[3]That is, the range of each curve is contained in Ω; otherwise replace the curves with $A_n \cap \Omega$.

The proof will proceed by measuring the size of the shadows of the $\{S_n\}_{n=1}^{N_\Omega}$ when projected onto the face of the cube $\mathbb{T}^3$ which is perpendicular to e_3, so it will be helpful to set some notation related to this projection. Let F_{e_3} be the face of the torus $\mathbb{T}^3$ which is perpendicular to e_3. For the sake of concreteness, we will occasionally identify F_{e_3} with the set of points $x = (x_1, x_2, x_3) \in \mathbb{T}^3$ such that $x_3 = 0$, or use that F_{e_3} is isomorphic to $\mathbb{T}^2$. Let A_n^p be the projection of A_n onto F_{e_3} defined by

$$A_n^p := \{(x_1, x_2) \in F_{e_3} : (x_1, x_2, x_3) \in A_n\}, \tag{4.33}$$

and let S_n^p be defined similarly as the projection of S_n onto F_{e_3}. For $x = (x_1, x_2, x_3) \in \mathbb{T}^3$ and $x' = (x_1', x_2', x_3') \in \mathbb{T}^3$ we let $P(x) = (x_1, x_2) \in F_{e_3}$ and $P(x') = (x_1', x_2') \in F_{e_3}$ be the projection of these points onto F_{e_3}. Since projections do not increase distances, we have that

$$|P(x) - P(x')| \le |x - x'|. \tag{4.34}$$

Since both A_n and A_n^p are smooth curves[4] and can be approximated by piecewise linear polygonal paths, (4.34), (4.28), and (4.30) imply that if $L_{n,x,x'}^p$ is the length of the projected curve A_n^p in between the points $P(x)$ and $P(x')$, then

$$L_{n,x,x'}^p \le 2|x - x'| \le 32 \left(\lambda_{q+1} r_1\right)^{-1}. \tag{4.35}$$

In particular, taking x and x' to be the endpoints of the curve A_n, we obtain a bound for the total length of A_n^p. Additionally, (4.34) and the third assumption of the lemma imply that S_n^p is contained inside a $2\pi(1 + \Gamma_{q+1}^{-1})(4n_* \lambda_{q+1})^{-1}$-neighborhood of A_n^p. Finally, since $\mathbb{W}_{\lambda_{q+1}, r_2, e_3}^k$ is independent of x_3 for all $k \in \{0, ..., r_2^{-1} - 1\}^2$, it is clear that the conclusion (4.32) will be achieved if we can show that there exists a shift k_0 such that

$$S_n^p \cap \left(\operatorname{supp} \mathbb{W}_{\lambda_{q+1}, r_2, e_3}^{k_0} \cap \{x_3 = 0\}\right) = \emptyset, \tag{4.36}$$

for all $1 \le n \le N_\Omega$. To prove (4.36), we will apply a covering argument to each S_n^p.

Let $\mathbb{S}_{\lambda_{q+1}}$ be the grid of $(\lambda_{q+1} n_*)^2$-many open squares contained in F_{e_3}, evenly centered around a grid of $(\lambda_{q+1} n_*)^2$-many points $\mathbb{G}_{\lambda_{q+1}}$ which contains the origin. By Proposition 4.3, for each choice of $k = (k_1, k_2) \in \{0, ..., r_2^{-1} - 1\}^2$, the support of the shifted pipe $\mathbb{W}_{\lambda_{q+1}, r_2, e_3}^k$ intersects F_{e_3} in a $\frac{2\pi}{4\lambda_{q+1} n_*}$-neighborhood of a finite subcollection of grid points from $\mathbb{G}_{\lambda_{q+1}}$, which we call $\mathbb{G}_{\lambda_{q+1}}^k$, and which by construction is $\frac{\mathbb{T}^3}{\lambda_{q+1} r_2 n_*}$-periodic. Furthermore, two subcollections for $k \ne k'$ contain no grid points in common. Let $\mathbb{S}_{\lambda_{q+1}}^k$ be the set of

---

[4]Technically, the proof still applies if A_n^p is self-intersecting, but the conclusions of Lemma 4.7 eliminate this possibility, so we shall ignore this issue and use the word "smooth."

open squares centered around grid points in $\mathbb{G}^k_{\lambda_{q+1}}$, so that $\mathbb{S}^k_{\lambda_{q+1}}$ and $\mathbb{S}^{k'}_{\lambda_{q+1}}$ are disjoint if $k \neq k'$. To prove (4.36), we will identify a shift k_0 such that the set of squares $\mathbb{S}^{k_0}_{\lambda_{q+1}}$ has empty intersection with S^p_n for all n. Then by Proposition 4.3, we have that the pipe flow $\mathbb{W}^{k_0}_{\lambda_{q+1},r_2,e_3}$ intersects F_{e_3} inside of $\mathbb{S}^{k_0}_{\lambda_{q+1}}$, and so we will have verified (4.36).

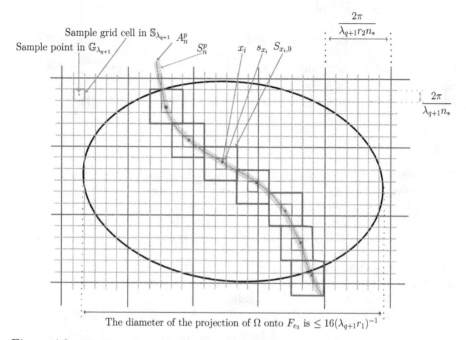

Figure 4.3: The boundary of the projection of Ω onto the face F_{e_3} is represented by the large dark oval. The small grid cells of sidelength $2\pi(\lambda_{q+1}n_*)^{-1}$ represent the elements of $\mathbb{S}_{\lambda_{q+1}}$, while the center points are the elements of $\mathbb{G}_{\lambda_{q+1}}$. A projected pipe S^p_n with axis A^p_n is represented in gray shading. A point $x_i \in A^p_n$, its associated grid cell s_{x_i}, and its 3×3 cluster $S_{x_i,9}$ are represented in the center of the image. The union of the 3×3 clusters, $\cup_i S_{x_i,9}$, generously covers the gray shaded projected pipe S^p_n.

In order to identify a suitable shift k_0 such that $\mathbb{S}^{k_0}_{\lambda_{q+1}}$ has empty intersection with S^p_n, we first present a generous cover for S^p_n; see Figure 4.3. Let $x_1 \in A^p_n$ be arbitrary. Set $s_{x_1} \in \mathbb{S}_{\lambda_{q+1}}$ to be the grid square of sidelength $\frac{2\pi}{\lambda_{q+1}n_*}$ containing x_1,[5] and let $S_{x_1,9}$ be the 3×3 cluster of squares surrounding s_{x_1}. Then either x_1 is within distance $\frac{2\pi}{\lambda_{q+1}n_*}$ of an endpoint of A^p_n, or the length of $A^p_n \cap S_{x_1,9}$ is at least $\frac{2\pi}{n_*\lambda_{q+1}}$. If possible, choose $x_2 \in A^p_n$ so that $S_{x_2,9}$ is disjoint from $S_{x_1,9}$, and iteratively continue choosing $x_i \in A^p_n$ with $S_{x_i,9}$ disjoint from $S_{x_j,9}$ with

---

[5]If x_1 is on the boundary of more than one square, any choice of s_{x_1} will work.

$1 \leq j \leq i - 1$. Due to aforementioned observation about the lower bound on the length of A_n^p in each $S_{x_i,9}$, after a finite number of steps, which we denote by i_n, one cannot choose $x_{i_{n+1}} \in A_n^p$ so that $S_{x_{i_{n+1}},9}$ is disjoint from previous clusters; see Figure 4.3. By the length constraint on A_n^p and the observations on the length of $A_n^p \cap S_{x_i,9}$ for each i, we obtain the bound

$$32(\lambda_{q+1}r_1)^{-1} \geq |A_n^p| \geq (i_n - 2)2\pi \left(n_*\lambda_{q+1}\right)^{-1},$$

which implies that i_n may be bounded from above as

$$i_n \leq \frac{32r_1^{-1}n_*}{2\pi} + 2 \leq 6n_*r_1^{-1} + 2 \leq 8n_*r_1^{-1} \tag{4.37}$$

since $r_1^{-1} \geq 1$. By the definition of i_n, any point $x \in A_n^p$ which does not belong to any of the clusters $\{S_{x_i,9}\}_{i=1}^{i_n}$ must be such that $S_{x,9}$ has non-empty intersection with $S_{x_j,9}$ for some $j \leq i_n$. Thus, if we denote by $S_{x_j,81}$ the cluster of 9×9 grid squares centered at x_j, it follows that x belongs to $S_{x_j,81}$, and thus $A_n^p \subset \cup_{i \leq i_n} S_{x_i,81}$. Furthermore, since it was observed earlier that S_n^p is contained inside a $2\pi(1 + \Gamma_{q+1}^{-1})(4n_*\lambda_{q+1})^{-1}$-neighborhood of A_n^p, we have in addition that

$$S_n^p \subset \bigcup_{i=1}^{i_n} S_{x_i,81}.$$

Thus, we have covered S_n^p using no more than

$$81i_n \leq 81 \cdot 8n_*r_1^{-1} = 648n_*r_1^{-1}$$

grid squares. Set $C_* = 1300n_*$. Repeating this argument for every $1 \leq n \leq N_\Omega$ and taking the union over n, we have thus covered $\cup_{n \leq N_\Omega} S_n^p$ using no more than

$$\frac{1}{2}C_*\mathcal{C}_A \cdot r_2^2 r_1^{-2} \cdot r_1^{-1} < r_2^{-2} \tag{4.38}$$

grid squares of sidelength $\frac{2\pi}{\lambda_{q+1}n_*}$; the strict inequality in (4.38) follows from the assumption (4.31).

In order to conclude the proof, we appeal to a pigeonhole argument, made possible by the bound (4.38). Indeed, the left side of (4.38) represents as an upper bound on the number of grid cells in $\mathbb{S}_{\lambda_{q+1}}$ which are deemed "occupied" by $\cup_{n \leq N_\Omega} S_n^p$, while the right side of (4.38) represents the number of possible choices for the shifts $k_0 \in \{0, ..., r_2^{-1} - 1\}^2$ belonging to the $\frac{2\pi}{\lambda_{q+1}r_2n_*}$-periodic subcollection $\mathbb{S}_{\lambda_{q+1}}^{k_0}$. See Figure 4.4 for details. We conclude by (4.38) and the pigeonhole principle that there exists a "free" shift $k_0 \in \{0, ..., r_2^{-1} - 1\}^2$ such that *none* of the squares in $\mathbb{S}_{\lambda_{q+1}}^{k_0}$ intersect the covering $\cup_{i \leq i_n} S_{x_i,81}$ of $\cup_{n \leq N_\Omega} S_n^p$. Choosing the pipe flow $\mathbb{W}_{\lambda_{q+1},r_2,e_3}^{k_0}$, we have proven (4.36), concluding the proof of the lemma when $\xi = e_3$.

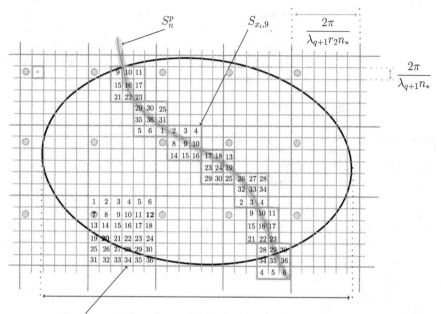

In this periodic cell, we check which grid cells are available.

Figure 4.4: We revisit Figure 4.3. Each cluster $S_{x_i,9}$ of nine cells covers a portion of S_n^p; their union covers the entirety. We would like to determine which set $\mathbb{S}_{\lambda_{q+1}}^{k_0}$ of $\frac{2\pi}{\lambda_{q+1}r_2n_*}$-periodic grid cells is free (we index these cells by the shift parameter k_0), so that we can place a $\frac{2\pi}{\lambda_{q+1}r_2n_*}$-periodic pipe flow $\mathbb{W}_{\lambda_{q+1},r_2,e_3}^{k_0}$ at the centers of the cells. This pipe flow then will not intersect the cells taken up by the union of the clusters $\cup_i S_{x_i,9}$. Towards this purpose, consider one of the periodic cells of sidelength $\frac{2\pi}{\lambda_{q+1}r_2n_*}$, e.g., bottom row, second from left. This cell contains r_2^{-2}-many sub-cells of sidelength $\frac{2\pi}{\lambda_{q+1}n_*}$, which in the figure we index by an integer $k \in \{1,\ldots,36\}$. In order to determine which of these sub-cells are "free," we verify for every k whether a periodic copy of the k-cell lies in the union of the clusters $\cup_i S_{x_i,9}$; if yes, we may not place a pipe in any periodic copies of this sub-cell. For instance, the cell with label 9 appears three times within the union of the clusters; the cell with label 3 appears twice; while the cell with label 36 appears just once. In the above figure we discover that there are only three "free" sub-cells, corresponding to the indices 7, 12, and 20. Any of these indices indicates a location where we may place a new pipe flow $\mathbb{W}_{\lambda_{q+1},r_2,e_3}^{k_0}$; in the figure, we have chosen k_0 to correspond to the label 7, and have represented by a $\frac{2\pi}{\lambda_{q+1}r_2n_*}$-periodic array of circles the intersections of the pipes in $\mathbb{W}_{\lambda_{q+1},r_2,e_3}^{k_0}$ with F_{e_3}.

To prove the proposition when $\xi \neq e_3$, first consider the portion[6] of $\Omega \subset \mathbb{R}^3$

---

[6]Recall that Ω is a $\mathbb{T}^3$-periodic set but can be considered as a subset of $\mathbb{R}^3$; cf. Definition 4.2.

restricted to the cube $[-\pi, \pi]^3$, denoted by $\Omega|_{[-\pi,\pi]^3}$, and consider similarly $S_n|_{[-\pi,\pi]^3}$ and $A_n|_{[-\pi,\pi]^3}$. Let $3\mathbb{T}_\xi^3$ be the $3 \times 3 \times 3$ cluster of periodic cells for $\mathbb{T}_\xi^3$ centered at the origin. Then $[-\pi, \pi]^3$ is contained in this cluster, and in particular $[-\pi, \pi]^3$ has *empty* intersection with the boundary of $3\mathbb{T}_\xi^3$ (understood as the boundary of the $3\mathbb{T}_\xi^3$-periodic cell centered at the origin when simply viewed as a subset of $\mathbb{R}^3$). Thus $\Omega|_{[0,2\pi]^3}$, $S_n|_{[-\pi,\pi]^3}$, and $A_n|_{[-\pi,\pi]^3}$ also have empty intersection with the boundary of $3\mathbb{T}_\xi^3$ and may be viewed as $3\mathbb{T}_\xi^3$-periodic sets. Up to a dilation which replaces $3\mathbb{T}_\xi^3$ with $\mathbb{T}_\xi^3$, we have exactly satisfied the assumptions of the proposition, but with $\mathbb{T}^3$-periodicity replaced by $\mathbb{T}_\xi^3$-periodicity. This dilation will shrink everything by a factor of 3, which we may compensate for by choosing a pipe flow $\mathbb{W}_{3\lambda_{q+1}, r_2, \xi}$, and then undoing the dilation at the end. Any constants related to this dilation are q-independent and may be absorbed into the geometric constant C_* at the end of the proof. At this point we may then redo the proof of the proposition with minimal adjustments. In particular, we replace the projection of S_n and A_n onto the face F_{e_3} of the box $\mathbb{T}^3$ with the projection of the restricted and dilated versions of S_n and A_n onto the face F_ξ of the box $\mathbb{T}_\xi^3$. We similarly replace the grids and squares on F_{e_3} with grids and squares on F_ξ, analogously to (4.3). The covering argument then proceeds exactly as before. The proof produces pipes belonging to the intermittent pipe flow $\mathbb{W}_{3\lambda_{q+1}, r_2, \xi}^{k_0}$ which are $\frac{\mathbb{T}^3}{3\lambda_{q+1}n_*r_2}$-periodic and disjoint from the dilated and restricted versions of the S_n's. Undoing the dilation, we find that $\mathbb{W}_{\lambda_{q+1}, r_2, \xi}^{k_0}$ is $\frac{\mathbb{T}^3}{\lambda_{q+1}r_2}$-periodic and disjoint from each S_n. Then all the conclusions of Proposition 4.8 have been achieved, finishing the proof. $\qquad\square$

Chapter Five

Mollification

Because the principal inductive assumptions for the velocity increments (3.13) and the Reynolds stress (3.15) are assumed to hold only for a limited number of space and material derivatives ($\leq 7\mathsf{N}_{\text{ind},v}$ and $\leq 3\mathsf{N}_{\text{ind},v}$ respectively), and because in our proof we need to appeal to derivative bounds of much higher orders, it is customary to employ a *mollification step* prior to adding the convex integration perturbation. This mollification step is discussed in Lemma 5.1. Note that the mollification step is only employed once (for every inductive step $q \mapsto q+1$), and is not repeated for the higher order stresses $R_{q,n,p}$. In particular, Lemma 5.1 already shows that the inductive assumption (3.12) holds for $q' = q$.

Lemma 5.1 (Mollifying the Euler-Reynolds system). *Let $(v_q, \mathring{R}_q)$ solve the Euler-Reynolds system (3.1), and assume that $\psi_{i,q'}, u_{q'}$ for $q' < q$, w_q, and $\mathring{R}_q$ satisfy (3.12)–(3.25b). Then, we mollify $(v_q, \mathring{R}_q)$ at spatial scale λ_q^{-1} and temporal scale $\widetilde{\tau}_{q-1}$ (cf. the notation in (9.64)), and accordingly define*

$$v_{\ell_q} := \mathcal{P}_{q,x,t} v_q \qquad \text{and} \qquad \mathring{R}_{\ell_q} := \mathcal{P}_{q,x,t} \mathring{R}_q. \tag{5.1}$$

The mollified pair $(v_{\ell_q}, \mathring{R}_{\ell_q})$ satisfies

$$\partial_t v_{\ell_q} + \text{div}\,(v_{\ell_q} \otimes v_{\ell_q}) + \nabla p_{\ell_q} = \text{div}\,\mathring{R}_{\ell_q} + \text{div}\,\mathring{R}_q^{\text{comm}}, \tag{5.2a}$$

$$\text{div}\,v_{\ell_q} = 0. \tag{5.2b}$$

The commutator stress $\mathring{R}_q^{\text{comm}}$ satisfies the estimate (consistent with (3.15) at level $q + 1$)

$$\left\| D^n D_{t,q}^m \mathring{R}_q^{\text{comm}} \right\|_{L^\infty} \leq \Gamma_{q+1}^{-1} \Gamma_{q+1}^{-\mathsf{C}_{\mathsf{R}}} \delta_{q+2} \lambda_{q+1}^n \mathcal{M}\left(m, \mathsf{N}_{\text{ind},t}, \tau_q^{-1}, \Gamma_q^{-1} \widetilde{\tau}_q^{-1}\right) \tag{5.3}$$

for all $n, m \leq 3\mathsf{N}_{\text{ind},v}$, and then we have that

$$\left\| D^n D_{t,q-1}^m (v_{\ell_q} - v_q) \right\|_{L^\infty} \leq \lambda_q^{-2} \delta_q^{1/2} \mathcal{M}\left(n, 2\mathsf{N}_{\text{ind},v}, \lambda_q, \widetilde{\lambda}_q\right)$$
$$\times \mathcal{M}\left(m, \mathsf{N}_{\text{ind},t}, \tau_{q-1}^{-1} \Gamma_q^{i-1}, \widetilde{\tau}_{q-1}^{-1} \Gamma_q^{-1}\right) \tag{5.4}$$

for all $n, m \leq 3\mathsf{N}_{\text{ind},v}$. Furthermore,

$$u_q = v_{\ell_q} - v_{\ell_{q-1}}$$

satisfies the bound (3.12) *with* q' *replaced by* q, *namely*

$$\left\|\psi_{i,q-1}D^n D_{t,q-1}^m u_q\right\|_{L^2} \leq \delta_q^{1/2} \mathcal{M}\left(n, 2\mathsf{N}_{\mathrm{ind},v}, \lambda_q, \widetilde{\lambda}_q\right)$$
$$\times \mathcal{M}\left(m, \mathsf{N}_{\mathrm{ind},t}, \Gamma_q^i \tau_{q-1}^{-1}, \widetilde{\tau}_{q-1}^{-1}\right). \qquad (5.5)$$

for all $n + m \leq 2\mathsf{N}_{\mathrm{fin}}$. *In fact, when either* $n \geq 3\mathsf{N}_{\mathrm{ind},v}$ *or* $m \geq 3\mathsf{N}_{\mathrm{ind},v}$ *are such that* $n + m \leq 2\mathsf{N}_{\mathrm{fin}}$, *then the above estimate holds uniformly:*

$$\left\|D^n D_{t,q-1}^m u_q\right\|_{L^\infty} \leq \Gamma_q^{-1} \delta_q^{1/2} \mathcal{M}\left(n, 2\mathsf{N}_{\mathrm{ind},v}, \lambda_q, \widetilde{\lambda}_q\right)$$
$$\times \mathcal{M}\left(m, \mathsf{N}_{\mathrm{ind},t}, \tau_{q-1}^{-1}, \widetilde{\tau}_{q-1}^{-1}\right) \qquad (5.6)$$

Finally, $\mathring{R}_{\ell_q}$ *satisfies bounds which extend* (3.15) *to*

$$\left\|\psi_{i,q-1}D^n D_{t,q-1}^m \mathring{R}_{\ell_q}\right\|_{L^1} \leq \Gamma_q^{-\mathsf{C_R}} \delta_{q+1} \mathcal{M}\left(n, 2\mathsf{N}_{\mathrm{ind},v}, \lambda_q, \widetilde{\lambda}_q\right)$$
$$\times \mathcal{M}\left(m, \mathsf{N}_{\mathrm{ind},t}, \Gamma_q^{i+2}\tau_{q-1}^{-1}, \widetilde{\tau}_{q-1}^{-1}\right) \qquad (5.7)$$

for all $n + m \leq 2\mathsf{N}_{\mathrm{fin}}$. *In fact, the above estimate holds uniformly,*

$$\left\|D^n D_{t,q-1}^m \mathring{R}_{\ell_q}\right\|_{L^\infty} \leq \Gamma_q^{-1} \Gamma_q^{-\mathsf{C_R}} \delta_{q+1} \mathcal{M}\left(n, 2\mathsf{N}_{\mathrm{ind},v}, \lambda_q, \widetilde{\lambda}_q\right)$$
$$\times \mathcal{M}\left(m, \mathsf{N}_{\mathrm{ind},t}, \tau_{q-1}^{-1}, \widetilde{\tau}_{q-1}^{-1}\right), \qquad (5.8)$$

whenever either $n \geq 3\mathsf{N}_{\mathrm{ind},v}$ *or* $m \geq 3\mathsf{N}_{\mathrm{ind},v}$ *are such that* $n + m \leq 2\mathsf{N}_{\mathrm{fin}}$.

Remark 5.2. The bounds (5.6) and (5.8) provide L^∞ estimates for $D^n D_{t,q-1}^m$ applied to u_q and $\mathring{R}_{\ell_q}$, respectively, but only when either n or m is sufficiently large. In the remaining cases, we note that (5.5), combined with the partition of unity property (3.16), and the inductive assumption (3.19) (with $M = 0$ and $K = 4$), implies the bound

$$\left\|D^n D_{t,q-1}^m u_q\right\|_{L^\infty(\mathrm{supp}\,\psi_{i,q-1})} \lesssim \delta_q^{1/2} \widetilde{\lambda}_q^{3/2} \mathcal{M}\left(n, 2\mathsf{N}_{\mathrm{ind},v}, \lambda_q, \widetilde{\lambda}_q\right)$$
$$\times \mathcal{M}\left(m, \mathsf{N}_{\mathrm{ind},t}, \tau_{q-1}^{-1}\Gamma_q^{i+1}, \widetilde{\tau}_{q-1}^{-1}\right) \qquad (5.9)$$

for all $n, m \leq 3\mathsf{N}_{\mathrm{ind},v}$. Indeed, we may apply Lemma A.3 (estimate (A.18b)) with $\psi_i = \psi_{i,q-1}$, $f = u_q$, $C_f = \delta_q^{1/2}$, $\rho = \lambda_{q-1}\Gamma_{q-1} \leq \lambda_q$ (cf. (9.38)), $\lambda = \lambda_q$, $\widetilde{\lambda} = \widetilde{\lambda}_q$, $\mu_i = \tau_q^{-1}\Gamma_q^i$, $\widetilde{\mu}_i = \widetilde{\tau}_{q-1}^{-1}$, $N_x = 2\mathsf{N}_{\mathrm{ind},v}$, $N_t = \mathsf{N}_{\mathrm{ind},t}$, and $N_\circ = 2\mathsf{N}_{\mathrm{fin}}$, to conclude that (5.9) holds for all $n + m \leq 2\mathsf{N}_{\mathrm{fin}} - 2$, and in particular for $n, m \leq 3\mathsf{N}_{\mathrm{ind},v}$.

A similar argument, shows that estimate (5.7) and Lemma A.3 imply

$$\left\|D^n D_{t,q-1}^m \mathring{R}_{\ell_q}\right\|_{L^\infty(\mathrm{supp}\,\psi_{i,q-1})} \lesssim \Gamma_q^{-\mathsf{C_R}} \delta_{q+1} \widetilde{\lambda}_q^3 \mathcal{M}\left(n, 2\mathsf{N}_{\mathrm{ind},v}, \lambda_q, \widetilde{\lambda}_q\right)$$
$$\times \mathcal{M}\left(m, \mathsf{N}_{\mathrm{ind},t}, \Gamma_q^{i+3}\tau_{q-1}^{-1}, \widetilde{\tau}_{q-1}^{-1}\right) \qquad (5.10)$$

for $n + m \leq 2N_{\mathrm{fin}} - 4$, and in particular for $n, m \leq 3N_{\mathrm{ind},v}$.

Proof of Lemma 5.1. The bound (5.3) requires a different proof than (5.5) and (5.7), so that we start with the former.

Proof of (5.3). Recall that

$$\mathring{R}_q^{\mathrm{comm}} = \mathcal{P}_{q,x,t} v_q \mathring{\otimes} \mathcal{P}_{q,x,t} v_q - \mathcal{P}_{q,x,t}(v_q \mathring{\otimes} v_q). \tag{5.11}$$

We note—cf. (9.64)—that $\mathcal{P}_{q,x,t}$ mollifies in space at length scale $\widetilde{\lambda}_q$, and in time at timescale $\widetilde{\tau}_{q-1}^{-1}$. Let us denote by K_q the space-time mollification kernel for $\mathcal{P}_{q,x,t}$, which thus equals the product of the bump functions $\phi_{\widetilde{\lambda}_q}^{(x)} \phi_{\widetilde{\tau}_{q-1}^{-1}}^{(t)}$. For brevity of notation (locally in this proof) it is convenient to denote space-time points as $(x,t), (y,s), (z,r) \in \mathbb{T}^3 \times \mathbb{R}$

$$(x,t) = \theta, \qquad (y,s) = \kappa, \qquad (z,r) = \zeta. \tag{5.12}$$

Using this notation we may write out the commutator stress $\mathring{R}_q^{\mathrm{comm}}$ explicitly, and symmetrizing the resulting expression leads to the formula

$$\mathring{R}_q^{\mathrm{comm}}(\theta) = \frac{-1}{2} \iint_{(\mathbb{T}^3 \times \mathbb{R})^2} (v_q(\theta - \kappa) - v_q(\theta - \zeta))^{\circ}$$
$$\otimes (v_q(\theta - \kappa) - v_q(\theta - \zeta)) K_q(\kappa) K_q(\zeta) \, d\kappa \, d\zeta. \tag{5.13}$$

Expanding v_q in a Taylor series in space and time around θ yields the formula

$$v_q(\theta - \kappa) = v_q(\theta) + \sum_{\substack{|\alpha|+m=1}}^{N_c-1} \frac{1}{\alpha! m!} D^\alpha \partial_t^m v_q(\theta)(-\kappa)^{(\alpha,m)} + R_{N_c}(\theta, \kappa), \tag{5.14}$$

where the remainder term with N_c derivatives is given by

$$R_{N_c}(\theta, \kappa) = \sum_{|\alpha|+m=N_c} \frac{N_c}{\alpha! m!} (-\kappa)^{(\alpha,m)} \int_0^1 (1-\eta)^{N_c-1} D^\alpha \partial_t^m v_q(\theta - \eta\kappa) \, d\eta. \tag{5.15}$$

The value of N_c will be chosen later so that $N_{\mathrm{ind},t} \ll N_c = N_{\mathrm{ind},v} - 2$, more precisely, such that conditions (5.24) and (9.50a) hold.

Using the fact that by (9.62) all moments of K_q vanish up to order N_c, we rewrite (5.13) as

$$\mathring{R}_q^{\mathrm{comm}}(\theta) = \int_{\mathbb{T}^3 \times \mathbb{R}} \sum_{\substack{|\alpha|+m=1}}^{N_c-1} \frac{(-\kappa)^{(\alpha,m)}}{\alpha! m!} D^\alpha \partial_t^m v_q(\theta) \mathring{\otimes}_s R_{N_c}(\theta, \kappa) K_q(\kappa) \, d\kappa$$

$$- \int_{\mathbb{T}^3 \times \mathbb{R}} R_{N_c}(\theta, \kappa) \mathring{\otimes} R_{N_c}(\theta, \kappa) K_q(\kappa) \, d\kappa$$

$$-\iint_{(\mathbb{T}^3\times\mathbb{R})^2} R_{N_c}(\theta,\kappa)\overset{\circ}{\otimes}_{\mathrm{sym}} R_{N_c}(\theta,\zeta)K_q(\kappa)K_q(\zeta)\,d\kappa\,d\zeta$$

$$=: \overset{\circ}{R}{}^{\mathrm{comm}}_{q,1}(\theta) + \overset{\circ}{R}{}^{\mathrm{comm}}_{q,2}(\theta) + \overset{\circ}{R}{}^{\mathrm{comm}}_{q,3}(\theta)\,, \tag{5.16}$$

where we have used the notation (9.66).

In order to prove (5.3), we first show that every term in $D^n D^m_{t,q}\overset{\circ}{R}{}^{\mathrm{comm}}_q$ can be decomposed into products of pure space and time differential operators applied to products of v_{ℓ_q} and v_q. More generally, for any sufficiently smooth function $F = F(x,t)$ and for any $n, m \geq 0$, the Leibniz rule implies that

$$D^n D^m_{t,q}F = D^n(\partial_t + v_{\ell_q}\cdot\nabla_x)^m F = \sum_{\substack{m'\leq m \\ n'+m'\leq n+m}} d_{n,m,n',m'}(x,t)D^{n'}\partial^{m'}_t F \tag{5.17a}$$

$$d_{n,m,n',m'}(x,t) = \sum_{k=0}^{m-m'}\ \sum_{\substack{\{\gamma\in\mathbb{N}^k\,:\,|\gamma|=n-n'+k, \\ \beta\in\mathbb{N}^k\,:\,|\beta|=m-m'-k\}}} c(m,n,k,\gamma,\beta)\prod_{\ell=1}^{k}\left(D^{\gamma_\ell}\partial^{\beta_\ell}_t v_{\ell_q}(x,t)\right), \tag{5.17b}$$

where $c(m,n,k,\gamma,\beta)$ denotes explicitly computable combinatorial coefficients which depend only on the factors inside the parentheses, which are in particular independent of q (which is why we do not carefully track these coefficients). Identity (5.17a)–(5.17b) holds because D and ∂_t commute; the proof is based on induction on n and m. Clearly, if $D_{t,q}$ in (5.17a) is replaced by $D_{t,q-1}$, then the same formula holds, with the v_{ℓ_q} factors in (5.17b) being replaced by $v_{\ell_{q-1}}$.

In order to prove (5.3) we consider (5.17a)–(5.17b) for $n, m \leq 3N_{\mathrm{ind},v}$, with $F = \overset{\circ}{R}{}^{\mathrm{comm}}_q$. In order to estimate the factors $d_{n,m,n',m'}$ in (5.17b), we need to bound $D^n\partial^m_t v_q$ for $n \leq 6N_{\mathrm{ind},v} + N_c$ and $m \leq 3N_{\mathrm{ind},v} + N_c$, with $n+m \leq 6N_{\mathrm{ind},v} + N_c$. Recall that $v_q = w_q + v_{\ell_{q-1}}$, and thus we will obtain the needed estimate from bounds on $D^n\partial^m_t w_q$ and $D^n\partial^m_t v_{\ell_{q-1}}$. We start with the latter.

We recall that $v_{\ell_{q-1}} = w_{q-1} + v_{\ell_{q-2}}$. Using (3.16) with $q' = q - 2$ and the inductive assumption (3.13) with q replaced with $q-1$, we obtain from Sobolev interpolation that $\|w_{q-1}\|_{L^\infty} \lesssim \|w_{q-1}\|^{1/4}_{L^2}\|D^2 w_{q-1}\|^{3/4}_{L^2} \lesssim \delta^{1/2}_{q-1}\lambda^{3/2}_{q-1}$. Additionally, combining (3.24) with $q' = q - 2$ and (3.18) with $q' = q - 2$, we obtain $\|v_{\ell_{q-2}}\|_{L^\infty} \lesssim \lambda^2_{q-2}\Gamma^{i_{\max}+1}_{q-1}\delta^{1/2}_{q-1} \lesssim \lambda^4_{q-2}\delta^{1/2}_{q-1}$. Jointly, these two estimates imply

$$\|v_{q-1}\|_{L^\infty} \lesssim \|w_{q-1}\|_{L^\infty} + \|v_{\ell_{q-2}}\|_{L^\infty} \lesssim \delta^{1/2}_{q-1}\lambda^4_{q-1}\,.$$

Now, using that $v_{\ell_{q-1}} = \mathcal{P}_{q-1,x,t}v_{q-1}$, and that the mollifier operator $\mathcal{P}_{q-1,x,t}$ localizes at scale $\widetilde{\lambda}_{q-1}$ in space and $\widetilde{\tau}^{-1}_{q-2}$ in time, we deduce the global estimate

$$\|D^n\partial^m_t v_{\ell_{q-1}}\|_{L^\infty} \lesssim (\lambda^4_{q-1}\delta^{1/2}_{q-1})\widetilde{\lambda}^n_{q-1}\widetilde{\tau}^{-m}_{q-2} \tag{5.18}$$

for $n + m \leq 2N_{\mathrm{fin}}$. Note that from the definitions (9.19) and (9.20), it is immediate that $\widetilde{\tau}_{q-2}^{-1} \ll \Gamma_q^{-1}\widetilde{\tau}_{q-1}^{-1}$.

As mentioned earlier, the bound for the space-time derivatives of $v_{\ell_{q-1}}$ needs to be combined with similar estimates for w_q in order to yield a control of v_q. For this purpose, we appeal to the Sobolev embedding $H^2 \subset L^\infty$ and the bound (3.13) (in which we take a supremum over $0 \leq i \leq i_{\max}$ and use (9.43)) to deduce

$$\left\| D^n D_{t,q-1}^m w_q \right\|_{L^\infty} \lesssim \left\| D^n D_{t,q-1}^m w_q \right\|_{H^2} \lesssim (\delta_q^{1/2}\lambda_q^2)\lambda_q^n(\widetilde{\tau}_{q-1}^{-1}\Gamma_q^{-1})^m \qquad (5.19)$$

for all $n \leq 7N_{\mathrm{ind},v} - 2$ and $m \leq 7N_{\mathrm{ind},v}$. Using the above estimate we may apply Lemma A.10 with the decomposition $\partial_t = -v_{\ell_{q-1}} \cdot \nabla + D_{t,q-1} = A + B$, $v = -v_{\ell_{q-1}}$ and $f = w_q$. The conditions (A.40) in Lemma A.10 holds in view of the inductive estimate (3.24) at level $q - 1$, with the following choice of parameters: $p = \infty$, $\Omega = \mathbb{T}^3$, $C_v = \lambda_{q-1}^4\delta_{q-1}^{1/2}$, $N_x = N_{\mathrm{ind},v} - 2$, $\lambda_v = \Gamma_{q-1}\lambda_{q-1}$, $\widetilde{\lambda}_v = \widetilde{\lambda}_{q-1}$, $N_t = N_{\mathrm{ind},t}$, $\mu_v = \lambda_{q-1}^2\tau_{q-1}^{-1}$, $\widetilde{\mu}_v = \Gamma_q^{-1}\widetilde{\tau}_{q-1}^{-1}$, and $N_* = {}^{3N_{\mathrm{fin}}}/_2$. On the other hand, using (5.19) we have that condition (A.41) holds with the parameters: $p = \infty$, $\Omega = \mathbb{T}^3$, $C_f = \delta_q^{1/2}\lambda_q^2$, $\lambda_f = \widetilde{\lambda}_f = \lambda_q$, $\mu_f = \widetilde{\mu}_f = \Gamma_q^{-1}\widetilde{\tau}_{q-1}^{-1}$, and $N_* = 7N_{\mathrm{ind},v} - 2$. We deduce from (A.44) and the inequalities $\widetilde{\lambda}_{q-1} \leq \lambda_q$ and $\lambda_{q-1}^4\delta_{q-1}^{1/2}\lambda_q \leq \Gamma_q^{-1}\widetilde{\tau}_{q-1}^{-1}$ (cf. (9.39), (9.43), and (9.20)) that

$$\left\| D^n \partial_t^m w_q \right\|_{L^\infty} \lesssim (\delta_q^{1/2}\lambda_q^2)\lambda_q^n(\widetilde{\tau}_{q-1}^{-1}\Gamma_q^{-1})^m \qquad (5.20)$$

holds for $n + m \leq 7N_{\mathrm{ind},v} - 2$.

By combining (5.18) and (5.20) with the definition (3.3) we thus deduce

$$\left\| D^n \partial_t^m v_q \right\|_{L^\infty} \lesssim (\lambda_{q-1}^4\delta_{q-1}^{1/2})\lambda_q^n(\widetilde{\tau}_{q-1}^{-1}\Gamma_q^{-1})^m \qquad (5.21)$$

for all $n + m \leq 7N_{\mathrm{ind},v} - 2$, where we have used that $\lambda_{q-1}^4\delta_{q-1}^{1/2} \geq \delta_q^{1/2}\lambda_q^2$ and that $\widetilde{\tau}_{q-2}^{-1} \leq \Gamma_q^{-1}\widetilde{\tau}_{q-1}^{-1}$. By the definition of v_{ℓ_q} in (5.1) we thus also deduce that

$$\left\| D^n \partial_t^m v_{\ell_q} \right\|_{L^\infty} \lesssim (\lambda_{q-1}^4\delta_{q-1}^{1/2})\lambda_q^n(\widetilde{\tau}_{q-1}^{-1}\Gamma_q^{-1})^m \qquad (5.22)$$

for all $n + m \leq 7N_{\mathrm{ind},v} - 2$. Note that by the definition of the mollifier operator $\mathbb{P}_{q,x,t}$, any further space derivative on v_{ℓ_q} costs a factor of $\widetilde{\lambda}_q$, while additional temporal derivatives cost $\widetilde{\tau}_{q-1}$, up to a $2N_{\mathrm{fin}}$ total number of derivatives.

With (5.22) in hand, we may return to (5.17b) and deduce that for $n, m \leq 3N_{\mathrm{ind},v}$, we have

$$\begin{aligned}
\left\| d_{n,m,n',m'} \right\|_{L^\infty} &\lesssim \sum_{k=0}^{m-m'} \lambda_q^{n-n'+k}(\widetilde{\tau}_{q-1}^{-1}\Gamma_q^{-1})^{m-m'-k}(\lambda_{q-1}^4\Gamma_q\delta_{q-1}^{1/2})^k \\
&\lesssim \lambda_q^{n-n'}(\widetilde{\tau}_{q-1}^{-1}\Gamma_q^{-1})^{m-m'}.
\end{aligned} \qquad (5.23)$$

In the last inequality above we have used that $\lambda_q \lambda_{q-1}^4 \Gamma_q \delta_{q-1}^{1/2} \leq \widetilde{\tau}_{q-1}^{-1} \Gamma_q^{-1}$, which is a consequence of (9.39), (9.43), and (9.20).

Returning to (5.17a) with $F = \mathring{R}_q^{\mathrm{comm}}$, we use the expansion in (5.16), the definition (5.15), and the bound (5.21) to estimate $D^{n'} \partial_t^{m'} \mathring{R}_q^{\mathrm{comm}}$ when $n', m' \leq 3\mathsf{N}_{\mathrm{ind},v}$. Using (5.21) and the choice

$$N_{\mathrm{c}} = \mathsf{N}_{\mathrm{ind},v} - 2, \tag{5.24}$$

which is required in order to ensure that $n' + m' + N_{\mathrm{c}} \leq 7\mathsf{N}_{\mathrm{ind},v} - 2$, we first obtain the pointwise estimate

$$\left| D^{n''} \partial_t^{m''} R_{N_{\mathrm{c}}}(\theta, \kappa) \right| \lesssim (\lambda_{q-1}^4 \delta_{q-1}^{1/2}) \sum_{|\alpha| + m_1 = N_{\mathrm{c}}} \left| \kappa^{(\alpha, m_1)} \right| \lambda_q^{n'' + |\alpha|} (\widetilde{\tau}_{q-1}^{-1} \Gamma_q^{-1})^{m'' + m_1}, \tag{5.25}$$

where we recall the notation in (5.12). Using (5.25), the Leibniz rule, and the fact that $\lambda_q \Gamma_q \leq \widetilde{\lambda}_q$, we may estimate

$$
\begin{aligned}
&\left\| D^{n'} \partial_t^{m'} \mathring{R}_{q,2}^{\mathrm{comm}} \right\|_{L^\infty} \\
&\lesssim (\lambda_{q-1}^4 \delta_{q-1}^{1/2})^2 \sum_{|\alpha| + m_1 = N_{\mathrm{c}}} \sum_{|\alpha'| + m_2 = N_{\mathrm{c}}} \lambda_q^{n' + |\alpha| + |\alpha'|} (\widetilde{\tau}_{q-1}^{-1} \Gamma_q^{-1})^{m' + m_1 + m_2} \\
&\qquad\qquad\qquad \times \int_{\mathbb{T}^3 \times \mathbb{R}} |\kappa^{(\alpha + \alpha', m_1 + m_2)}| |K_q(\kappa)| d\kappa \\
&\lesssim (\lambda_{q-1}^4 \delta_{q-1}^{1/2})^2 \sum_{|\alpha| + m_1 = N_{\mathrm{c}}} \sum_{|\alpha'| + m_2 = N_{\mathrm{c}}} \lambda_q^{n' + |\alpha| + |\alpha'|} (\widetilde{\tau}_{q-1}^{-1} \Gamma_q^{-1})^{m' + m_1 + m_2} \\
&\qquad\qquad\qquad \times \widetilde{\lambda}_q^{-|\alpha| - |\alpha'|} \widetilde{\tau}_{q-1}^{m_1 + m_2} \\
&\lesssim (\lambda_{q-1}^4 \delta_{q-1}^{1/2})^2 \lambda_q^{n'} (\widetilde{\tau}_{q-1}^{-1} \Gamma_q^{-1})^{m'} \Gamma_q^{-2N_{\mathrm{c}}}
\end{aligned}
$$

whenever $n', m' \leq 3\mathsf{N}_{\mathrm{ind},v}$. It is clear that a very similar argument also gives the bound

$$\left\| D^{n'} \partial_t^{m'} \mathring{R}_{q,3}^{\mathrm{comm}} \right\|_{L^\infty} \lesssim (\lambda_{q-1}^4 \delta_{q-1}^{1/2})^2 \lambda_q^{n'} (\widetilde{\tau}_{q-1}^{-1} \Gamma_q^{-1})^{m'} \Gamma_q^{-2N_{\mathrm{c}}}$$

for the same range of n' and m'. Lastly, by combining (5.25), (5.21), and the Leibniz rule, we similarly deduce

$$
\begin{aligned}
&\left\| D^{n'} \partial_t^{m'} \mathring{R}_{q,1}^{\mathrm{comm}} \right\|_{L^\infty} \\
&\lesssim (\lambda_{q-1}^4 \delta_{q-1}^{1/2})^2 \sum_{|\alpha| + m_1 = 1}^{N_{\mathrm{c}} - 1} \sum_{|\alpha'| + m_2 = N_{\mathrm{c}}} \lambda_q^{n' + |\alpha| + |\alpha'|} (\widetilde{\tau}_{q-1}^{-1} \Gamma_q^{-1})^{m' + m_1 + m_2} \\
&\qquad\qquad\qquad \times \int_{\mathbb{T}^3 \times \mathbb{R}} \left| \kappa^{(\alpha + \alpha', m_1 + m_2)} \right| |K_q(\kappa)| d\kappa
\end{aligned}
$$

$$\lesssim (\lambda_{q-1}^4 \delta_{q-1}^{1/2})^2 \sum_{|\alpha|+m_1=1}^{N_c-1} \sum_{|\alpha'|+m_2=N_c} \lambda_q^{n'+|\alpha|+|\alpha'|} (\widetilde{\tau}_{q-1}^{-1}\Gamma_q^{-1})^{m'+m_1+m_2}$$
$$\times \widetilde{\lambda}_q^{-|\alpha|-|\alpha'|} \widetilde{\tau}_{q-1}^{m_1+m_2}$$
$$\lesssim (\lambda_{q-1}^4 \delta_{q-1}^{1/2})^2 \lambda_q^{n'} (\widetilde{\tau}_{q-1}^{-1}\Gamma_q^{-1})^{m'} \Gamma_q^{-N_c-1}.$$

Combining the above three bounds, identity (5.16) yields

$$\left\| D^{n'} \partial_t^{m'} \mathring{R}_q^{\mathrm{comm}} \right\|_{L^\infty} \lesssim (\lambda_{q-1}^4 \delta_{q-1}^{1/2})^2 \lambda_q^{n'} (\widetilde{\tau}_{q-1}^{-1}\Gamma_q^{-1})^{m'} \Gamma_q^{-N_c-1} \qquad (5.26)$$

whenever $n', m' \leq 3\mathsf{N}_{\mathrm{ind},v}$.

Lastly, by combining (5.17a) with (5.23) and (5.26) we obtain

$$\left\| D^n D_{t,q}^m \mathring{R}_q^{\mathrm{comm}} \right\|_{L^\infty} \lesssim (\lambda_{q-1}^4 \delta_{q-1}^{1/2})^2 \lambda_q^n (\widetilde{\tau}_{q-1}^{-1}\Gamma_q^{-1})^m \Gamma_q^{-N_c-1}$$

for all $n, m \leq 3\mathsf{N}_{\mathrm{ind},v}$. Therefore, in order to verify (5.3), we need to verify that

$$(\lambda_{q-1}^4 \delta_{q-1}^{1/2})^2 \lambda_q^n (\widetilde{\tau}_{q-1}^{-1}\Gamma_q^{-1})^m \Gamma_q^{-N_c}$$
$$\leq \Gamma_{q+1}^{-1} \Gamma_{q+1}^{-C_R} \delta_{q+2} \lambda_{q+1}^n \mathcal{M}\left(m, \mathsf{N}_{\mathrm{ind},t}, \tau_q^{-1}, \widetilde{\tau}_q^{-1}\Gamma_q^{-1}\right)$$

for $0 \leq n, m \leq 3\mathsf{N}_{\mathrm{ind},v}$. Since $\lambda_q \leq \lambda_{q+1}$, $\widetilde{\tau}_{q-1}^{-1} \leq \widetilde{\tau}_q^{-1}$, and $\widetilde{\tau}_{q-1}^{-1}\Gamma_q^{-1} \geq \tau_q^{-1} \geq \tau_{q-1}^{-1}$, the above condition is ensured by the more restrictive condition

$$\lambda_{q-1}^8 \Gamma_{q+1}^{1+C_R} \frac{\delta_{q-1}}{\delta_{q+2}} \left(\frac{\widetilde{\tau}_{q-1}^{-1}\Gamma_q^{-1}}{\tau_q^{-1}}\right)^{\mathsf{N}_{\mathrm{ind},t}} \leq \lambda_{q-1}^8 \Gamma_{q+1}^{1+C_R} \frac{\delta_{q-1}}{\delta_{q+2}} \left(\frac{\widetilde{\tau}_{q-1}^{-1}}{\tau_{q-1}^{-1}}\right)^{\mathsf{N}_{\mathrm{ind},t}}$$
$$\leq \Gamma_q^{N_c} = \Gamma_q^{\mathsf{N}_{\mathrm{ind},v}-2}, \qquad (5.27)$$

which holds as soon as $\mathsf{N}_{\mathrm{ind},v}$ is chosen sufficiently large with respect to $\mathsf{N}_{\mathrm{ind},t}$; see (9.50a) below. This completes the proof of (5.3).

Proof of (5.5) and (5.6). Using Hölder's inequality and the extra factor of Γ_q^{-1} present in (5.6), it is clear that for all n, m such that (5.6) holds, the estimate (5.5) is also true. The proof is thus split in three parts: first we consider $n, m \leq 3\mathsf{N}_{\mathrm{ind},v}$, then we consider $m > 3\mathsf{N}_{\mathrm{ind},v}$, and lastly we consider $n > 3\mathsf{N}_{\mathrm{ind},v}$.

We start with the proof of (5.5). In view of (3.4), we first bound the main term, $\mathcal{P}_{q,x,t} w_q$, which we claim may be estimated as

$$\left\| \psi_{i,q-1} D^n D_{t,q-1}^m \mathcal{P}_{q,x,t} w_q \right\|_{L^2}$$
$$\leq \frac{1}{2} \delta_q^{1/2} \mathcal{M}\left(n, 2\mathsf{N}_{\mathrm{ind},v}, \lambda_q, \widetilde{\lambda}_q\right) \mathcal{M}\left(m, \mathsf{N}_{\mathrm{ind},t}, \tau_{q-1}^{-1}\Gamma_q^i, \widetilde{\tau}_{q-1}^{-1}\right). \qquad (5.28)$$

for all $n, m \leq 3\mathsf{N}_{\mathrm{ind},v}$, and as

$$
\begin{aligned}
\left\| D^n D^m_{t,q-1} \mathcal{P}_{q,x,t} w_q \right\|_{L^\infty} & \\
& \leq \Gamma_q^{-2} \delta_q^{1/2} \mathcal{M}\left(n, 2\mathsf{N}_{\mathrm{ind},v}, \lambda_q, \widetilde{\lambda}_q\right) \mathcal{M}\left(m, \mathsf{N}_{\mathrm{ind},t}, \tau_{q-1}^{-1}, \widetilde{\tau}_{q-1}^{-1}\right).
\end{aligned} \tag{5.29}
$$

when $n + m \leq 2\mathsf{N}_{\mathrm{fin}}$, and either $n > 3\mathsf{N}_{\mathrm{ind},v}$ or $m > 3\mathsf{N}_{\mathrm{ind},v}$. By the definition of $\mathcal{P}_{q,x,t}$ in (9.64), in view of the moment condition (9.62) for the associated mollifier kernel, we have that

$$
\begin{aligned}
\mathcal{P}_{q,x,t} w_q(\theta) - w_q(\theta) = & \sum_{|\alpha|+m''=N_c} \frac{N_c}{\alpha! m''!} \iint_{\mathbb{T}^3 \times \mathbb{R}} K_q(\kappa)(-\kappa)^{(\alpha,m'')} \\
& \times \int_0^1 (1-\eta)^{N_c-1} D^\alpha \partial_t^{m''} w_q(\theta - \eta\kappa)\, d\eta d\kappa,
\end{aligned} \tag{5.30}
$$

where we have appealed to the notation in (5.12), and $N_c = \mathsf{N}_{\mathrm{ind},v} - 2$. For $n, m \leq 3\mathsf{N}_{\mathrm{ind},v}$, we appeal to the identity (5.17a) with $F = \mathcal{P}_{q,x,t} w_q - w_q$, and with $D_{t,q}$ replaced by $D_{t,q-1}$, to obtain

$$
\begin{aligned}
\left\| D^n D^m_{t,q-1} (\mathcal{P}_{q,x,t} w_q - w_q) \right\|_{L^\infty} & \\
& \lesssim \sum_{\substack{m' \leq m \\ n'+m' \leq n+m}} \left\| d_{n,m,n',m'} \right\|_{L^\infty} \left\| D^{n'} \partial_t^{m'} (\mathcal{P}_{q,x,t} w_q - w_q) \right\|_{L^\infty},
\end{aligned} \tag{5.31}
$$

where

$$
d_{n,m,n',m'} = \sum_{k=0}^{m-m'} \sum_{\substack{\{\gamma \in \mathbb{N}^k : |\gamma|=n-n'+k, \\ \beta \in \mathbb{N}^k : |\beta|=m-m'-k\}}} c(m,n,k,\gamma,\beta) \prod_{\ell=1}^{k} \left(D^{\gamma_\ell} \partial_t^{\beta_\ell} v_{\ell_{q-1}}(x,t) \right).
$$

From (5.18), and the parameter inequality $\lambda_{q-1}^4 \delta_{q-1}^{1/2} \widetilde{\lambda}_{q-1} \leq \Gamma_q^{-1} \widetilde{\tau}_{q-1}^{-1}$ we deduce the bound

$$
\left\| D^{n''} \partial_t^{m''} v_{\ell_{q-1}} \right\|_{L^\infty} \lesssim \widetilde{\lambda}_{q-1}^{n''-1} (\Gamma_q^{-1} \widetilde{\tau}_{q-1}^{-1})^{m''+1}
$$

for $n'' + m'' \leq 2\mathsf{N}_{\mathrm{fin}}$, and therefore

$$
\left\| d_{n,m,n',m'} \right\|_{L^\infty} \lesssim \lambda_q^{n-n'} (\widetilde{\tau}_{q-1}^{-1} \Gamma_q^{-1})^{m-m'}. \tag{5.32}
$$

Combining this estimate with the bound (5.20), we deduce that

$$
\begin{aligned}
\left\| D^n D^m_{t,q-1} (\mathcal{P}_{q,x,t} w_q - w_q) \right\|_{L^\infty} & \\
& \lesssim \sum_{\substack{m' \leq m \\ n'+m' \leq n+m}} \lambda_q^{n-n'} (\widetilde{\tau}_{q-1}^{-1} \Gamma_q^{-1})^{m-m'} \left\| D^{n'} \partial_t^{m'} (\mathcal{P}_{q,x,t} w_q - w_q) \right\|_{L^\infty}
\end{aligned}
$$

$$\lesssim \sum_{\substack{m' \leq m \\ n'+m' \leq n+m}} \sum_{|\alpha|+m''=N_c} \lambda_q^{n-n'} (\widetilde{\tau}_{q-1}^{-1} \Gamma_q^{-1})^{m-m'}$$

$$\times (\delta_q^{1/2} \lambda_q^2) \lambda_q^{n'+|\alpha|} (\widetilde{\tau}_{q-1}^{-1} \Gamma_q^{-1})^{m'+m''} \int_{\mathbb{T}^3 \times \mathbb{R}} \left| \kappa^{(\alpha,m'')} \right| |K_q(\kappa)| d\kappa$$

$$\lesssim (\delta_q^{1/2} \lambda_q^2) \sum_{|\alpha|+m''=N_c} \lambda_q^{n+|\alpha|} (\widetilde{\tau}_{q-1}^{-1} \Gamma_q^{-1})^{m+m''} \widetilde{\lambda}_q^{-|\alpha|} \widetilde{\tau}_{q-1}^{m''}$$

$$\lesssim (\delta_q^{1/2} \lambda_q^2) \lambda_q^n (\widetilde{\tau}_{q-1}^{-1} \Gamma_q^{-1})^m \Gamma_q^{-N_c} . \tag{5.33}$$

Next, we claim that the above estimate is consistent with (5.28): for $n, m \leq 3N_{\text{ind},v}$ we have

$$(\delta_q^{1/2} \lambda_q^2) \lambda_q^n (\widetilde{\tau}_{q-1}^{-1} \Gamma_q^{-1})^m \Gamma_q^{-N_c} \lesssim \Gamma_q^{-1} \delta_q^{1/2} \lambda_q^n \mathcal{M} \left(m, N_{\text{ind},t}, \tau_{q-1}^{-1} \Gamma_q^{i-1}, \widetilde{\tau}_{q-1}^{-1} \Gamma_q^{-1} \right) . \tag{5.34}$$

Recalling the definition of N_c in (5.24), the above bound is in turn implied by the estimate

$$\Gamma_q^3 \lambda_q^2 \left(\frac{\widetilde{\tau}_{q-1}^{-1}}{\tau_{q-1}^{-1}} \right)^{N_{\text{ind},t}} \leq \Gamma_q^{N_{\text{ind},v}},$$

which holds since $N_{\text{ind},v} \gg N_{\text{ind},t}$; in fact, it is easy to see that the above condition is less stringent than (5.27). Summarizing (5.33)–(5.34), and appealing to the inductive assumption (3.13), we deduce that

$$\left\| \psi_{i,q-1} D^n D_{t,q-1}^m \mathcal{P}_{q,x,t} w_q \right\|_{L^2}$$
$$\lesssim \left\| \psi_{i,q-1} D^n D_{t,q-1}^m w_q \right\|_{L^2} + \left\| D^n D_{t,q-1}^m (\mathcal{P}_{q,x,t} w_q - w_q) \right\|_{L^\infty}$$
$$\lesssim \Gamma_q^{-1} \delta_q^{1/2} \lambda_q^n \mathcal{M} \left(m, N_{\text{ind},t}, \tau_{q-1}^{-1} \Gamma_q^{i-1}, \widetilde{\tau}_{q-1}^{-1} \Gamma_q^{-1} \right) \tag{5.35}$$

for all $0 \leq n, m \leq 3N_{\text{ind},v}$. The above estimate verifies (5.28).

We next turn to the proof of (5.29). The key observation is that when establishing (5.35), the two main properties of the mollification kernel $K_q(\kappa)$ which we have used are: the vanishing of the moments $\iint_{\mathbb{T}^3 \times \mathbb{R}} K_q(\kappa)(-\kappa)^{(\alpha,m'')} d\kappa = 0$ for $1 \leq |\alpha| + m'' \leq N_{\text{ind},v}$ and the fact that $\|K_q(\kappa)(-\kappa)^{(\alpha,m'')}\|_{L^1(d\kappa)} \lesssim \widetilde{\lambda}_q^{-|\alpha|} \widetilde{\tau}_{q-1}^{m''}$ for all $|\alpha| + m'' \leq N_{\text{ind},v}$. We claim that, for any $\widetilde{n} + \widetilde{m} \leq 2N_{\text{fin}}$, the kernel

$$K_q^{(\widetilde{n},\widetilde{m})}(y,s) := D_y^{\widetilde{n}} \partial_s^{\widetilde{m}} K_q(y,s) \widetilde{\lambda}_q^{-\widetilde{n}} \widetilde{\tau}_{q-1}^{\widetilde{m}}$$

satisfies exactly the same two properties. The second property, about the L^1 norm, is immediate by scaling and the above definition, from the properties of the Friedrichs mollifier densities ϕ and $\widetilde{\phi}$ from (9.62). Concerning the vanishing moment condition, we note that $K_q^{(n,m)}$ has in fact more vanishing moments than K_q, as is easily seen from integration by parts in κ. The upshot of this

observation is that in precisely the same way that (5.35) was proven, we may
show that

$$\left\| D^n D^m_{t,q-1} D^{\tilde{n}} \partial_t^{\tilde{m}} \mathcal{P}_{q,x,t} w_q \right\|_{L^2}$$

$$\lesssim \sum_{i=0}^{i_{\max}} \left\| \psi_{i,q-1} D^n D^m_{t,q-1} w_q \right\|_{L^2} + \left\| D^n D^m_{t,q-1} (D^{\tilde{n}} \partial_t^{\tilde{m}} \mathcal{P}_{q,x,t} w_q - w_q) \right\|_{L^\infty}$$

$$\lesssim \Gamma_q^{-1} \delta_q^{1/2} \lambda_q^n \tilde{\lambda}_q^{\tilde{n}} (\tilde{\tau}_{q-1}^{-1} \Gamma_q^{-1})^m (\tilde{\tau}_{q-1}^{-1})^{\tilde{m}} \tag{5.36}$$

for all $0 \le n, m \le 3N_{\mathrm{ind},v}$, and for all $0 \le \tilde{n} + \tilde{m} \le 2N_{\mathrm{fin}}$. Here we have used
(3.16) and (3.18) with $q' = q - 1$, and the parameter inequality $\tau_{q-1}^{-1} \Gamma_q^{i_{\max}-1} \le$
$\tau_{q-1}^{-1} \lambda_{q-1}^2 \le \tilde{\tau}_{q-1}^{-1} \Gamma_q^{-1}$.

Next, consider $n + m \le 2N_{\mathrm{fin}}$ such that $n \le 3N_{\mathrm{ind},v}$ and $m > 3N_{\mathrm{ind},v}$. Define
$\bar{m} = m - 3N_{\mathrm{ind},v} > 0$, which is the number of excess material derivatives not
covered by the bound (5.35). We rewrite the term which we need to estimate
in (5.29) as

$$\left\| D^n D^m_{t,q-1} \mathcal{P}_{q,x,t} w_q \right\|_{L^\infty} = \left\| D^n D^{3N_{\mathrm{ind},v}}_{t,q-1} D^{\bar{m}}_{t,q-1} \mathcal{P}_{q,x,t} w_q \right\|_{L^\infty} . \tag{5.37}$$

Using (5.17a)–(5.17b) we expand $D^{\bar{m}}_{t,q-1}$ into space and time derivatives and
apply the Leibniz rule to deduce

$$D^{\bar{m}}_{t,q-1} \mathcal{P}_{q,x,t} w_q = \sum_{\substack{\bar{m}' \le \bar{m} \\ \bar{n}' + \bar{m}' \le \bar{m}}} d_{\bar{m},\bar{n}',\bar{m}'} D^{\bar{n}'} \partial_t^{\bar{m}'} \mathcal{P}_{q,x,t} w_q \tag{5.38a}$$

$$d_{\bar{m},\bar{n}',\bar{m}'}(x,t) = \sum_{k=0}^{\bar{m}-\bar{m}'} \sum_{\substack{\{\gamma \in \mathbb{N}^k : |\gamma| = -\bar{n}'+k, \\ \beta \in \mathbb{N}^k : |\beta| = \bar{m}-\bar{m}'-k\}}} c(\bar{m},k,\gamma,\beta) \prod_{\ell=1}^k \left(D^{\gamma_\ell} \partial_t^{\beta_\ell} v_{\ell_{q-1}}(x,t) \right) . \tag{5.38b}$$

Using the Leibniz rule, the previously established bound (5.36), and the Sobolev
embedding $H^2 \subset L^\infty$, we deduce that

$$\left\| D^n D^{3N_{\mathrm{ind},v}}_{t,q-1} D^{\bar{m}}_{t,q-1} \mathcal{P}_{q,x,t} w_q \right\|_{L^\infty}$$

$$\lesssim \sum_{a=0}^n \sum_{b=0}^{3N_{\mathrm{ind},v}} \sum_{\substack{\bar{m}' \le \bar{m} \\ \bar{n}' + \bar{m}' \le \bar{m}}} \left\| D^a D^b_{t,q-1} d_{\bar{m},\bar{n}',\bar{m}'} \right\|_{L^\infty}$$

$$\times \left\| D^{n-a} D^{3N_{\mathrm{ind},v}-b}_{t,q-1} D^{\bar{n}'} \partial_t^{\bar{m}'} \mathcal{P}_{q,x,t} w_q \right\|_{L^\infty}$$

$$\lesssim \sum_{a=0}^{n} \sum_{b=0}^{3N_{\mathrm{ind},v}} \sum_{\substack{\bar{m}' \leq \bar{m} \\ \bar{n}' + \bar{m}' \leq \bar{m}}} \left\| D^a D_{t,q-1}^b d_{\bar{m},\bar{n}',\bar{m}'} \right\|_{L^\infty}$$

$$\times \, \Gamma_q^{-1} \delta_q^{1/2} \lambda_q^{n-a} \widetilde{\lambda}_q^{\bar{n}'+2} (\widetilde{\tau}_{q-1}^{-1} \Gamma_q^{-1})^{3N_{\mathrm{ind},v}-b} (\widetilde{\tau}_{q-1}^{-1})^{\bar{m}'} . \qquad (5.39)$$

Thus, in order to obtain the desired bound on (5.37), we need to estimate space and material derivatives $D^a D_{t,q-1}^b$ of the term defined in (5.38b), and in particular $D^{\gamma_\ell} \partial_t^{\beta_\ell} v_{\ell_{q-1}}$. We may, however, appeal to (5.31)–(5.32) with $(\mathcal{P}_{q,x,t} w_q - w_q)$ replaced by $D^{\gamma_\ell} \partial_t^{\beta_\ell} v_{\ell_{q-1}}$ and to the bound (5.18) to deduce that

$$\left\| D^{a'} D_{t,q-1}^{b'} D^{\gamma_\ell} \partial_t^{\beta_\ell} v_{\ell_{q-1}} \right\|_{L^\infty}$$

$$\lesssim \sum_{\substack{b'' \leq b' \\ a''+b'' \leq a'+b'}} \lambda_q^{a'-a''} (\Gamma_q^{-1} \widetilde{\tau}_{q-1}^{-1})^{b'-b''} \left\| D^{a''} \partial_t^{b''} D^{\gamma_\ell} \partial_t^{\beta_\ell} v_{\ell_{q-1}} \right\|_{L^\infty}$$

$$\lesssim (\lambda_{q-1}^4 \delta_{q-1}^{1/2}) \lambda_q^{a'} \widetilde{\lambda}_{q-1}^{\gamma_\ell} (\Gamma_q^{-1} \widetilde{\tau}_{q-1}^{-1})^{b'+\beta_\ell}$$

$$\lesssim \lambda_q^{a'} \widetilde{\lambda}_{q-1}^{\gamma_\ell-1} (\Gamma_q^{-1} \widetilde{\tau}_{q-1}^{-1})^{b'+\beta_\ell+1} .$$

In the last estimate we have used the parameter inequality $\lambda_{q-1}^4 \delta_{q-1}^{1/2} \widetilde{\lambda}_{q-1} \leq \Gamma_q^{-1} \widetilde{\tau}_{q-1}^{-1}$. Using the above bound and the definition (5.38b) we deduce that

$$\left\| D^a D_{t,q-1}^b d_{\bar{m},\bar{n}',\bar{m}'} \right\|_{L^\infty} \lesssim \lambda_q^a \widetilde{\lambda}_{q-1}^{-\bar{n}'} (\Gamma_q^{-1} \widetilde{\tau}_{q-1}^{-1})^{b+\bar{m}-\bar{m}'} . \qquad (5.40)$$

The above display may be combined with (5.39) and yields

$$\left\| D^n D_{t,q-1}^{3N_{\mathrm{ind},v}} D_{t,q-1}^{\bar{m}} \mathcal{P}_{q,x,t} w_q \right\|_{L^\infty}$$

$$\lesssim \Gamma_q^{-1} \delta_q^{1/2} \lambda_q^n \widetilde{\lambda}_q^2 \sum_{b=0}^{3N_{\mathrm{ind},v}} \sum_{\substack{\bar{m}' \leq \bar{m} \\ \bar{n}' + \bar{m}' \leq \bar{n}+\bar{m}}} \widetilde{\lambda}_{q-1}^{-\bar{n}'} (\Gamma_q^{-1} \widetilde{\tau}_{q-1}^{-1})^{b+\bar{m}-\bar{m}'} (\widetilde{\tau}_{q-1}^{-1} \Gamma_q^{-1})^{3N_{\mathrm{ind},v}-b} \widetilde{\tau}_{q-1}^{-\bar{m}'}$$

$$\lesssim \Gamma_q^{-1} \delta_q^{1/2} \lambda_q^n \widetilde{\lambda}_q^2 \sum_{\bar{m}' \leq \bar{m}} (\Gamma_q^{-1} \widetilde{\tau}_{q-1}^{-1})^{m-\bar{m}'} (\widetilde{\tau}_{q-1}^{-1})^{\bar{m}'} , \qquad (5.41)$$

where we have recalled that $3N_{\mathrm{ind},v} + \bar{m} = m$. The above estimate has to be compared with the right side of (5.29), and for this purpose we note that for $\bar{m}' \leq \bar{m} = m - 3N_{\mathrm{ind},v}$ we have

$$\lambda_q^n (\Gamma_q^{-1} \widetilde{\tau}_{q-1}^{-1})^{m-\bar{m}'} (\widetilde{\tau}_{q-1}^{-1})^{\bar{m}'}$$

$$\lesssim \mathcal{M} \left(n, 2N_{\mathrm{ind},v}, \lambda_q, \widetilde{\lambda}_q \right) \Gamma_q^{-(m-\bar{m}')} (\widetilde{\tau}_{q-1}^{-1})^{-m}$$

$$\lesssim \Gamma_q^{-3N_{\mathrm{ind},v}} (\widetilde{\tau}_{q-1}^{-1} \tau_{q-1})^{N_{\mathrm{ind},t}} \mathcal{M} \left(n, 2N_{\mathrm{ind},v}, \lambda_q, \widetilde{\lambda}_q \right) \mathcal{M} \left(m, N_{\mathrm{ind},t}, \tau_{q-1}^{-1}, \widetilde{\tau}_{q-1}^{-1} \right) ,$$

where we have used the fact that $m - \bar{m}' \geq m - \bar{m} = 3N_{\text{ind},v}$. Taking $N_{\text{ind},v} \gg N_{\text{ind},t}$ such that

$$\widetilde{\lambda}_q^2 (\widetilde{\tau}_{q-1}^{-1} \tau_{q-1})^{N_{\text{ind},t}} \leq \Gamma_q^{3N_{\text{ind},v}-2}, \tag{5.42}$$

a condition which is satisfied due to (9.50c), it follows from (5.41) that (5.29) holds whenever $m > 3N_{\text{ind},v}$, $n \leq 3N_{\text{ind},v}$, and $m + n \leq 2N_{\text{fin}}$.

It remains to consider the case $n > 3N_{\text{ind},v}$, $n + m \leq 2N_{\text{fin}}$. In this case we still use (5.38a)–(5.38b), but with $\bar{m}$ replaced by m, and similarly to (5.39), but by appealing to the bounds (5.18) and (5.32) instead of (5.40), we obtain

$$\left\| D^n D_{t,q-1}^m \mathcal{P}_{q,x,t} w_q \right\|_{L^\infty}$$

$$\lesssim \sum_{a=0}^{n} \sum_{\substack{\bar{m}' \leq m \\ \bar{n}' + \bar{m}' \leq m}} \left\| D^a d_{m,\bar{n}',\bar{m}'} \right\|_{L^\infty} \left\| D^{n-a+\bar{n}'} \partial_t^{\bar{m}'} \mathcal{P}_{q,x,t} w_q \right\|_{L^\infty}$$

$$\lesssim \sum_{a=0}^{n} \sum_{\substack{\bar{m}' \leq m \\ \bar{n}' + \bar{m}' \leq m}} \lambda_q^{a-\bar{n}'} (\Gamma_q^{-1} \widetilde{\tau}_{q-1}^{-1})^{m-\bar{m}'}$$

$$\times \Gamma_q^{-1} \delta_q^{1/2} \widetilde{\lambda}_q^2 \mathcal{M} \left(n - a + \bar{n}', 3N_{\text{ind},v}, \lambda_q, \widetilde{\lambda}_q \right) (\widetilde{\tau}_{q-1}^{-1})^{\bar{m}'}$$

$$\lesssim \Gamma_q^{-1} \delta_q^{1/2} \widetilde{\lambda}_q^2 \mathcal{M} \left(n, 3N_{\text{ind},v}, \lambda_q, \widetilde{\lambda}_q \right) (\widetilde{\tau}_{q-1}^{-1})^m.$$

To conclude the proof of (5.29) in this case, we note that for $n \geq 3N_{\text{ind},v}$ the definition (9.19) implies

$$\mathcal{M} \left(n, 3N_{\text{ind},v}, \lambda_q, \widetilde{\lambda}_q \right) \leq \Gamma_{q+1}^{-5N_{\text{ind},v}} \mathcal{M} \left(n, 2N_{\text{ind},v}, \lambda_q, \widetilde{\lambda}_q \right),$$

and this factor is sufficiently small to absorb losses due to bad material derivative estimates. Indeed, we have that

$$\Gamma_q^{-1} \delta_q^{1/2} \widetilde{\lambda}_q^2 \mathcal{M} \left(n, 3N_{\text{ind},v}, \lambda_q, \widetilde{\lambda}_q \right) (\widetilde{\tau}_{q-1}^{-1})^m$$

$$\lesssim \Gamma_q^{-3} \delta_q^{1/2} \mathcal{M} \left(n, 2N_{\text{ind},v}, \lambda_q, \widetilde{\lambda}_q \right)$$

$$\times \mathcal{M} \left(m, N_{\text{ind},t}, \tau_{q-1}^{-1}, \widetilde{\tau}_{q-1}^{-1} \right) \Gamma_q^2 \widetilde{\lambda}_q^2 \left(\frac{\widetilde{\tau}_{q-1}^{-1}}{\tau_{q-1}^{-1}} \right)^{N_{\text{ind},t}} \Gamma_{q+1}^{-5N_{\text{ind},v}}$$

$$\lesssim \Gamma_q^{-1} \delta_q^{1/2} \mathcal{M} \left(n, 2N_{\text{ind},v}, \lambda_q, \widetilde{\lambda}_q \right) \mathcal{M} \left(m, N_{\text{ind},t}, \tau_{q-1}^{-1}, \widetilde{\tau}_{q-1}^{-1} \right)$$

by appealing to the condition $N_{\text{ind},v} \gg N_{\text{ind},t}$ given in (9.50b). This concludes the proof of (5.29) for all $n + m \leq 2N_{\text{fin}}$ if either n or m is larger than $3N_{\text{ind},v}$.

The bounds (5.28)–(5.29) estimate the leading order contribution to u_q. According to the decomposition (3.4), the proofs of (5.5) and (5.6) are completed if we are able to verify that

$$\left\| D^n D_{t,q-1}^m (\mathcal{P}_{q,x,t} - \mathrm{Id}) v_{\ell_{q-1}} \right\|_{L^\infty}$$
$$\leq \Gamma_q^{-2} \delta_q^{1/2} \mathcal{M}\left(n, 2\mathsf{N}_{\mathrm{ind},v}, \lambda_q, \widetilde{\lambda}_q\right) \mathcal{M}\left(m, \mathsf{N}_{\mathrm{ind},t}, \tau_{q-1}^{-1}, \widetilde{\tau}_{q-1}^{-1}\right) \qquad (5.43)$$

holds for all $n + m \leq 2\mathsf{N}_{\mathrm{fin}}$.

In order to establish this bound, we appeal to (5.31)–(5.32) and obtain

$$\left\| D^n D_{t,q-1}^m (\mathcal{P}_{q,x,t} - \mathrm{Id}) v_{\ell_{q-1}} \right\|_{L^\infty}$$
$$\lesssim \sum_{\substack{m' \leq m \\ n' + m' \leq n+m}} \lambda_q^{n-n'} (\widetilde{\tau}_{q-1}^{-1} \Gamma_q^{-1})^{m-m'} \left\| D^{n'} \partial_t^{m'} (\mathcal{P}_{q,x,t} - \mathrm{Id}) v_{\ell_{q-1}} \right\|_{L^\infty} \qquad (5.44)$$

for $n, m \geq 0$ such that $n + m \leq 2\mathsf{N}_{\mathrm{fin}}$. Here we distinguish two cases. If either $n > 3\mathsf{N}_{\mathrm{ind},v}$ or $m > 3\mathsf{N}_{\mathrm{ind},v}$, then we simply appeal to (5.18), use that $\mathcal{P}_{q,x,t}$ commutes with D and ∂_t, and obtain from the above display that

$$\left\| D^n D_{t,q-1}^m (\mathcal{P}_{q,x,t} - \mathrm{Id}) v_{\ell_{q-1}} \right\|_{L^\infty}$$
$$\lesssim \sum_{\substack{m' \leq m \\ n' + m' \leq n+m}} \lambda_q^{n-n'} (\widetilde{\tau}_{q-1}^{-1} \Gamma_q^{-1})^{m-m'} (\lambda_{q-1}^4 \delta_{q-1}^{1/2}) \widetilde{\lambda}_{q-1}^{n'} \widetilde{\tau}_{q-2}^{-m'}$$
$$\lesssim (\lambda_{q-1}^4 \delta_{q-1}^{1/2}) \lambda_q^n (\widetilde{\tau}_{q-1}^{-1} \Gamma_q^{-1})^m$$
$$\lesssim (\lambda_{q-1}^4 \delta_{q-1}^{1/2}) (\tau_{q-1} \widetilde{\tau}_{q-1}^{-1})^{\mathsf{N}_{\mathrm{ind},t}} \Gamma_q^{-3\mathsf{N}_{\mathrm{ind},v}}$$
$$\qquad \times \mathcal{M}\left(n, 2\mathsf{N}_{\mathrm{ind},v}, \lambda_q, \widetilde{\lambda}_q\right) \mathcal{M}\left(m, \mathsf{N}_{\mathrm{ind},t}, \tau_{q-1}^{-1}, \widetilde{\tau}_{q-1}^{-1}\right)$$
$$\lesssim \left(\lambda_{q-1}^4 \delta_{q-1}^{1/2} \Gamma_q^2 \delta_q^{-1/2} (\tau_{q-1} \widetilde{\tau}_{q-1}^{-1})^{\mathsf{N}_{\mathrm{ind},t}} \Gamma_q^{-3\mathsf{N}_{\mathrm{ind},v}}\right) \Gamma_q^{-2} \delta_q^{1/2}$$
$$\qquad \times \mathcal{M}\left(n, 2\mathsf{N}_{\mathrm{ind},v}, \lambda_q, \widetilde{\lambda}_q\right) \mathcal{M}\left(m, \mathsf{N}_{\mathrm{ind},t}, \tau_{q-1}^{-1}, \widetilde{\tau}_{q-1}^{-1}\right).$$

Using that $\mathsf{N}_{\mathrm{ind},v} \gg \mathsf{N}_{\mathrm{ind},t}$, as described in (9.50c), the above estimate then readily implies (5.43).

We are thus left to consider (5.44) for $n, m \leq 3\mathsf{N}_{\mathrm{ind},v}$. In this case, the bound for the term $\| D^{n'} \partial_t^{m'} (\mathcal{P}_{q,x,t} - \mathrm{Id}) v_{\ell_{q-1}} \|_{L^\infty}$ present in (5.44) is different. Similarly to (5.30) we use the fact that the kernel K_q has vanishing moments of orders between 1 and $\mathsf{N}_{\mathrm{ind},v}$, and thus we have

$$\mathcal{P}_{q,x,t} v_{\ell_{q-1}}(\theta) - v_{\ell_{q-1}}(\theta)$$
$$= \sum_{|\alpha| + m'' = \mathsf{N}_{\mathrm{ind},v}} \frac{\mathsf{N}_{\mathrm{ind},v}}{\alpha! m''!} \iint_{\mathbb{T}^3 \times \mathbb{R}} K_q(\kappa)(-\kappa)^{(\alpha, m'')}$$
$$\qquad \times \int_0^1 (1-\eta)^{\mathsf{N}_{\mathrm{ind},v}-1} D^\alpha \partial_t^{m''} v_{\ell_{q-1}}(\theta - \eta\kappa) \, d\eta \, d\kappa. \qquad (5.45)$$

Using (5.18) and (5.45), we may then estimate

$$\left\| D^{n'}\partial_t^{m'}(\mathcal{P}_{q,x,t} - \mathrm{Id})v_{\ell_{q-1}} \right\|_{L^\infty}$$
$$\lesssim (\lambda_{q-1}^4 \delta_{q-1}^{1/2}) \sum_{|\alpha|+m''=\mathsf{N}_{\mathrm{ind},v}} \widetilde{\lambda}_q^{-|\alpha|} \widetilde{\tau}_{q-1}^{m''} \widetilde{\lambda}_{q-1}^{n'+|\alpha|}(\Gamma_q^{-1}\widetilde{\tau}_q^{-1})^{m'+m''}$$
$$\lesssim (\lambda_{q-1}^4 \delta_{q-1}^{1/2})\Gamma_q^{-\mathsf{N}_{\mathrm{ind},v}} \lambda_q^{n'}(\Gamma_q^{-1}\widetilde{\tau}_q^{-1})^{m'} \, .$$

Combining the above display with (5.44) we arrive at

$$\left\| D^n D_{t,q-1}^m (\mathcal{P}_{q,x,t} - \mathrm{Id})v_{\ell_{q-1}} \right\|_{L^\infty}$$
$$\lesssim (\lambda_{q-1}^4 \delta_{q-1}^{1/2})\Gamma_q^{-\mathsf{N}_{\mathrm{ind},v}} \lambda_q^n (\Gamma_q^{-1}\widetilde{\tau}_{q-1}^{-1})^m$$
$$\lesssim (\lambda_{q-1}^4 \delta_{q-1}^{1/2})(\widetilde{\tau}_{q-1}^{-1}\tau_{q-1})^{\mathsf{N}_{\mathrm{ind},t}}\Gamma_q^{-\mathsf{N}_{\mathrm{ind},v}}$$
$$\times \mathcal{M}\left(n, 2\mathsf{N}_{\mathrm{ind},v}, \lambda_q, \widetilde{\lambda}_q\right) \mathcal{M}\left(m, \mathsf{N}_{\mathrm{ind},t}, \tau_{q-1}^{-1}, \widetilde{\tau}_{q-1}^{-1}\right) \, . \qquad (5.46)$$

Using the fact that $\mathsf{N}_{\mathrm{ind},v} \gg \mathsf{N}_{\mathrm{ind},t}$—see condition (9.50c)—the above estimate concludes the proof of (5.43).

Combining the bounds (5.28), (5.29), and (5.43) concludes the proofs of (5.5) and (5.6).

Proof of (5.4). By (3.3) we have that

$$v_{\ell_q} - v_q = (\mathcal{P}_{q,x,t} - \mathrm{Id})v_q = (\mathcal{P}_{q,x,t} - \mathrm{Id})w_q + (\mathcal{P}_{q,x,t} - \mathrm{Id})v_{\ell_{q-1}} \, .$$

From (5.33) and (5.34) we deduce that the first term on the right side of the above display is bounded as

$$\left\| D^n D_{t,q-1}^m (\mathcal{P}_{q,x,t} - \mathrm{Id})w_q \right\|_{L^\infty}$$
$$\lesssim \left(\delta_q^{1/2}\Gamma_q^2 \lambda_q^2 (\widetilde{\tau}_{q-1}^{-1}\tau_{q-1})^{\mathsf{N}_{\mathrm{ind},t}}\Gamma_q^{-\mathsf{N}_{\mathrm{ind},v}} \right) \lambda_q^n \mathcal{M}\left(m, \mathsf{N}_{\mathrm{ind},t}, \tau_{q-1}^{-1}\Gamma_q^{i-1}, \widetilde{\tau}_{q-1}^{-1}\Gamma_q^{-1}\right) \, ,$$

while the second term is estimated from (5.46) as

$$\left\| D^n D_{t,q-1}^m (\mathcal{P}_{q,x,t} - \mathrm{Id})v_{\ell_{q-1}} \right\|_{L^\infty}$$
$$\lesssim \left(\delta_{q-1}^{1/2}\lambda_{q-1}^4 (\widetilde{\tau}_{q-1}^{-1}\tau_{q-1})^{\mathsf{N}_{\mathrm{ind},t}}\Gamma_q^{-\mathsf{N}_{\mathrm{ind},v}} \right)$$
$$\times \mathcal{M}\left(n, 2\mathsf{N}_{\mathrm{ind},v}, \lambda_q, \widetilde{\lambda}_q\right) \mathcal{M}\left(m, \mathsf{N}_{\mathrm{ind},t}, \tau_{q-1}^{-1}, \widetilde{\tau}_{q-1}^{-1}\right) \, ,$$

for $n, m \leq 3\mathsf{N}_{\mathrm{ind},v}$. Since $\mathsf{N}_{\mathrm{ind},v} \gg \mathsf{N}_{\mathrm{ind},t}$—see, e.g., the parameter inequality (9.50a)—the above two displays directly imply (5.4).

Proof of (5.7) **and** (5.8). The argument is nearly identical to how the inductive bounds on w_q in (3.13) were shown earlier to imply bounds for $\mathcal{P}_{q,x,t}w_q$ as in (5.28). The crucial ingredients in this proof were that for each material derivative the bound on the mollified function $\mathcal{P}_{q,x,t}w_q$ is relaxed by a factor of

Γ_q, the cost of space derivatives is relaxed from λ_q to $\widetilde{\lambda}_q$ when $n \geq \mathsf{N}_{\text{ind},v}$, and the available number of estimates on the unmollified function w_q is much larger than $\mathsf{N}_{\text{ind},v}$ (more precisely, $7\mathsf{N}_{\text{ind},v}$). But the same ingredients are available for the transfer of estimates from $\mathring{R}_q$ to $\mathring{R}_{\ell_q} = \mathcal{P}_{q,x,t}\mathring{R}_q$. Indeed, the derivatives available in (3.15) extend significantly past $\mathsf{N}_{\text{ind},v}$ (this time up to $3\mathsf{N}_{\text{ind},v}$). When comparing the desired bound on $\mathring{R}_{\ell_q}$ in (5.7) with the available inductive bound in (3.15) we note that the cost of each material derivative is relaxed by a factor of Γ_q, and that the cost of each additional space derivative is relaxed from λ_q to $\widetilde{\lambda}_q$ when n is sufficiently large. To avoid redundancy, we omit these details. $\qquad\square$

Chapter Six

Cutoffs

This section is dedicated to the construction of the cutoff functions described in Section 2.5, which play the role of a joint Eulerian-and-Lagrangian Littlewood-Paley frequency decompositon, which in addition keeps track of the size of objects in physical space. During a first pass at the book, the reader may skip this technical section—if the Lemmas 6.8, 6.14, 6.18, 6.21, 6.35, 6.36, 6.38, 6.40, and 6.41, and Corollaries 6.27 and 6.33 are taken for granted.

This section is organized as follows. In Section 6.1 we define the velocity cutoff functions $\psi_{i,q}$, recursively in terms of the previous level (meaning $q-1$) velocity cutoff functions $\psi_{i',q-1}$, which are assumed to satisfy the inductive bounds and properties mentioned in Section 3.2.3. In Section 6.2 we then verify that the velocity cutoff functions at level q, and the velocity fields u_q and v_{ℓ_q}, satisfy all the inductive estimates claimed in Sections 3.2.3 and 3.2.4, for $q' = q$. This section is the bulk of Chapter 6; and it is here that the various commutators between Eulerian (space and time) derivatives and Lagrangian derivatives cause a plethora of difficulties.

Remark 6.1. We note that by the conclusion of Section 6.2 we have verified all the inductive assumptions from Section 3.2, except for (3.13)–(3.14) for the new velocity increment w_{q+1}, and (3.15) for the new stress $\mathring{R}_{q+1}$. These three inductive assumptions will be revisited, broken down, and restated in Chapter 7 and proven in Chapter 8.

Next, in Section 6.3 we introduce the temporal cutoffs $\chi_{i,k,q}$, indexed by k, which are meant to subdivide the support of the velocity cutoff $\psi_{i,q}$ into time slices of width inversely to the *local Lipschitz norm* of v_{ℓ_q}. This allows us in Section 6.4 to properly define and estimate the Lagrangian flow maps induced by the incompressible vector field v_{ℓ_q}, on the support of $\psi_{i,q}\chi_{i,k,q}$. We next turn to defining the stress cutoff functions $\omega_{i,j,q,n,p}$, indexed by j, for the stress $\mathring{R}_{q,n,p}$, on the support of $\psi_{i,q}$. Coupling the stress and velocity cutoffs in this way allows us in Section 6.7 to sharply estimate spatial and material derivatives of these higher order stresses, but also to estimate the derivatives of the stress cutoffs themselves. At last, we define in Section 6.8 the checkerboard cutoffs $\zeta_{q,i,k,n,\vec{l}}$, indexed by an address $\vec{l} = (l, w, h)$ which identifies a specific cube of sidelength $2\pi/\lambda_{q,n,0}$ within $\mathbb{T}^3$. This specific size of the support of $\zeta_{q,i,k,n,\vec{l}}$ is important for ensuring that Oscillation Type 2 errors vanish (see Lemmas 8.11 and 8.12). These cutoff functions are flowed by the backward Lagrangian flows $\Phi_{i,k,q}$ defined earlier, explaining their dependence on the indices q, i, k. Lastly, the cumulative cutoff function $\eta_{i,j,k,q,n,p,\vec{l}}$ is defined in Section 6.9, along with some of its principal properties. We emphasize that this cumulative cutoff has embedded into it information about the local size and cost of space/Lagrangian derivatives of the velocity, the stress, and the Lagrangian maps.

6.1 DEFINITION OF THE VELOCITY CUTOFF FUNCTIONS

For all $q \geq 1$ and $0 \leq m \leq \mathsf{N}_{\mathrm{cut},t}$, we construct the following cutoff functions. The proof is contained in Appendix A.2.

Lemma 6.2. *For all $q \geq 1$ and $0 \leq m \leq \mathsf{N}_{\mathrm{cut},t}$, there exist smooth cutoff functions $\widetilde{\psi}_{m,q}, \psi_{m,q} : [0,\infty) \to [0,1]$ which satisfy the following.*

1. *The support of $\widetilde{\psi}_{m,q}$ is precisely the set $\left[0, \Gamma_q^{2(m+1)}\right]$, and furthermore*

   a) *on the interval $\left[0, \frac{1}{4}\Gamma_q^{2(m+1)}\right]$, $\widetilde{\psi}_{m,q} \equiv 1$;*

   b) *on the interval $\left[\frac{1}{4}\Gamma_q^{2(m+1)}, \Gamma_q^{2(m+1)}\right]$, $\widetilde{\psi}_{m,q}$ decreases from 1 to 0.*

2. *The support of $\psi_{m,q}$ is precisely the set $\left[\frac{1}{4}, \Gamma_q^{2(m+1)}\right]$, and furthermore*

   a) *on the interval $\left[\frac{1}{4}, 1\right]$, $\psi_{m,q}$ increases from 0 to 1;*

   b) *on the interval $\left[1, \frac{1}{4}\Gamma_q^{2(m+1)}\right]$, $\psi_{m,q} \equiv 1$;*

   c) *on the interval $\left[\frac{1}{4}\Gamma_q^{2(m+1)}, \Gamma_q^{2(m+1)}\right]$, $\psi_{m,q}$ decreases from 1 to 0.*

3. *For all $y \geq 0$, a partition of unity is formed as*

$$\widetilde{\psi}_{m,q}^2(y) + \sum_{i \geq 1} \psi_{m,q}^2\left(\Gamma_q^{-2i(m+1)}y\right) = 1. \tag{6.1}$$

4. *$\widetilde{\psi}_{m,q}$ and $\psi_{m,q}\left(\Gamma_q^{-2i(m+1)}\cdot\right)$ satisfy*

$$\operatorname{supp}\widetilde{\psi}_{m,q}(\cdot) \cap \operatorname{supp}\psi_{m,q}\left(\Gamma_q^{-2i(m+1)}\cdot\right) = \emptyset \quad \text{if} \quad i \geq 2,$$
$$\operatorname{supp}\psi_{m,q}\left(\Gamma_q^{-2i(m+1)}\cdot\right) \cap \operatorname{supp}\psi_{m,q}\left(\Gamma_q^{-2i'(m+1)}\cdot\right) = \emptyset \quad \text{if} \quad |i - i'| \geq 2. \tag{6.2}$$

5. *For $0 \leq N \leq \mathsf{N}_{\mathrm{fin}}$, when $0 \leq y < \Gamma_q^{2(m+1)}$ we have*

$$\frac{|D^N \widetilde{\psi}_{m,q}(y)|}{(\widetilde{\psi}_{m,q}(y))^{1-N/\mathsf{N}_{\mathrm{fin}}}} \lesssim \Gamma_q^{-2N(m+1)}. \tag{6.3}$$

For $\frac{1}{4} < y < 1$ we have

$$\frac{|D^N \psi_{m,q}(y)|}{(\psi_{m,q}(y))^{1-N/N_{\text{fin}}}} \lesssim 1, \tag{6.4}$$

while for $\frac{1}{4}\Gamma_q^{2(m+1)} < y < \Gamma_q^{2(m+1)}$ we have

$$\frac{|D^N \psi_{m,q}(y)|}{(\psi_{m,q}(y))^{1-N/N_{\text{fin}}}} \lesssim \Gamma_q^{-2N(m+1)}. \tag{6.5}$$

In each of the above inequalities, the implicit constants depend on N but not m or q.

Definition 6.3. *Given $i, j, q \geq 0$, we define*

$$i_* = i_*(j, q) = i_*(j) = \min\{i \geq 0 : \Gamma_{q+1}^i \geq \Gamma_q^j\}.$$

In view of the definition (3.10), we see that

$$i_*(j) = \left\lceil j \frac{\log(\lambda_q) - \log(\lambda_{q-1})}{\log(\lambda_{q+1}) - \log(\lambda_q)} \right\rceil = \left\lceil j \frac{\log\left(\lceil a^{b^q}\rceil\right) - \log\left(\lceil a^{b^{q-1}}\rceil\right)}{\log\left(\lceil a^{b^{q+1}}\rceil\right) - \log\left(\lceil a^{b^q}\rceil\right)} \right\rceil.$$

One may check that as $q \to \infty$ or $a \to \infty$, $i_*(j)$ converges to $\lceil \frac{j}{b} \rceil$ for any j, and so if a is sufficiently large, $i_*(j)$ is bounded from above and below independently of q for each j. Note that in particular, for $j = 0$ we have that $i_*(j) = 0$.

At stage $q \geq 1$ of the iteration (by convention $w_0 = u_0 = 0$) and for $m \leq N_{\text{cut},t}$ and $j_m \geq 0$, we can now define

$$h_{m,j_m,q}^2(x,t) := \sum_{n=0}^{N_{\text{cut},x}} \Gamma_{q+1}^{-2i_*(j_m)} \delta_q^{-1} (\lambda_q \Gamma_q)^{-2n} \left(\tau_{q-1}^{-1} \Gamma_{q+1}^{i_*(j_m)+2}\right)^{-2m}$$

$$\times |D^n D_{t,q-1}^m u_q(x,t)|^2. \tag{6.6}$$

Definition 6.4 (Intermediate cutoff functions). *Given $q \geq 1$, $m \leq N_{\text{cut},t}$, and $j_m \geq 0$ we define $\psi_{m,i_m,j_m,q}$ by*

$$\psi_{m,i_m,j_m,q}(x,t) = \psi_{m,q+1}\left(\Gamma_{q+1}^{-2(i_m-i_*(j_m))(m+1)} h_{m,j_m,q}^2(x,t)\right) \tag{6.7}$$

for $i_m > i_*(j_m)$, while for $i_m = i_*(j_m)$,

$$\psi_{m,i_*(j_m),j_m,q}(x,t) = \tilde{\psi}_{m,q+1}\left(h_{m,j_m,q}^2(x,t)\right). \tag{6.8}$$

The intermediate cutoff functions $\psi_{m,i_m,j_m,q}$ are equal to zero for $i_m < i_*(j_m)$.

The indices i_m and j_m will be shown to run up to some maximal values $i_{\max}$

and $\tilde{\imath}_{\max}$ to be determined in the proof (see Lemma 6.14 and (6.27)). With this notation and in view of (6.1) and (6.2), it immediately follows that

$$\sum_{i_m \geq 0} \psi^2_{m,i_m,j_m,q} = \sum_{i_m \geq i_*(j_m)} \psi^2_{m,i_m,j_m,q} = \sum_{\{i_m : \, \Gamma^{i_m}_{q+1} \geq \Gamma^{j_m}_q\}} \psi^2_{m,i_m,j_m,q} \equiv 1 \quad (6.9)$$

for any m, and for $|i_m - i'_m| \geq 2$

$$\psi_{m,i_m,j_m,q}\psi_{m,i'_m,j_m,q} = 0. \tag{6.10}$$

Definition 6.5 (m^{th} velocity cutoff function). *For $q \geq 1$ and $i_m \geq 0$,[1] we inductively define the mth velocity cutoff function*

$$\psi^2_{m,i_m,q} = \sum_{\{j_m : \, i_m \geq i_*(j_m)\}} \psi^2_{j_m,q-1}\psi^2_{m,i_m,j_m,q}. \tag{6.11}$$

In order to define the full velocity cutoff function, we use the notation

$$\vec{\imath} = \{i_m\}^{N_{\text{cut},t}}_{m=0} = (i_0, ..., i_{N_{\text{cut},t}}) \in \mathbb{N}_0^{N_{\text{cut},t}+1} \tag{6.12}$$

to denote a tuple of non-negative integers of length $N_{\text{cut},t} + 1$.

Definition 6.6 (Velocity cutoff function). *For $0 \leq i \leq i_{max}(q)$ and $q \geq 0$, we inductively define the velocity cutoff function $\psi_{i,q}$ as follows. When $q = 0$, we let*

$$\psi_{i,0} = \begin{cases} 1 & \text{if } i = 0 \\ 0 & \text{otherwise.} \end{cases} \tag{6.13}$$

Then, we inductively on q define

$$\psi^2_{i,q} = \sum_{\left\{\vec{\imath} : \, \max\limits_{0 \leq m \leq N_{\text{cut},t}} i_m = i\right\}} \prod_{m=0}^{N_{\text{cut},t}} \psi^2_{m,i_m,q} \tag{6.14}$$

for all $q \geq 1$.

The sum used to define $\psi_{i,q}$ for $q \geq 1$ is over all tuples with a maximum entry of i. The number of such tuples is clearly q-independent once it is demonstrated in Lemma 6.14 that $i_m \leq i_{\max}(q)$ (which implies $i \leq i_{\max}(q)$), and $i_{\max}(q)$ is bounded above independently of q.

---

[1] Later we will show that $\psi_{m,i_m,q} \equiv 0$ if $i \geq i_{\max}$.

For notational convenience, given an $\vec{i}$ as in the sum of (6.14), we shall denote

$$\text{supp} \left(\prod_{m=0}^{N_{\text{cut},t}} \psi_{m,i_m,q} \right) = \bigcap_{m=0}^{N_{\text{cut},t}} \text{supp}\,(\psi_{m,i_m,q}) =: \text{supp}\,(\psi_{\vec{i},q}). \tag{6.15}$$

In particular, we will frequently use that $(x,t) \in \text{supp}\,(\psi_{i,q})$ if and only if there exists $\vec{i} \in \mathbb{N}_0^{N_{\text{cut},t}+1}$ such that $\max_{0 \le m \le N_{\text{cut},t}} i_m = i$, and $(x,t) \in \text{supp}\,(\psi_{\vec{i},q})$.

6.2 PROPERTIES OF THE VELOCITY CUTOFF FUNCTIONS

6.2.1 Partitions of unity

Lemma 6.7 ($\psi_{m,i_m,q}$–Partition of unity). *For all m, we have that*

$$\sum_{i_m \ge 0} \psi_{m,i_m,q}^2 \equiv 1, \qquad \psi_{m,i_m,q}\psi_{m,i_m',q} = 0 \quad \text{for} \quad |i_m - i_m'| \ge 2. \tag{6.16}$$

Proof of Lemma 6.7. The proof proceeds inductively. When $q = 0$ there is nothing to prove as $\psi_{m,i_m,q}$ is not defined. Thus we assume $q \ge 1$. From (6.13) for $q = 0$ and (3.16) for $q \ge 1$, we assume that the functions $\{\psi_{j,q-1}^2\}_{j \ge 0}$ form a partition of unity. To show the first part of (6.16), we may use (6.9) and (6.11) and reorder the summation to obtain

$$\sum_{i_m \ge 0} \psi_{m,i_m,q}^2 = \sum_{i_m \ge 0} \sum_{\{j_m\,:\,i_*(j_m) \le i_m\}} \psi_{j_m,q-1}^2 \psi_{m,i_m,j_m,q}^2(x,t)$$

$$= \sum_{j_m \ge 0} \psi_{j_m,q-1}^2 \underbrace{\sum_{\{i_m\,:\,i_m \ge i_*(j_m)\}} \psi_{m,i_m,j_m,q}^2}_{\equiv 1 \text{ by } (6.9)} = \sum_{j_m \ge 0} \psi_{j_m,q-1}^2 \equiv 1.$$

The last equality follows from the inductive assumption (3.16).

The proof of the second claim is more involved and will be split into cases. Using the definition in (6.11), we have that

$$\psi_{m,i_m,q}\psi_{m,i_m',q}$$

$$= \sum_{\{j_m:i_m \ge i_*(j_m)\}} \sum_{\{j_m':i_m' \ge i_*(j_m')\}} \psi_{j_m,q-1}^2 \psi_{j_m',q-1}^2 \psi_{m,i_m,j_m,q}^2 \psi_{m,i_m',j_m',q}^2.$$

Recalling the inductive assumption (3.16), we have that the above sum only includes pairs of indices j_m and j_m' such that $|j_m - j_m'| \le 1$. So we may assume that

$$(x,t) \in \text{supp}\,\psi_{m,i_m,j_m,q} \cap \text{supp}\,\psi_{m,i_m',j_m',q}, \tag{6.17}$$

where $|j_m - j'_m| \leq 1$. The first and simplest case is the case $j_m = j'_m$. We then appeal to (6.10) to deduce that it must be the case that $|i_m - i'_m| \leq 1$ in order for (6.17) to be true.

Before moving to the second and third cases, we first show that by symmetry it will suffice to prove that $\psi_{m,i_m,q}\psi_{m,i'_m,q} = 0$ when $i'_m \leq i_m - 2$. Assuming this has been proven, let i_{m_1}, i_{m_2} be given with $|i_{m_1} - i_{m_2}| \geq 2$. Without loss of generality we may assume that $i_{m_1} \geq i_{m_2}$, which implies that $i_{m_1} \geq i_{m_2} + 2$. Using the assumption and setting $i_{m_2} = i'_m$ and $i_{m_1} = i_m$, we deduce that $\psi_{m,i_{m_1},q}\psi_{m,i_{m_2},q} = 0$. Thus, we have reduced the proof to showing that $\psi_{m,i_m,q}\psi_{m,i'_m,q} = 0$ when $i'_m \leq i_m - 2$, which we will show next by contradiction.

Let us consider the second case, $j'_m = j_m + 1$. When $i_m = i_*(j_m)$, using that $i_*(j_m) \leq i_*(j_m + 1)$, we obtain

$$i'_m \leq i_m - 2 = i_*(j_m) - 2 < i_*(j_m + 1) = i_*(j'_m),$$

and so by Definition 6.4, we have that $\psi_{m,i'_m,j'_m,q} = 0$. Thus, in this case there is nothing to prove, and we need to only consider the case $i_m > i_*(j_m)$. From (6.17), points 1 and 2 from Lemma 6.2, and Definition 6.4, we have that

$$h_{m,j_m,q}(x,t) \in \left[\frac{1}{2}\Gamma_{q+1}^{(m+1)(i_m - i_*(j_m))}, \Gamma_{q+1}^{(m+1)(i_m+1-i_*(j_m))}\right], \qquad (6.18a)$$

$$h_{m,j_m+1,q}(x,t) \leq \Gamma_{q+1}^{(m+1)(i'_m+1-i_*(j_m+1))}. \qquad (6.18b)$$

Note that from the definition of $h_{m,j_m,q}$ in (6.6), we have that

$$\Gamma_{q+1}^{(m+1)(i_*(j_m+1)-i_*(j_m))} h_{m,j_m+1,q} = h_{m,j_m,q}.$$

Then, since $i'_m \leq i_m - 2$, from (6.18b) we have that

$$\Gamma_{q+1}^{-(m+1)(i_m-i_*(j_m))} h_{m,j_m,q}$$
$$= \Gamma_{q+1}^{-(m+1)(i_m-i_*(j_m))} h_{m,j_m+1,q} \Gamma_{q+1}^{(m+1)(i_*(j_m+1)-i_*(j_m))}$$
$$\leq \Gamma_{q+1}^{-(m+1)(i_m-i_*(j_m))} \Gamma_{q+1}^{(m+1)(i'_m+1-i_*(j_m+1))} \Gamma_{q+1}^{(m+1)(i_*(j_m+1)-i_*(j_m))}$$
$$= \Gamma_{q+1}^{(m+1)(i'_m+1-i_m)}$$
$$\leq \Gamma_{q+1}^{-(m+1)}.$$

Since $m \geq 0$, the above estimate contradicts the lower bound on $h_{m,j_m,q}$ in (6.18a) because $\Gamma_{q+1}^{-1} \ll 1/2$ for a sufficiently large.

We move to the third and final case, $j'_m = j_m - 1$. As before, if $i_m = i_*(j_m)$, then since $i_*(j_m) \leq i_*(j_m - 1) + 1$, we have that

$$i'_m \leq i_m - 2 = i_*(j_m) - 2 \leq i_*(j_m - 1) - 1 < i_*(j_m - 1) = i_*(j'_m),$$

which by Definition 6.4 implies that $\psi_{m,i'_m,j'_m,q} = 0$, and there is nothing to prove. Thus, we only must consider the case $i_m > i_*(j_m)$. Using the definition (6.6) we have that

$$h_{m,j_m,q} = \Gamma_{q+1}^{(m+1)(i_*(j_m-1)-i_*(j_m))} h_{m,j_m-1,q} \,.$$

On the other hand, for $i'_m \leq i_m - 2$ we have from (6.18b) that

$$h_{m,j_m-1,q} \leq \Gamma_{q+1}^{(m+1)(i'_m+1-i_*(j_m-1))} \leq \Gamma_{q+1}^{(m+1)(i_m-1-i_*(j_m-1))} \,.$$

Therefore, combining the above two displays and the inequality $-i_*(j_m) \geq -i_*(j_m - 1) - 1$, we obtain the bound

$$\Gamma_{q+1}^{-(m+1)(i_m-i_*(j_m))} h_{m,j_m,q}$$
$$\leq \Gamma_{q+1}^{-(m+1)(i_m-i_*(j_m))} \Gamma_{q+1}^{(m+1)(i_*(j_m-1)-i_*(j_m))} \Gamma_{q+1}^{(m+1)(i_m-1-i_*(j_m-1))}$$
$$= \Gamma_{q+1}^{-(m+1)} \,.$$

As before, since $m \geq 0$ this produces a contradiction with the lower bound on $h_{m,j_m,q}$ given in (6.18a), since $\Gamma_{q+1}^{-1} \ll 1/2$. $\qquad\square$

With Lemma 6.7 in hand, we can now verify the inductive assumption (3.16) at level q.

Lemma 6.8 ($\psi_{i,q}$ is a partition of unity). *We have that for $q \geq 0$,*

$$\sum_{i\geq 0} \psi_{i,q}^2 \equiv 1 \,, \qquad \psi_{i,q}\psi_{i',q} = 0 \quad \text{for} \quad |i-i'| \geq 2. \qquad (6.19)$$

Proof of Lemma 6.8. When $q = 0$, both statements are immediate from (6.13). To prove the first claim for $q \geq 1$, let us introduce the notation

$$\Lambda_i = \left\{ \vec{i} = (i_0, ..., i_{\mathsf{N}_{\text{cut},t}}) \colon \max_{0 \leq m \leq \mathsf{N}_{\text{cut},t}} i_m = i \right\}. \qquad (6.20)$$

Then

$$\psi_{i,q}^2 = \sum_{\vec{i}\in\Lambda_i} \prod_{m=0}^{\mathsf{N}_{\text{cut},t}} \psi_{m,i_m,q}^2,$$

and thus

$$\sum_{i\geq 0} \psi_{i,q}^2 = \sum_{i\geq 0}\sum_{\vec{i}\in\Lambda_i} \prod_{m=0}^{\mathsf{N}_{\text{cut},t}} \psi_{m,i_m,q}^2 = \sum_{\vec{i}\in\mathsf{N}_0^{\mathsf{N}_{\text{cut},t}+1}} \left(\prod_{m=0}^{\mathsf{N}_{\text{cut},t}} \psi_{m,i_m,q}^2 \right)$$
$$= \prod_{m=0}^{\mathsf{N}_{\text{cut},t}} \left(\sum_{i_m\geq 0} \psi_{m,i_m,q}^2 \right) = \prod_{m=0}^{\mathsf{N}_{\text{cut},t}} 1 = 1$$

after using (6.16).

To prove the second claim, assume towards a contradiction that there exists $|i - i'| \geq 2$ such that $\psi_{i,q}\psi_{i',q} \geq 0$. Then

$$0 \neq \psi_{i,q}^2 \psi_{i',q}^2 = \sum_{\vec{i} \in \Lambda_i} \sum_{\vec{i}' \in \Lambda_{i'}} \prod_{m=0}^{N_{cut,t}} \psi_{m,i_m,q}^2 \psi_{m,i'_m,q}^2. \tag{6.21}$$

In order for (6.21) to be non-vanishing, by (6.16), there must exist indexes $\vec{i} = (i_0, ..., i_{N_{cut,t}}) \in \Lambda_i$ and $\vec{i}' = (i'_0, ..., i'_{N_{cut,t}}) \in \Lambda_{i'}$ such that $|i_m - i'_m| \leq 1$ for all $0 \leq m \leq N_{cut,t}$. By the definition of i and i', there exist m_* and m'_* such that

$$i_{m_*} = \max_m i_m = i, \qquad i'_{m'_*} = \max_m i'_m = i'.$$

But then

$$i = i_{m_*} \leq i'_{m_*} + 1 \leq i'_{m'_*} + 1 = i' + 1$$
$$i' = i'_{m'_*} \leq i_{m'_*} + 1 \leq i_{m_*} + 1 = i + 1,$$

implying that $|i - i'| \leq 1$, a contradiction. □

In view of the preceding two lemmas and (6.10), and for convenience of notation, we define

$$\psi_{i\pm,q}(x,t) = \left(\psi_{i-1,q}^2(x,t) + \psi_{i,q}^2(x,t) + \psi_{i+1,q}^2(x,t)\right)^{1/2}, \tag{6.22}$$

which are cutoffs with the property that

$$\psi_{i\pm,q} \equiv 1 \quad \text{on} \quad \text{supp}\,(\psi_{i,q}). \tag{6.23}$$

Remark 6.9. The definition (6.14) is not convenient to use directly for estimating material derivatives of the $\psi_{i,q}$ cutoffs, because differentiating the terms $\psi_{m,i_m,q}$ *individually* ignores certain cancellations which arise due to the fact that $\{\psi_{m,i_m,q}\}_{i_m \geq 0}$ is a partition of unity (as was shown above in Lemma 6.7). For this purpose, we *re-sum* the terms in the definition (6.14) as follows. For any given $0 \leq m \leq N_{cut,t}$, we introduce the summed cutoff function

$$\Psi_{m,i,q}^2 = \sum_{i_m=0}^{i} \psi_{m,i_m,q}^2 \tag{6.24}$$

and note via Lemma 6.7 its chief property:

$$D(\Psi_{m,i,q}^2) = D(\psi_{m,i,q}^2)\mathbf{1}_{\text{supp}\,(\psi_{m,i+1,q})} = D(\psi_{m,i,q}^2)\mathbf{1}_{\text{supp}\,(\psi_{m,i+1,q})}. \tag{6.25}$$

The above inclusion holds because on the support of $\psi_{m,i_m,q}$ with $i_m < i$, we have that $\Psi_{m,i,q} \equiv 1$. With the notation (6.24) we return to the definition (6.14) and note that

$$\psi_{i,q}^2 = \sum_{m=0}^{N_{\text{cut},t}} \psi_{m,i,q}^2 \prod_{m'=0}^{m-1} \Psi_{m',i,q}^2 \prod_{m''=m+1}^{N_{\text{cut},t}} (\Psi_{m'',i,q}^2 - \psi_{m'',i,q}^2)$$

$$= \sum_{m=0}^{N_{\text{cut},t}} \psi_{m,i,q}^2 \prod_{m'=0}^{m-1} \Psi_{m',i,q}^2 \prod_{m''=m+1}^{N_{\text{cut},t}} \Psi_{m'',i-1,q}^2 . \qquad (6.26)$$

Remark 6.10. Define $j_*(i,q) = \max\{j : i_*(j) \leq i\}$ to be the largest index of j_m appearing in the sum in (6.11). We note here that

$$\Gamma_{q+1}^{i-1} < \Gamma_q^{j_*(i,q)} \leq \Gamma_{q+1}^i \qquad (6.27)$$

holds. This fact will be used later on in the proof in conjunction with Lemma 6.14 to bound the maximal values of j_m.

The following lemma is a direct consequence of the definitions of the cutoffs.

Lemma 6.11. *If* $(x,t) \in \text{supp}\,(\psi_{m,i_m,j_m,q})$ *then*

$$h_{m,j_m,q} \leq \Gamma_{q+1}^{(m+1)(i_m+1-i_*(j_m))} . \qquad (6.28)$$

Moreover, if $i_m > i_*(j_m)$ *we have*

$$h_{m,j_m,q} \geq (1/2)\Gamma_{q+1}^{(m+1)(i_m-i_*(j_m))} \qquad (6.29)$$

on the support of $\psi_{m,i_m,j_m,q}$. *As a consequence, we have*

$$\left\| D^N D_{t,q-1}^m u_q \right\|_{L^\infty(\text{supp}\,\psi_{m,i_m,q})} \leq \delta_q^{1/2} \Gamma_{q+1}^{i_m+1} (\lambda_q \Gamma_q)^N (\tau_{q-1}^{-1} \Gamma_{q+1}^{i_m+3})^m \qquad (6.30)$$

$$\left\| D^N D_{t,q-1}^M u_q \right\|_{L^\infty(\text{supp}\,\psi_{i,q})} \leq \delta_q^{1/2} \Gamma_{q+1}^{i+1} (\lambda_q \Gamma_q)^N (\tau_{q-1}^{-1} \Gamma_{q+1}^{i+3})^M \qquad (6.31)$$

for all $0 \leq m, M \leq N_{\text{cut},t}$, *and* $0 \leq N \leq N_{\text{cut},x}$.

Proof of Lemma 6.11. Estimates (6.28) and (6.29) follow directly from the definitions of $\widetilde{\psi}_{m,q+1}$ and $\psi_{m,q+1}$. In order to prove (6.30), we note that for $(x,t) \in \text{supp}\,(\psi_{m,i_m,q})$, by (6.11) there must exist a j_m with $i_*(j_m) \leq i_m$ such that $(x,t) \in \text{supp}\,(\psi_{m,i_m,j_m,q})$. Using (6.28), we conclude that

$$\left\| D^N D_{t,q-1}^m u_q \right\|_{L^\infty(\text{supp}\,\psi_{m,i_m,j_m,q})}$$
$$\leq \Gamma_{q+1}^{(m+1)(i_m+1-i_*(j_m))} \Gamma_{q+1}^{i_*(j_m)} (\Gamma_q \lambda_q)^N (\Gamma_{q+1}^{i_*(j_m)+2} \tau_{q-1}^{-1})^m \delta_q^{1/2}$$
$$= \delta_q^{1/2} \Gamma_{q+1}^{i_m+1} (\lambda_q \Gamma_q)^N (\tau_{q-1}^{-1} \Gamma_{q+1}^{i_m+3})^m , \qquad (6.32)$$

which completes the proof of (6.30). The proof of (6.31) follows from the fact that we have employed the *maximum* over m of i_m to define $\psi_{i,q}$ in (6.6). $\qquad \square$

An immediate corollary of the bound (5.9) and of the previous lemma is that estimates for the derivatives of u_q are also available on the support of $\psi_{i,q}$, instead of $\psi_{i,q-1}$.

Corollary 6.12. *For $N, M \leq 3\mathsf{N}_{\mathrm{ind},v}$, and $i \geq 0$, we have the bound*

$$\left\| D^N D^M_{t,q-1} u_q \right\|_{L^\infty(\mathrm{supp}\,\psi_{i,q})}$$
$$\lesssim \Gamma^{i+1}_{q+1} \delta^{1/2}_q \mathcal{M}\left(N, 2\mathsf{N}_{\mathrm{ind},v}, \Gamma_q \lambda_q, \widetilde{\lambda}_q\right) \mathcal{M}\left(M, \mathsf{N}_{\mathrm{ind},t}, \Gamma^{i+3}_{q+1} \tau^{-1}_{q-1}, \widetilde{\tau}^{-1}_{q-1}\right). \quad (6.33)$$

Recall that if either $N > 3\mathsf{N}_{\mathrm{ind},v}$ or $M > 3\mathsf{N}_{\mathrm{ind},v}$ are such that $N + M \leq 2\mathsf{N}_{\mathrm{fin}}$, suitable estimates for $D^N D^M_{t,q-1} u_q$ are already provided by (5.6).

Proof of Corollary 6.12. When $0 \leq N \leq \mathsf{N}_{\mathrm{cut},x}$ and $0 \leq M \leq \mathsf{N}_{\mathrm{cut},t} \leq \mathsf{N}_{\mathrm{ind},t}$, the desired bound was already established in (6.31).

For the remaining cases, note that if $0 \leq m \leq \mathsf{N}_{\mathrm{cut},t}$ and $(x,t) \in \mathrm{supp}\,\psi_{m,i_m,q}$, there exists $j_m \geq 0$ with $i_*(j_m) \leq i_m$, such that $(x,t) \in \mathrm{supp}\,\psi_{j_m,q-1}$. Thus, we may appeal to (5.9) and deduce that

$$\left| D^N D^M_{t,q-1} u_q \right| \lesssim \delta^{1/2}_q \widetilde{\lambda}^{3/2}_q \mathcal{M}\left(N, 2\mathsf{N}_{\mathrm{ind},v}, \lambda_q, \widetilde{\lambda}_q\right) \mathcal{M}\left(M, \mathsf{N}_{\mathrm{ind},t}, \Gamma^{j_m+1}_q \tau^{-1}_{q-1}, \widetilde{\tau}^{-1}_{q-1}\right).$$

Since $i_*(j_m) \leq i_m$ implies $\Gamma^{j_m}_q \leq \Gamma^{i_m}_{q+1}$, we deduce that

$$\left\| D^N D^M_{t,q-1} u_q \right\|_{L^\infty(\mathrm{supp}\,\psi_{m,i_m,q})}$$
$$\lesssim \delta^{1/2}_q \widetilde{\lambda}^{3/2}_q \mathcal{M}\left(N, 2\mathsf{N}_{\mathrm{ind},v}, \lambda_q, \widetilde{\lambda}_q\right) \mathcal{M}\left(M, \mathsf{N}_{\mathrm{ind},t}, \Gamma^{i_m+1}_{q+1} \tau^{-1}_{q-1}, \widetilde{\tau}^{-1}_{q-1}\right).$$

Note that the above estimate does not have a factor of $\Gamma^{i_m+1}_{q+1}$ next to the $\delta^{1/2}_q$ at the amplitude.

We now consider two cases. If $\mathsf{N}_{\mathrm{cut},x} < N \leq 3\mathsf{N}_{\mathrm{ind},v}$, then

$$\mathcal{M}\left(N, 2\mathsf{N}_{\mathrm{ind},v}, \lambda_q, \widetilde{\lambda}_q\right) \lesssim \Gamma^{-\mathsf{N}_{\mathrm{cut},x}}_q \mathcal{M}\left(N, 2\mathsf{N}_{\mathrm{ind},v}, \Gamma_q \lambda_q, \widetilde{\lambda}_q\right).$$

On the other hand, if $\mathsf{N}_{\mathrm{cut},t} < M \leq 3\mathsf{N}_{\mathrm{ind},v}$, then

$$\mathcal{M}\left(M, \mathsf{N}_{\mathrm{ind},t}, \Gamma^{i_m+1}_{q+1} \tau^{-1}_{q-1}, \widetilde{\tau}^{-1}_{q-1}\right) \lesssim \Gamma^{-2\mathsf{N}_{\mathrm{cut},t}}_{q+1} \mathcal{M}\left(M, \mathsf{N}_{\mathrm{ind},t}, \Gamma^{i_m+3}_{q+1} \tau^{-1}_{q-1}, \widetilde{\tau}^{-1}_{q-1}\right).$$

Combining the above three displays, and recalling the definition of $\psi_{i,q}$ in (6.14), we deduce that if either $N > \mathsf{N}_{\mathrm{cut},x}$ or $M > \mathsf{N}_{\mathrm{cut},t}$, we have

$$\left\| D^N D^M_{t,q-1} u_q \right\|_{L^\infty(\mathrm{supp}\,\psi_{i,q})}$$
$$\lesssim \delta^{1/2}_q \widetilde{\lambda}^{3/2}_q \max\{\Gamma^{-\mathsf{N}_{\mathrm{cut},x}}_q, \Gamma^{-2\mathsf{N}_{\mathrm{cut},t}}_{q+1}\}$$
$$\times \mathcal{M}\left(N, 2\mathsf{N}_{\mathrm{ind},v}, \Gamma_q \lambda_q, \widetilde{\lambda}_q\right) \mathcal{M}\left(M, \mathsf{N}_{\mathrm{ind},t}, \Gamma^{i+1}_{q+1} \Gamma^2_q \tau^{-1}_{q-1}, \widetilde{\tau}^{-1}_{q-1}\right),$$

and the proof of (6.33) is completed by taking $N_{\text{cut},x}$ and $N_{\text{cut},t}$ sufficiently large to ensure that

$$\widetilde{\lambda}_q^{3/2} \max\{\Gamma_q^{-N_{\text{cut},x}}, \Gamma_{q+1}^{-2N_{\text{cut},t}}\} \leq 1. \qquad (6.34)$$

This condition holds by (9.51). $\qquad\qquad\qquad\qquad\qquad\qquad\qquad\qquad\qquad\qquad$ □

6.2.2 Pure spatial derivatives

In this section we prove that the cutoff functions $\psi_{i,q}$ satisfy sharp spatial derivative estimates, which are consistent with (3.19) for $q' = q$.

Lemma 6.13 (Spatial derivatives for the cutoffs). *Fix $q \geq 1$, $0 \leq m \leq N_{\text{cut},t}$, and $i_m \geq 0$. For all $j_m \geq 0$ such that $i_m \geq i_*(j_m)$ and all $N \leq N_{\text{fin}}$, we have*

$$\mathbf{1}_{\text{supp}\,(\psi_{j_m,q-1})} \frac{|D^N \psi_{m,i_m,j_m,q}|}{\psi_{m,i_m,j_m,q}^{1-N/N_{\text{fin}}}} \lesssim \mathcal{M}\left(N, N_{\text{ind},v}, \lambda_q \Gamma_q, \widetilde{\lambda}_q \Gamma_q\right), \qquad (6.35)$$

which in turn implies

$$\frac{|D^N \psi_{i,q}|}{\psi_{i,q}^{1-N/N_{\text{fin}}}} \lesssim \mathcal{M}\left(N, N_{\text{ind},v}, \lambda_q \Gamma_q, \widetilde{\lambda}_q \Gamma_q\right) \qquad (6.36)$$

for all $i \geq 0$, all $N \leq N_{\text{fin}}$.

Proof of Lemma 6.13. We first show that (5.9) implies (6.35). We distinguish between two cases. The first case is when $\psi = \widetilde{\psi}_{m,q+1}$ or $\psi = \psi_{m,q+1}$ and we have the lower bound

$$h_{m,j_m,q}^2 \Gamma_{q+1}^{-2(i_m - i_*(j_m))(m+1)} \geq \frac{1}{4}\Gamma_{q+1}^{2(m+1)} \qquad (6.37)$$

so that (6.5) applies. The goal is then to apply Lemma A.4 to the function $\psi = \widetilde{\psi}_{m,q+1}$ or $\psi = \psi_{m,q+1}$ as described above in conjunction with $\Gamma_\psi = \Gamma_{q+1}^{m+1}$, $\Gamma = \Gamma_{q+1}^{(m+1)(i_m - i_*(j_m))}$, and $h(x,t) = (h_{m,j_m,q}(x,t))^2$. The assumption (A.21) holds by (6.3) or (6.5) for all $N \leq N_{\text{fin}}$, and so we need to obtain bounds on the derivatives of $h_{m,j_m,q}^2$, which are consistent with assumption (A.22) of Lemma A.4. For $B \leq N_{\text{fin}}$, the Leibniz rule gives

$$|D^B h_{m,j_m,q}^2| \lesssim (\lambda_q \Gamma_q)^B \sum_{B'=0}^{B} \sum_{n=0}^{N_{\text{cut},x}} \Gamma_{q+1}^{-i_*(j_m)} (\tau_{q-1}^{-1} \Gamma_{q+1}^{i_*(j_m)+2})^{-m} (\lambda_q \Gamma_q)^{-n-B'} \delta_q^{-1/2}$$

$$\times |D^{n+B'} D_{t,q-1}^m u_q|$$

$$\times \Gamma_{q+1}^{-i_*(j_m)} (\tau_{q-1}^{-1} \Gamma_{q+1}^{i_*(j_m)+2})^{-m} (\lambda_q \Gamma_q)^{-n-B+B'} \delta_q^{-1/2}$$

$$\times \left| D^{n+B-B'} D^m_{t,q-1} u_q \right|. \tag{6.38}$$

For the terms with $L \in \{n+B', n+B-B'\} \le \mathsf{N}_{\mathrm{cut},x}$ we may appeal to estimate (6.28), which gives

$$\Gamma_{q+1}^{-i_*(j_m)} (\tau_{q-1}^{-1} \Gamma_{q+1}^{i_*(j_m)+2})^{-m} (\lambda_q \Gamma_q)^{-L} \delta_q^{-1/2} \left\| D^L D^m_{t,q-1} u_q \right\|_{L^\infty(\operatorname{supp}\psi_{m,i_m,j_m,q})}$$
$$\le \Gamma_{q+1}^{(m+1)(i_m+1-i_*(j_m))}. \tag{6.39}$$

On the other hand, for $\mathsf{N}_{\mathrm{cut},x} < L \in \{n+B', n+B-B'\} \le \mathsf{N}_{\mathrm{cut},x}+B \le 2\mathsf{N}_{\mathrm{fin}} - \mathsf{N}_{\mathrm{ind},t}$, we may appeal to estimates (5.6) and (5.9), and since $m \le \mathsf{N}_{\mathrm{cut},t} < \mathsf{N}_{\mathrm{ind},t}$, we deduce that

$$\Gamma_{q+1}^{-i_*(j_m)} (\tau_{q-1}^{-1} \Gamma_{q+1}^{i_*(j_m)+2})^{-m} (\lambda_q \Gamma_q)^{-L} \delta_q^{-1/2} \left\| D^L D^m_{t,q-1} u_q \right\|_{L^\infty(\operatorname{supp}\psi_{j_m,q-1})}$$
$$\lesssim (\Gamma_q^{j_m+1} \Gamma_{q+1}^{-i_*(j_m)-2})^m (\Gamma_q^{-L} \widetilde{\lambda}_q^{3/2}) \lambda_q^{-L} \mathcal{M}\left(L, 2\mathsf{N}_{\mathrm{ind},v}, \lambda_q, \widetilde{\lambda}_q\right)$$
$$\lesssim \mathcal{M}\left(L, 2\mathsf{N}_{\mathrm{ind},v}, 1, \lambda_q^{-1}\widetilde{\lambda}_q\right)$$
$$\le \Gamma_{q+1}^{(m+1)(i_m+1-i_*(j_m))} \mathcal{M}\left(L, 2\mathsf{N}_{\mathrm{ind},v}, 1, \lambda_q^{-1}\widetilde{\lambda}_q\right). \tag{6.40}$$

In the last inequality we have used that $i_m \ge i_*(j_m)$, while in the second to last inequality we have used that if $L \ge \mathsf{N}_{\mathrm{cut},x}$ then $\Gamma_q^L \ge \widetilde{\lambda}_q^{3/2}$, which follows once $\mathsf{N}_{\mathrm{cut},x}$ is chosen to be sufficiently large, as in (9.51). Summarizing the bounds (6.38)–(6.40), since $n \le \mathsf{N}_{\mathrm{cut},x}$, we arrive at

$$\mathbf{1}_{\operatorname{supp}(\psi_{j_m,q-1}\psi_{m,i_m,j_m,q})} \left| D^B h^2_{m,j_m,q} \right|$$
$$\lesssim (\lambda_q \Gamma_q)^B \mathcal{M}\left(2\mathsf{N}_{\mathrm{cut},x}+B, 2\mathsf{N}_{\mathrm{ind},v}, 1, \lambda_q^{-1}\widetilde{\lambda}_q\right) \Gamma_{q+1}^{2(m+1)(i_m+1-i_*(j_m))}$$
$$\lesssim \mathcal{M}\left(B, \mathsf{N}_{\mathrm{ind},v}, \lambda_q \Gamma_q, \widetilde{\lambda}_q \Gamma_q\right) \Gamma_{q+1}^{2(m+1)(i_m+1-i_*(j_m))}$$

whenever $B \le \mathsf{N}_{\mathrm{fin}}$. Here we have used that $2\mathsf{N}_{\mathrm{cut},x} \le \mathsf{N}_{\mathrm{ind},v}$. Thus, assumption (A.22) holds with $C_h = \Gamma_{q+1}^{2(m+1)(i_m+1-i_*(j_m))}$, $\lambda = \Gamma_q \lambda_q$, $\Lambda = \widetilde{\lambda}_q \Gamma_q$, $N_* = \mathsf{N}_{\mathrm{ind},v}$. Note that with these choices of parameters, we have $C_h \Gamma_\psi^{-2} \Gamma^{-2} = 1$. We may thus apply Lemma A.4 and conclude that

$$\mathbf{1}_{\operatorname{supp}(\psi_{j_m,q-1})} \frac{\left| D^N \psi_{m,i_m,j_m,q} \right|}{\psi_{m,i_m,j_m,q}^{1-N/\mathsf{N}_{\mathrm{fin}}}} \lesssim \mathcal{M}\left(N, \mathsf{N}_{\mathrm{ind},v}, \lambda_q \Gamma_q, \widetilde{\lambda}_q \Gamma_q\right)$$

for all $N \le \mathsf{N}_{\mathrm{fin}}$, proving (6.35) in the first case.

Recalling the inequality (6.37), the second case is when $\psi = \psi_{m,q+1}$ and

$$h^2_{m,j_m,q} \Gamma_{q+1}^{-2(i_m-i_*(j_m))(m+1)} \le \frac{1}{4} \Gamma_{q+1}^{2(m+1)}. \tag{6.41}$$

However, since $\psi_{m,q+1}$ is uniformly equal to 1 when the left hand side of the above display takes values in $\left[1, \frac{1}{4}\Gamma_{q+1}^{2(m+1)}\right]$, (6.35) is trivially satisfied. Thus we may reduce to the case that

$$h_{m,j_m,q}^2 \Gamma_{q+1}^{-2(i_m - i_*(j_m))(m+1)} \leq 1. \tag{6.42}$$

As in the first case, we aim to apply Lemma A.4 with $h = h_{m,j_m,q}^2$, but now with $\Gamma_\psi = 1$ and $\Gamma = \Gamma_{q+1}^{(m+1)(i_m - i_*(j_m))}$. From (6.4), the assumption (A.21) holds. Towards estimating derivatives of h, for the terms with $L \in \{n+B', n+B-B'\} \leq N_{cut,x}$, (6.42) gives immediately that

$$\Gamma_{q+1}^{-i_*(j_m)}(\tau_{q-1}^{-1}\Gamma_{q+1}^{i_*(j_m)+2})^{-m}(\lambda_q\Gamma_q)^{-L}\delta_q^{-1/2}\left\|D^L D_{t,q-1}^m u_q\right\|_{L^\infty(\text{supp}\,\psi_{m,i_m,j_m,q})}$$
$$\leq \Gamma_{q+1}^{(m+1)(i_m - i_*(j_m))}. \tag{6.43}$$

Conversely, when $N_{cut,x} > L$, we may argue as in the estimates which gave (6.40), only this time using the fact that since $i_m \geq i_*(j_m)$, we can achieve the slightly improved bound[2]

$$\Gamma_{q+1}^{(m+1)(i_m - i_*(j_m))}\mathcal{M}\left(L, 2N_{ind,v}, 1, \lambda_q^{-1}\widetilde{\lambda}_q\right). \tag{6.44}$$

We then arrive at

$$\mathbf{1}_{\text{supp}\,(\psi_{j_m,q-1}\psi_{m,i_m,j_m,q})}\left|D^B h_{m,j_m,q}^2\right|$$
$$\lesssim (\lambda_q\Gamma_q)^B \mathcal{M}\left(2N_{cut,x}+B, 2N_{ind,v}, 1, \lambda_q^{-1}\widetilde{\lambda}_q\right)\Gamma_{q+1}^{2(m+1)(i_m - i_*(j_m))}$$
$$\lesssim \mathcal{M}\left(B, N_{ind,v}, \lambda_q\Gamma_q, \widetilde{\lambda}_q\Gamma_q\right)\Gamma_{q+1}^{2(m+1)(i_m - i_*(j_m))}$$

whenever $B \leq N_{fin}$, again using that $2N_{cut,x} \leq N_{ind,v}$. Thus, assumption (A.22) now holds with $\mathcal{C}_h = \Gamma_{q+1}^{2(m+1)(i_m - i_*(j_m))}$, $\lambda = \Gamma_q\lambda_q$, $\Lambda = \widetilde{\lambda}_q\Gamma_q$, $N_* = N_{ind,v}$. Note that with these new choices of parameters, we still have $\mathcal{C}_h\Gamma_\psi^{-2}\Gamma^{-2} = 1$. We may thus apply Lemma A.4 and conclude that

$$\mathbf{1}_{\text{supp}\,(\psi_{j_m,q-1})}\frac{\left|D^N \psi_{m,i_m,j_m,q}\right|}{\psi_{m,i_m,j_m,q}^{1-N/N_{fin}}} \lesssim \mathcal{M}\left(N, N_{ind,v}, \lambda_q\Gamma_q, \widetilde{\lambda}_q\Gamma_q\right)$$

for all $N \leq N_{fin}$, proving (6.35) in the second case.

From the definition (6.11), and the bound (6.35), we next estimate deriva-

---

[2]This bound was also available in (6.40), but we wrote the worse bound there to match the chosen value of $\mathcal{C}_h$.

tives of the mth velocity cutoff function $\psi_{m,i_m,q}$, and claim that

$$\frac{|D^N \psi_{m,i_m,q}|}{\psi_{m,i_m,q}^{1-N/\mathsf{N}_{\text{fin}}}} \lesssim \mathcal{M}\left(N, \mathsf{N}_{\text{ind},v}, \lambda_q \Gamma_q, \widetilde{\lambda}_q \Gamma_q\right) \tag{6.45}$$

for all $i_m \geq 0$, all $N \leq \mathsf{N}_{\text{fin}}$. We prove (6.45) by induction on N. When $N = 0$ the bound trivially holds, which gives the induction base. For the induction step, assume that (6.45) holds for all $N' \leq N-1$. By the Leibniz rule we obtain

$$D^N(\psi_{m,i_m,q}^2) = 2\psi_{m,i_m,q} D^N \psi_{m,i_m,q} + \sum_{N'=1}^{N-1} \binom{N}{N'} D^{N'} \psi_{m,i_m,q} \, D^{N-N'} \psi_{m,i_m,q} \tag{6.46}$$

and thus

$$\frac{D^N \psi_{m,i_m,q}}{\psi_{m,i_m,q}^{1-N/\mathsf{N}_{\text{fin}}}} = \frac{D^N(\psi_{m,i_m,q}^2)}{2\psi_{m,i_m,q}^{2-N/\mathsf{N}_{\text{fin}}}} - \frac{1}{2}\sum_{N'=1}^{N-1} \binom{N}{N'} \frac{D^{N'} \psi_{m,i_m,q}}{\psi_{m,i_m,q}^{1-N'/\mathsf{N}_{\text{fin}}}} \frac{D^{N-N'} \psi_{m,i_m,q}}{\psi_{m,i_m,q}^{1-(N-N')/\mathsf{N}_{\text{fin}}}}.$$

Since $N', N-N' \leq N-1$ by the induction assumption (6.45) we obtain

$$\frac{|D^N \psi_{m,i_m,q}|}{\psi_{m,i_m,q}^{1-N/\mathsf{N}_{\text{fin}}}} \lesssim \frac{|D^N(\psi_{m,i_m,q}^2)|}{\psi_{m,i_m,q}^{2-N/\mathsf{N}_{\text{fin}}}} + \mathcal{M}\left(N, \mathsf{N}_{\text{ind},v}, \lambda_q \Gamma_q, \widetilde{\lambda}_q \Gamma_q\right). \tag{6.47}$$

Thus, establishing (6.45) for the Nth derivative reduces to bounding the first term on the right side of the above. For this purpose we recall (6.11) and compute

$$\frac{|D^N(\psi_{m,i_m,q}^2)|}{\psi_{m,i_m,q}^{2-N/\mathsf{N}_{\text{fin}}}}$$

$$= \frac{1}{\psi_{m,i_m,q}^{2-N/\mathsf{N}_{\text{fin}}}} \sum_{\{j_m:\, i_*(j_m) \leq i_m\}} \sum_{K=0}^{N} \binom{N}{K} D^K(\psi_{j_m,q-1}^2) D^{N-K}(\psi_{m,i_m,j_m,q}^2)$$

$$= \sum_{\{j_m:\, i_*(j_m) \leq i_m\}} \sum_{K=0}^{N} \sum_{L_1=0}^{K} \sum_{L_2=0}^{N-K} \binom{N}{K}\binom{K}{L_1}\binom{N-K}{L_2} \frac{\psi_{j_m,q-1}^{2-K/\mathsf{N}_{\text{fin}}} \psi_{m,i_m,j_m,q}^{2-(N-K)/\mathsf{N}_{\text{fin}}}}{\psi_{m,i_m,q}^{2-N/\mathsf{N}_{\text{fin}}}}$$

$$\times \frac{D^{L_1}\psi_{j_m,q-1}}{\psi_{j_m,q-1}^{1-L_1/\mathsf{N}_{\text{fin}}}} \frac{D^{K-L_1}\psi_{j_m,q-1}}{\psi_{j_m,q-1}^{1-(K-L_1)/\mathsf{N}_{\text{fin}}}} \frac{D^{L_2}\psi_{m,i_m,j_m,q}}{\psi_{m,i_m,j_m,q}^{1-L_2/\mathsf{N}_{\text{fin}}}} \frac{D^{N-K-L_2}\psi_{m,i_m,j_m,q}}{\psi_{m,i_m,j_m,q}^{1-(N-K-L_2)/\mathsf{N}_{\text{fin}}}}.$$

Since $K, N-K \leq N$, and $\psi_{j_m,q-1}, \psi_{m,i_m,j,q} \leq 1$, we have by (6.14) that

$$\frac{\psi_{j_m,q-1}^{2-K/\mathsf{N}_{\text{fin}}} \psi_{m,i_m,j_m,q}^{2-(N-K)/\mathsf{N}_{\text{fin}}}}{\psi_{m,i_m,q}^{2-N/\mathsf{N}_{\text{fin}}}} \leq \frac{\psi_{j_m,q-1}^{2-N/\mathsf{N}_{\text{fin}}} \psi_{m,i_m,j_m,q}^{2-N/\mathsf{N}_{\text{fin}}}}{\psi_{m,i_m,q}^{2-N/\mathsf{N}_{\text{fin}}}} \leq 1.$$

Furthermore, the estimate (6.35) and the inductive assumption (3.19), combined with the parameter estimate $\Gamma_{q-1}\tilde{\lambda}_{q-1} \leq \Gamma_q \lambda_q$ (see (9.38)) and the previous three displays, conclude the proof of (6.45). In particular, note that this upper bound is independent of the value of i_m.

In order to conclude the proof of the lemma, we argue that (6.45) implies (6.36). Recalling (6.14), we have that $\psi_{i,q}^2$ is given as a sum of products of $\psi_{m,i_m,q}^2$, for which suitable derivative bounds are available (due to (6.45)). Thus, the proof of (6.36) is again done by induction on N, mutatis mutandi to the proof of (6.45): indeed, we note that $\psi_{m,i_m,q}^2$ was also given as a sum of squares of cutoff functions, for which derivative bounds were available. The proof of the induction step is thus again based on the application of the Leibniz rule for $\psi_{i,q}^2$; in order to avoid redundancy we omit these details.   □

6.2.3   Maximal indices appearing in the cutoff

A consequence of the inductive assumptions, Lemma 6.11, and of Lemma 6.13 above is that we may a priori estimate the maximal i appearing in $\psi_{i,q}$, labeled as $i_{\max}(q)$.

Lemma 6.14 (Maximal i index in the definition of the cutoff). *There exists $i_{\max} = i_{\max}(q) \geq 0$, determined by the formula (6.53) below, such that*

$$\psi_{i,q} \equiv 0 \quad \text{for all} \quad i > i_{\max} \tag{6.48}$$

and

$$\Gamma_{q+1}^{i_{\max}} \leq \lambda_q^{5/3} \tag{6.49}$$

holds for all $q \geq 0$, where the implicit constant is independent of q. Moreover $i_{\max}(q)$ is bounded uniformly in q as

$$i_{\max}(q) \leq \frac{4}{\varepsilon_\Gamma(b-1)}, \tag{6.50}$$

assuming λ_0 is sufficiently large.

Proof of Lemma 6.14. Assume $i \geq 0$ is such that $\text{supp}(\psi_{i,q}) \neq \emptyset$. Our goal is to prove that $\Gamma_{q+1}^i \leq \lambda_q^{5/3}$.

From (6.14) it follows that for any $(x,t) \in \text{supp}(\psi_{i,q})$, there must exist at least one $\vec{i} = (i_0, \ldots, i_{N_{\text{cut},t}})$ such that $\max_{0 \leq m \leq N_{\text{cut},t}} i_m = i$, and with $\psi_{m,i_m,q}(x,t) \neq 0$ for all $0 \leq m \leq N_{\text{cut},t}$. Therefore, in light of (6.11), for each such m there exists a maximal j_m such that $i_*(j_m) \leq i_m$, with $(x,t) \in \text{supp}(\psi_{j_m,q-1}) \cap \text{supp}(\psi_{m,i_m,j_m,q})$. In particular, this holds for any of the indices m such that $i_m = i$. For the remainder of the proof, we fix such an index $0 \leq m \leq N_{\text{cut},t}$.

If we have $i = i_m = i_*(j_m) = i_*(j_m, q)$, since $(x, t) \in \text{supp}(\psi_{j_m, q-1})$, by the inductive assumption (3.18) we have that $j_m \leq i_{\max}(q-1)$. Then, due to (6.27), we have $\Gamma_{q+1}^{i-1} < \Gamma_q^{j_m} \leq \Gamma_q^{i_{\max}(q-1)}$, and thus

$$\Gamma_{q+1}^i \leq \Gamma_{q+1} \Gamma_q^{i_{\max}(q-1)} \leq \Gamma_{q+1} \lambda_{q-1}^{5/3} < \lambda_q^{5/3}. \tag{6.51}$$

The last inequality above uses the fact that $\lambda_q^{(b+1)/2} \leq \lambda_{q+1}$ since $b > 1$ and a is taken sufficiently large.

On the other hand, if $i = i_m \geq i_*(j_m) + 1$, from (6.29) we have $|h_{m, j_m, q}(x, t)| \geq (1/2) \Gamma_{q+1}^{(m+1)(i_m - i_*(j_m))}$, and by the pigeonhole principle there exists $0 \leq n \leq N_{\text{cut}, x}$ with

$$|D^n D_{t, q-1}^m u_q(x, t)| \geq \frac{\Gamma_{q+1}^{i_*(j_m)}}{2N_{\text{cut}, x}} \Gamma_{q+1}^{(m+1)(i_m - i_*(j_m))} \delta_q^{1/2} (\lambda_q \Gamma_q)^n (\tau_{q-1}^{-1} \Gamma_{q+1}^{i_*(j_m)+2})^m$$

$$\geq \frac{1}{2N_{\text{cut}, x}} \Gamma_{q+1}^{i_m} \delta_q^{1/2} \lambda_q^n (\tau_{q-1}^{-1} \Gamma_{q+1}^{i_m+2})^m,$$

and we also know that $(x, t) \in \text{supp}(\psi_{j_m, q-1})$. By (5.9), the fact that $N_{\text{cut}, x} \leq 2N_{\text{ind}, v} - 2$, and $N_{\text{cut}, t} \leq N_{\text{ind}, t}$, we know that

$$|D^n D_{t, q-1}^m u_q(x, t)| \leq M_b \delta_q^{1/2} \lambda_q^n \tilde{\lambda}_q^{3/2} (\tau_{q-1}^{-1} \Gamma_q^{j_m+1})^m$$

$$\leq M_b \delta_q^{1/2} \lambda_q^n \tilde{\lambda}_q^{3/2} (\tau_{q-1}^{-1} \Gamma_{q+1}^{i_*(j_m)+1})^m$$

$$\leq M_b \delta_q^{1/2} \lambda_q^n \tilde{\lambda}_q^{3/2} (\tau_{q-1}^{-1} \Gamma_{q+1}^{i_m})^m$$

for some constant M_b which is the maximal constant appearing in the $\lesssim$ symbol of (5.9) with $n + m \leq N_{\text{fin}}$. In particular, M_b is independent of q. The proof is now completed, since the previous two inequalities and the assumption that $i_m = i \geq i_{\max}(q) + 1$ imply that

$$\Gamma_{q+1}^i \leq 2N_{\text{cut}, x} M_b \tilde{\lambda}_q^{3/2} \leq \lambda_q^{5/3}. \tag{6.52}$$

In view of (6.51) and (6.52), the value of $i_{\max}$ is chosen as

$$i_{\max}(q) = \sup \left\{ i' : \Gamma_{q+1}^{i'} \leq \lambda_q^{5/3} \right\}. \tag{6.53}$$

To show that $i_{\max}(q) < \infty$, and in particular that it is bounded independently of q, note that

$$\frac{\log(\lambda_q^{5/3})}{\log(\Gamma_{q+1})} \to \frac{5/3}{\varepsilon_\Gamma(b-1)}$$

as $q \to \infty$. Thus, assuming λ_0 is sufficiently large, since $(b-1)\varepsilon_\Gamma \leq 1/5$, the bound (6.50) holds. $\square$

6.2.4   Mixed derivative estimates

Recall from (3.7) the notation $D_q = u_q \cdot \nabla$ for the directional derivative in the direction of u_q. With this notation—cf. (3.6)—we have $D_{t,q} = D_{t,q-1} + D_q$. Thus, D_q derivatives are useful for transferring bounds on $D_{t,q-1}$ derivatives to bounds on $D_{t,q}$ derivatives.

From the Leibniz rule we have that

$$D_q^K = \sum_{j=1}^{K} f_{j,K} D^j, \tag{6.54}$$

where

$$f_{j,K} = \sum_{\{\gamma \in \mathbb{N}^K \,:\, |\gamma|=K-j\}} c_{j,K,\gamma} \prod_{\ell=1}^{K} D^{\gamma_\ell} u_q, \tag{6.55}$$

where $c_{j,K,\gamma}$ are explicitly computable coefficients that depend only on K, j, and γ. Similarly to the coefficients in (A.49), the precise value of these constants is not important, since all the indices appearing throughout the proof are taken to be less than $2N_{\text{fin}}$. The decomposition (6.54)–(6.55) will be used frequently in this section.

Remark 6.15. Since throughout the book the maximal number of spatial or material derivatives is bounded from above by $2N_{\text{fin}}$, which is a number that is independent of q, we have not explicitly stated the formula for the coefficients $c_{a,k,\beta}$ in (A.49), as all these constants will be absorbed in a $\lesssim$ symbol. We note, however, that the proof of Lemma A.13 does yield a recursion relation for the $c_{a,k,\beta}$, which may be used if desired to compute the $c_{a,k,\beta}$ explicitly.

With the notation in (6.55) we have the following bounds.

Lemma 6.16. *For $q \geq 1$ and $1 \leq K \leq 2N_{\text{fin}}$, the functions $\{f_{j,K}\}_{j=1}^{K}$ defined in (6.55) obey the estimate*

$$\|D^a f_{j,K}\|_{L^\infty(\mathrm{supp}\,\psi_{i,q})} \lesssim (\Gamma_{q+1}^{i+1} \delta_q^{1/2})^K \mathcal{M}\left(a + K - j, 2N_{\mathrm{ind},v}, \Gamma_q \lambda_q, \widetilde{\lambda}_q\right). \tag{6.56}$$

for any $a \leq 2N_{\text{fin}} - K + j$ and any $0 \leq i \leq i_{\max}(q)$.

Proof of Lemma 6.16. Note that no material derivative appears in (6.55), and thus to establish (6.56) we appeal to Corollary 6.12 with $M = 0$, and to the bound (5.6) with $m = 0$. From the product rule we obtain that

$$\|D^a f_j\|_{L^\infty(\mathrm{supp}\,\psi_{i,q})}$$

$$\lesssim \sum_{\{\gamma \in \mathbb{N}^K \,:\, |\gamma|=K-j\}} \sum_{\{\alpha \in \mathbb{N}^k \,:\, |\alpha|=a\}} \prod_{\ell=1}^{K} \|D^{\alpha_\ell + \gamma_\ell} u_q\|_{L^\infty(\mathrm{supp}\,\psi_{i,q})}$$

$$\lesssim \sum_{\{\gamma \in \mathbb{N}^K \,:\, |\gamma|=K-j\}} \sum_{\{\alpha \in \mathbb{N}^k \,:\, |\alpha|=a\}} \prod_{\ell=1}^{K} \Gamma_{q+1}^{i+1} \delta_q^{1/2} \mathcal{M}\left(\alpha_\ell + \gamma_\ell, 2\mathsf{N}_{\mathrm{ind},v}, \Gamma_q \lambda_q, \widetilde{\lambda}_q\right)$$

$$\lesssim (\Gamma_{q+1}^{i+1} \delta_q^{1/2})^K \mathcal{M}\left(a + K - j, 2\mathsf{N}_{\mathrm{ind},v}, \Gamma_q \lambda_q, \widetilde{\lambda}_q\right)$$

since $|\gamma| = K - j$. $\square$

Next, we supplement the space and material derivative estimates for u_q obtained in (5.6) and (6.33), with derivatives bounds that combine space, directional, and material derivatives.

Lemma 6.17. *For $q \geq 1$ and $0 \leq i \leq i_{\max}$, we have that*

$$\left\| D^N D_q^K D_{t,q-1}^M u_q \right\|_{L^\infty(\mathrm{supp}\,\psi_{i,q})}$$
$$\lesssim (\Gamma_{q+1}^{i+1} \delta_q^{1/2})^{K+1} \mathcal{M}(N + K, 2\mathsf{N}_{\mathrm{ind},v}, \Gamma_q \lambda_q, \widetilde{\lambda}_q) \mathcal{M}\left(M, \mathsf{N}_{\mathrm{ind},t}, \Gamma_{q+1}^{i+3} \tau_{q-1}^{-1}, \widetilde{\tau}_{q-1}\right)$$
$$\lesssim (\Gamma_{q+1}^{i+1} \delta_q^{1/2}) \mathcal{M}(N, 2\mathsf{N}_{\mathrm{ind},v}, \Gamma_q \lambda_q, \widetilde{\lambda}_q)(\Gamma_{q+1}^{i-c_0} \tau_q^{-1})^K \mathcal{M}\left(M, \mathsf{N}_{\mathrm{ind},t}, \Gamma_{q+1}^{i+3} \tau_{q-1}^{-1}, \widetilde{\tau}_{q-1}\right)$$

holds for $0 \leq K + N + M \leq 2\mathsf{N}_{\mathrm{fin}}$.

Proof of Lemma 6.17. The second estimate in the lemma follows from the parameter inequality $\Gamma_{q+1}^{1+c_0} \widetilde{\lambda}_q \delta_q^{1/2} \leq \tau_q^{-1}$, which is a consequence of (9.39). In order to prove the first statement, we let $0 \leq a \leq N$ and $1 \leq j \leq K$. From estimate (6.33) and (5.6) we obtain

$$\left\| D^{N-a+j} D_{t,q-1}^M u_q \right\|_{L^\infty(\mathrm{supp}\,\psi_{i,q})} \lesssim (\Gamma_{q+1}^{i+1} \delta_q^{1/2}) \mathcal{M}\left(N - a + j, 2\mathsf{N}_{\mathrm{ind},v}, \Gamma_q \lambda_q, \widetilde{\lambda}_q\right)$$
$$\times \mathcal{M}\left(M, \mathsf{N}_{\mathrm{ind},t}, \Gamma_{q+1}^{i+3} \tau_{q-1}^{-1}, \widetilde{\tau}_{q-1}\right),$$

which may be combined with (6.54)–(6.55), and the bound (6.56), to obtain that

$$\left\| D^N D_q^K D_{t,q-1}^M u_q \right\|_{L^\infty(\mathrm{supp}\,\psi_{i,q})}$$
$$\lesssim \sum_{a=0}^{N} \sum_{j=1}^{K} \left\| D^a f_{j,K} \right\|_{L^\infty(\mathrm{supp}\,\psi_{i,q})} \left\| D^{N-a+j} D_{t,q-1}^M w_q \right\|_{L^\infty(\mathrm{supp}\,\psi_{i,q})}$$
$$\lesssim (\Gamma_{q+1}^{i+1} \delta_q^{1/2})^{K+1} \mathcal{M}\left(N + K, 2\mathsf{N}_{\mathrm{ind},v}, \Gamma_q \lambda_q, \widetilde{\lambda}_q\right) \mathcal{M}\left(M, \mathsf{N}_{\mathrm{ind},t}, \Gamma_{q+1}^{i+3} \tau_{q-1}^{-1}, \widetilde{\tau}_{q-1}\right)$$

holds, concluding the proof of the lemma. $\square$

The next lemma shows that the inductive assumptions (3.22)–(3.25b) hold also for $q' = q$.

Lemma 6.18. *For $q \geq 1$, $k \geq 1$, $\alpha, \beta \in \mathbb{N}^k$ with $|\alpha| = K$ and $|\beta| = M$, we have*

$$\left\| \left(\prod_{i=1}^{k} D^{\alpha_i} D_{t,q-1}^{\beta_i} \right) u_q \right\|_{L^\infty(\operatorname{supp}\psi_{i,q})}$$
$$\lesssim (\Gamma_{q+1}^{i+1} \delta_q^{1/2}) \mathcal{M}\left(K, 2\mathsf{N}_{\mathrm{ind},v}, \Gamma_q \lambda_q, \widetilde{\lambda}_q \right) \mathcal{M}\left(M, \mathsf{N}_{\mathrm{ind},t}, \Gamma_{q+1}^{i+3} \tau_{q-1}^{-1}, \Gamma_{q+1}^{-1} \widetilde{\tau}_q^{-1} \right) \tag{6.57}$$

for all $K + M \leq 3\mathsf{N}_{\mathrm{fin}}/2 + 1$. Additionally, for $N \geq 0$, the bound

$$\left\| D^N \left(\prod_{i=1}^{k} D_q^{\alpha_i} D_{t,q-1}^{\beta_i} \right) u_q \right\|_{L^\infty(\operatorname{supp}\psi_{i,q})}$$
$$\lesssim (\Gamma_{q+1}^{i+1} \delta_q^{1/2})^{K+1} \mathcal{M}\left(N + K, 2\mathsf{N}_{\mathrm{ind},v}, \Gamma_q \lambda_q, \widetilde{\lambda}_q \right)$$
$$\times \mathcal{M}\left(M, \mathsf{N}_{\mathrm{ind},t}, \Gamma_{q+1}^{i+3} \tau_{q-1}^{-1}, \Gamma_{q+1}^{-1} \widetilde{\tau}_q^{-1} \right) \tag{6.58}$$
$$\lesssim (\Gamma_{q+1}^{i+1} \delta_q^{1/2}) \mathcal{M}\left(N, 2\mathsf{N}_{\mathrm{ind},v}, \Gamma_q \lambda_q, \widetilde{\lambda}_q \right) (\Gamma_{q+1}^{i-c_0} \tau_q^{-1})^K$$
$$\times \mathcal{M}\left(M, \mathsf{N}_{\mathrm{ind},t}, \Gamma_{q+1}^{i+3} \tau_{q-1}^{-1}, \Gamma_{q+1}^{-1} \widetilde{\tau}_q^{-1} \right) \tag{6.59}$$

holds for all $0 \leq K + M + N \leq 3\mathsf{N}_{\mathrm{fin}}/2 + 1$. Lastly, we have the estimate

$$\left\| \left(\prod_{i=1}^{k} D^{\alpha_i} D_{t,q}^{\beta_i} \right) Dv_{\ell_q} \right\|_{L^\infty(\operatorname{supp}\psi_{i,q})}$$
$$\lesssim (\Gamma_{q+1}^{i+1} \delta_q^{1/2} \widetilde{\lambda}_q) \mathcal{M}\left(K, 2\mathsf{N}_{\mathrm{ind},v}, \Gamma_q \lambda_q, \widetilde{\lambda}_q \right) \mathcal{M}\left(M, \mathsf{N}_{\mathrm{ind},t}, \Gamma_{q+1}^{i-c_0} \tau_q^{-1}, \Gamma_{q+1}^{-1} \widetilde{\tau}_q^{-1} \right) \tag{6.60}$$

for all $K + M \leq 3\mathsf{N}_{\mathrm{fin}}/2$, and

$$\left\| \left(\prod_{i=1}^{k} D^{\alpha_i} D_{t,q}^{\beta_i} \right) v_{\ell_q} \right\|_{L^\infty(\operatorname{supp}\psi_{i,q})}$$
$$\lesssim (\Gamma_{q+1}^{i+1} \delta_q^{1/2} \lambda_q^2) \mathcal{M}\left(K, 2\mathsf{N}_{\mathrm{ind},v}, \Gamma_q \lambda_q, \widetilde{\lambda}_q \right) \mathcal{M}\left(M, \mathsf{N}_{\mathrm{ind},t}, \Gamma_{q+1}^{i-c_0} \tau_q^{-1}, \Gamma_{q+1}^{-1} \widetilde{\tau}_q^{-1} \right) \tag{6.61}$$

for all $K + M \leq 3\mathsf{N}_{\mathrm{fin}}/2 + 1$.

Remark 6.19. As shown in Remark 3.4, the bound (6.59) and identity (A.39) imply that estimate (3.26) also holds with $q' = q$.

Proof of Lemma 6.18. We note that (6.59) follows directly from (6.58), by appealing to the parameter inequality $\Gamma_{q+1}^{1+c_0} \delta_q^{1/2} \widetilde{\lambda}_q \leq \tau_q^{-1}$, which is a consequence of (9.39). We first show that (6.57) holds, then establish (6.58), and lastly prove the bounds (6.60)–(6.61).

Proof of (6.57). The statement is proven by induction on k. For $k = 1$ the estimate is given by Corollary 6.12 and the bound (5.6); in fact, for $k = 1$ we have derivatives estimates up to level $2\mathsf{N}_{\mathrm{fin}}$, and not just $3\mathsf{N}_{\mathrm{fin}}/2 + 1$. For the induction step, assume that (6.57) holds for any $k' \leq k - 1$. We denote

$$P_{k'} = \Big(\prod_{i=1}^{k'} D^{\alpha_i} D_{t,q-1}^{\beta_i} \Big) u_q \tag{6.62}$$

and write

$$\Big(\prod_{i=1}^{k} D^{\alpha_i} D_{t,q-1}^{\beta_i} \Big) u_q$$

$$= (D^{\alpha_k} D_{t,q-1}^{\beta_k})(D^{\alpha_{k-1}} D_{t,q-1}^{\beta_{k-1}}) P_{k-2}$$

$$= (D^{\alpha_k + \alpha_{k-1}} D_{t,q-1}^{\beta_k + \beta_{k-1}}) P_{k-2} + D^{\alpha_k} \Big[D_{t,q-1}^{\beta_k}, D^{\alpha_{k-1}} \Big] D_{t,q-1}^{\beta_{k-1}} P_{k-2}. \tag{6.63}$$

The first term in (6.63) already obeys the correct bound, since we know that (6.57) holds for $k' = k - 1$. In order to treat the second term on the right side of (6.63), we use Lemma A.12 to write the commutator as

$$D^{\alpha_k} \Big[D_{t,q-1}^{\beta_k}, D^{\alpha_{k-1}} \Big] D_{t,q-1}^{\beta_{k-1}} P_{k-2}$$

$$= D^{\alpha_k} \sum_{1 \leq |\gamma| \leq \beta_k} \frac{\beta_k!}{\gamma!(\beta_k - |\gamma|)!} \Big(\prod_{\ell=1}^{\alpha_{k-1}} (\operatorname{ad} D_{t,q-1})^{\gamma_\ell}(D) \Big) D_{t,q-1}^{\beta_k + \beta_{k-1} - |\gamma|} P_{k-2}. \tag{6.64}$$

From Lemma A.13 and the Leibniz rule we claim that one may expand

$$\prod_{\ell=1}^{\alpha_{k-1}} (\operatorname{ad} D_{t,q-1})^{\gamma_\ell}(D) = \sum_{j=1}^{\alpha_{k-1}} g_j D^j \tag{6.65}$$

for some explicit functions g_j which obey the estimate

$$\| D^a g_j \|_{L^\infty(\operatorname{supp} \psi_{i,q})} \lesssim \widetilde{\lambda}_{q-1}^{a+\alpha_{k-1}-j} \mathcal{M}\big(|\gamma|, \mathsf{N}_{\mathrm{ind},t}, \Gamma_{q+1}^i \Gamma_q^{-\mathsf{c}_0} \tau_{q-1}^{-1}, \Gamma_q^{-1} \widetilde{\tau}_{q-1}^{-1} \big) \tag{6.66}$$

for all a such that $a + \alpha_{k-1} - j + |\gamma| \leq 3\mathsf{N}_{\mathrm{fin}}/2$. The claim (6.66) requires a proof, which we sketch next. Using the definition (6.11), the inductive estimate (3.23) at level $q' = q - 1$, $k = 1$, and the parameter inequality (9.39) at level $q - 1$, for any $0 \leq m \leq \mathsf{N}_{\mathrm{cut},t}$ we have that

$$\big\| D^a D_{t,q-1}^b D v_{\ell_{q-1}} \big\|_{L^\infty(\operatorname{supp} \psi_{m,i_m,q})}$$

$$\lesssim \sum_{\{j_m : \, \Gamma_q^{j_m} \leq \Gamma_{q+1}^{i_m}\}} \left\| D^a D_{t,q-1}^b D v_{\ell_{q-1}} \right\|_{L^\infty(\text{supp}\,\psi_{j_m,q-1})}$$

$$\lesssim \sum_{\{j_m : \, \Gamma_q^{j_m} \leq \Gamma_{q+1}^{i_m}\}} (\Gamma_q^{j_m+1} \delta_{q-1}^{1/2}) \widetilde{\lambda}_{q-1}^{a+1} \mathcal{M}\left(b, \mathsf{N}_{\text{ind},t}, \Gamma_q^{j_m-\mathsf{c}_0} \tau_{q-1}^{-1}, \Gamma_q^{-1} \widetilde{\tau}_{q-1}^{-1}\right)$$

$$\lesssim (\Gamma_{q+1}^{i_m} \Gamma_q \delta_{q-1}^{1/2}) \widetilde{\lambda}_{q-1}^{a+1} \mathcal{M}\left(b, \mathsf{N}_{\text{ind},t}, \Gamma_{q+1}^{i_m} \Gamma_q^{-\mathsf{c}_0} \tau_{q-1}^{-1}, \Gamma_q^{-1} \widetilde{\tau}_{q-1}^{-1}\right)$$

$$\lesssim \widetilde{\lambda}_{q-1}^a \mathcal{M}\left(b+1, \mathsf{N}_{\text{ind},t}, \Gamma_{q+1}^{i_m} \Gamma_q^{-\mathsf{c}_0} \tau_{q-1}^{-1}, \Gamma_q^{-1} \widetilde{\tau}_{q-1}^{-1}\right)$$

for all $a + b \leq 3\mathsf{N}_{\text{fin}}/2$. Thus, from the definition (6.14) we deduce that

$$\left\| D^a D_{t,q-1}^b D v_{\ell_{q-1}} \right\|_{L^\infty(\text{supp}\,\psi_{i,q})}$$
$$\lesssim \widetilde{\lambda}_{q-1}^a \mathcal{M}\left(b+1, \mathsf{N}_{\text{ind},t}, \Gamma_{q+1}^i \Gamma_q^{-\mathsf{c}_0} \tau_{q-1}^{-1}, \Gamma_q^{-1} \widetilde{\tau}_{q-1}^{-1}\right) \qquad (6.67)$$

for all $a + b \leq 3\mathsf{N}_{\text{fin}}/2$. When combined with the formula (A.49), which allows us to write

$$(\text{ad}\, D_{t,q-1})^\gamma (D) = f_{\gamma,q-1} \cdot \nabla \qquad (6.68)$$

for an explicit function $f_{\gamma,q-1}$ which is defined in terms of $v_{\ell_{q-1}}$, estimate (6.67) and the Leibniz rule gives the estimate

$$\left\| D^a f_{\gamma,q-1} \right\|_{L^\infty(\text{supp}\,\psi_{i,q})} \lesssim \widetilde{\lambda}_{q-1}^a \mathcal{M}\left(\gamma, \mathsf{N}_{\text{ind},t}, \Gamma_{q+1}^i \Gamma_q^{-\mathsf{c}_0} \tau_{q-1}^{-1}, \Gamma_q^{-1} \widetilde{\tau}_{q-1}^{-1}\right) \quad (6.69)$$

for all $a + \gamma \leq 3\mathsf{N}_{\text{fin}}/2$. In order to conclude the proof of (6.65)–(6.66), we use (6.68) to write

$$\prod_{\ell=1}^{\alpha_{k-1}} (\text{ad}\, D_{t,q-1})^{\gamma_\ell} (D) = \prod_{\ell=1}^{\alpha_{k-1}} (f_{\gamma_\ell,q-1} \cdot \nabla) = \sum_{j=1}^{\alpha_{k-1}} g_j D^j,$$

and now the claimed estimate for g_j follows from the previously established bound (6.69) for the $f_{\gamma_\ell,q-1}$'s and their derivatives, and the Leibniz rule.

With (6.65)–(6.66) in hand, and using estimate (6.57) with $k' = k - 1$, we return to (6.64) and obtain

$$\left\| D^{\alpha_k} \left[D_{t,q-1}^{\beta_k}, D^{\alpha_{k-1}} \right] D_{t,q-1}^{\beta_{k-1}} P_{k-2} \right\|_{L^\infty(\text{supp}\,\psi_{i,q})}$$

$$\lesssim \sum_{j=1}^{\alpha_{k-1}} \sum_{1 \leq |\gamma| \leq \beta_k} \left\| D^{\alpha_k} \left(g_j \, D^j D_{t,q-1}^{\beta_k+\beta_{k-1}-|\gamma|} P_{k-2} \right) \right\|_{L^\infty(\text{supp}\,\psi_{i,q})}$$

$$\lesssim \sum_{j=1}^{\alpha_{k-1}} \sum_{1 \leq |\gamma| \leq \beta_k} \sum_{a'=0}^{\alpha_k} \left\| D^{\alpha_k - a'} g_j \right\|_{L^\infty(\text{supp}\,\psi_{i,q})}$$
$$\times \left\| D^{a'+j} D_{t,q-1}^{\beta_k+\beta_{k-1}-|\gamma|} P_{k-2} \right\|_{L^\infty(\text{supp}\,\psi_{i,q})}$$

$$\lesssim \sum_{j=1}^{\alpha_{k-1}} \sum_{|\gamma|=1}^{\beta_k} \sum_{a'=0}^{\alpha_k} \lambda_q^{\alpha_k+\alpha_{k-1}-j-a'} \mathcal{M}\left(|\gamma|, \mathsf{N}_{\mathrm{ind},t}, \Gamma_{q+1}^i \Gamma_q^{-c_0} \tau_{q-1}^{-1}, \Gamma_q^{-1} \widetilde{\tau}_{q-1}^{-1}\right)$$

$$\times (\Gamma_{q+1}^{i+1} \delta_q^{1/2}) \mathcal{M}\left(K - \alpha_k - \alpha_{k-1} + j + a', 2\mathsf{N}_{\mathrm{ind},v}, \Gamma_q \lambda_q, \widetilde{\lambda}_q\right)$$

$$\times \mathcal{M}\left(M - |\gamma|, \mathsf{N}_{\mathrm{ind},t}, \Gamma_{q+3}^{i+1} \tau_{q-1}^{-1}, \Gamma_{q+1}^{-1} \widetilde{\tau}_q^{-1}\right)$$

$$\lesssim (\Gamma_{q+1}^{i+1} \delta_q^{1/2}) \mathcal{M}\left(K, 2\mathsf{N}_{\mathrm{ind},v}, \Gamma_q \lambda_q, \widetilde{\lambda}_q\right) \mathcal{M}\left(M, \mathsf{N}_{\mathrm{ind},t}, \Gamma_{q+1}^{i+3} \tau_{q-1}^{-1}, \Gamma_{q+1}^{-1} \widetilde{\tau}_q^{-1}\right)$$

$$\tag{6.70}$$

for $M \leq \mathsf{N}_{\mathrm{ind},t}$ and $K + M \leq {}^{3\mathsf{N}_{\mathrm{fin}}}/2 + 1$. The $+1$ in the range of derivatives is simply a consequence of the fact that the summand in the third line of the above display starts with $j \geq 1$ and with $|\gamma| \geq 1$. This concludes the proof of the inductive step for (6.57).

Proof of (6.58). This estimate follows from Lemma A.10. Indeed, letting $v = f = u_q$, $B = D_{t,q-1}$, $\Omega = \operatorname{supp} \psi_{i,q}$, and $p = \infty$, the previously established bound (6.57) allows us to verify conditions (A.40)–(A.41) of Lemma A.10 with $N_* = {}^{3\mathsf{N}_{\mathrm{fin}}}/2 + 1$, $\mathcal{C}_v = \mathcal{C}_f = \Gamma_{q+1}^{i+1} \delta_q^{1/2}$, $\lambda_v = \lambda_f = \Gamma_q \lambda_q$, $\widetilde{\lambda}_v = \widetilde{\lambda}_f = \widetilde{\lambda}_q$, $N_x = 2\mathsf{N}_{\mathrm{ind},v}$, $\mu_v = \mu_f = \Gamma_{q+1}^{i+3} \tau_{q-1}^{-1}$, $\widetilde{\mu}_v = \widetilde{\mu}_f = \Gamma_{q+1}^{-1} \widetilde{\tau}_q^{-1}$, and $N_t = \mathsf{N}_{\mathrm{ind},t}$. As $|\alpha| = K$ and $|\beta| = M$, the bound (6.58) now is a direct consequence of (A.42).

Proof of (6.60) **and** (6.61). First we consider the bound (6.60), inductively on k. For the case $k = 1$ the main idea is to appeal to estimate (A.44) in Lemma A.10 with the operators $A = D_q$ and $B = D_{t,q-1}$ and the functions $v = u_q$ and $f = Dv_{\ell_q}$, so that $D^n(A + B)^m f = D^n D_{t,q}^m Dv_{\ell_q}$. As before, the assumption (A.40) holds due to (6.57) with $\Omega = \operatorname{supp} \psi_{i,q}$, $N_* = {}^{3\mathsf{N}_{\mathrm{fin}}}/2 + 1$, $\mathcal{C}_v = \Gamma_{q+1}^{i+1} \delta_q^{1/2}$, $\lambda_v = \Gamma_q \lambda_q$, $\widetilde{\lambda}_v = \widetilde{\lambda}_q$, $N_x = 2\mathsf{N}_{\mathrm{ind},v}$, $\mu_v = \Gamma_{q+1}^{i+3} \tau_{q-1}^{-1}$, $\widetilde{\mu}_v = \Gamma_{q+1}^{-1} \widetilde{\tau}_q^{-1}$, and $N_t = \mathsf{N}_{\mathrm{ind},t}$. Verifying condition (A.41) is this time more involved, and follows by rewriting $f = Dv_{\ell_q} = Du_q + Dv_{\ell_{q-1}}$. By using (6.57), and the parameter inequality $\Gamma_{q+1}^3 \tau_{q-1}^{-1} \leq \Gamma_{q+1}^{-c_0} \tau_q^{-1}$ (cf. (9.40)), we conveniently obtain

$$\left\|\left(\prod_{i=1}^k D^{\alpha_i} D_{t,q-1}^{\beta_i}\right) Du_q\right\|_{L^\infty(\operatorname{supp} \psi_{i,q})}$$

$$\lesssim (\Gamma_{q+1}^{i+1} \delta_q^{1/2} \widetilde{\lambda}_q) \mathcal{M}\left(K, 2\mathsf{N}_{\mathrm{ind},v}, \Gamma_q \lambda_q, \widetilde{\lambda}_q\right) \mathcal{M}\left(M, \mathsf{N}_{\mathrm{ind},t}, \Gamma_{q+1}^{i-c_0} \tau_q^{-1}, \Gamma_{q+1}^{-1} \widetilde{\tau}_q^{-1}\right)$$

$$\tag{6.71}$$

for all $|\alpha| + |\beta| = K + M \leq {}^{3\mathsf{N}_{\mathrm{fin}}}/2$ (note that the maximal number of derivatives is not ${}^{3\mathsf{N}_{\mathrm{fin}}}/2 + 1$ anymore, but instead is just ${}^{3\mathsf{N}_{\mathrm{fin}}}/2$; the reason is that we are estimating Du_q and not u_q). On the other hand, from the inductive assumption (3.23) with $q' = q - 1$ we obtain that

$$\left\|\left(\prod_{i=1}^k D^{\alpha_i} D_{t,q-1}^{\beta_i}\right) Dv_{\ell_{q-1}}\right\|_{L^\infty(\operatorname{supp} \psi_{j,q-1})}$$

$$\lesssim (\Gamma_q^{j+1}\delta_{q-1}^{1/2})(\widetilde{\lambda}_{q-1})^{K+1}\mathcal{M}\left(M,\mathsf{N}_{\mathrm{ind},t},\Gamma_q^{j-\mathsf{c}_0}\tau_{q-1}^{-1},\widetilde{\tau}_{q-1}^{-1}\right)$$

for $K + M \leq 3\mathsf{N}_{\mathrm{fin}}/2$. Recalling the definitions (6.11)–(6.14) and the notation (6.15), we have that $(x,t) \in \mathrm{supp}\,(\psi_{i,q})$ if and only if $(x,t) \in \mathrm{supp}\,(\psi_{\widetilde{i},q})$, and thus for every $m \in \{0,\ldots,\mathsf{N}_{\mathrm{cut},t}\}$, there exists j_m with $\Gamma_q^{j_m} \leq \Gamma_{q+1}^{i_m} \leq \Gamma_{q+1}^{i}$ and $(x,t) \in \mathrm{supp}\,(\psi_{j_m,q-1})$. Thus, the above stated estimate and our usual parameter inequalities imply that

$$\left\|\left(\prod_{i=1}^{k}D^{\alpha_i}D_{t,q-1}^{\beta_i}\right)Dv_{\ell_{q-1}}\right\|_{L^{\infty}(\mathrm{supp}\,\psi_{i,q})}$$

$$\lesssim (\Gamma_{q+1}^{i+1}\delta_{q-1}^{1/2}\widetilde{\lambda}_{q-1})(\widetilde{\lambda}_{q-1})^{K}\mathcal{M}\left(M,\mathsf{N}_{\mathrm{ind},t},\Gamma_{q+1}^{i}\Gamma_q^{-\mathsf{c}_0}\tau_{q-1}^{-1},\widetilde{\tau}_{q-1}^{-1}\right)$$

$$\lesssim (\Gamma_{q+1}^{i+1}\delta_q^{1/2}\widetilde{\lambda}_q)(\Gamma_q\lambda_q)^{K}\mathcal{M}\left(M,\mathsf{N}_{\mathrm{ind},t},\Gamma_{q+1}^{i-\mathsf{c}_0}\tau_q^{-1},\Gamma_{q+1}^{-1}\widetilde{\tau}_q^{-1}\right) \qquad (6.72)$$

whenever $K + M \leq 3\mathsf{N}_{\mathrm{fin}}/2$. Here we have used that $\delta_{q-1}^{1/2}\widetilde{\lambda}_{q-1} \leq \delta_q^{1/2}\widetilde{\lambda}_q$ and that $\Gamma_{q+1}^{i}\Gamma_q^{-\mathsf{c}_0}\tau_{q-1}^{-1} \leq \Gamma_{q+1}^{i-\mathsf{c}_0}\tau_q^{-1} \leq \Gamma_{q+1}^{-1}\widetilde{\tau}_q^{-1}$, for all $i \leq i_{\max}$. In the last inequality, we have used (9.20) and (6.49). Combining (6.71) and (6.72) we may now verify condition (A.41) for $f = Dv_{\ell_q}$, with $p = \infty$, $\Omega = \mathrm{supp}\,(\psi_{i,q})$, $\mathcal{C}_f = \Gamma_{q+1}^{i+1}\delta_q^{1/2}\widetilde{\lambda}_q$, $\lambda_f = \Gamma_q\lambda_q, \widetilde{\lambda}_f = \widetilde{\lambda}_q, N_x = 2\mathsf{N}_{\mathrm{ind},v}, \mu_f = \Gamma_{q+1}^{i-\mathsf{c}_0}\tau_q^{-1}, \widetilde{\mu}_f = \Gamma_{q+1}^{-1}\widetilde{\tau}_q^{-1}, N_t = \mathsf{N}_{\mathrm{ind},t}$, and $N_* = 3\mathsf{N}_{\mathrm{fin}}/2$. We may thus appeal to (A.44) and obtain that

$$\left\|D^{K}D_{t,q}^{M}Dv_{\ell_q}\right\|_{L^{\infty}(\mathrm{supp}\,\psi_{i,q})}$$

$$\lesssim (\Gamma_{q+1}^{i+1}\delta_q^{1/2}\widetilde{\lambda}_q)\mathcal{M}\left(K,2\mathsf{N}_{\mathrm{ind},v},\Gamma_q\lambda_q,\widetilde{\lambda}_q\right)$$

$$\times \mathcal{M}\left(M,\mathsf{N}_{\mathrm{ind},t},\max\{\Gamma_{q+1}^{i-\mathsf{c}_0}\tau_q^{-1},\Gamma_{q+1}^{i+1}\delta_q^{1/2}\widetilde{\lambda}_q\},\max\{\Gamma_{q+1}^{-1}\widetilde{\tau}_q^{-1},\Gamma_{q+1}^{i+1}\delta_q^{1/2}\widetilde{\lambda}_q\}\right)$$

whenever $K + M \leq 3\mathsf{N}_{\mathrm{fin}}/2$. The parameter inequalities $\Gamma_{q+1}^{\mathsf{c}_0+1}\delta_q^{1/2}\widetilde{\lambda}_q \leq \tau_q^{-1}$ from (9.39) and $\Gamma_{q+1}^{i+2}\delta_q^{1/2}\widetilde{\lambda}_q \leq \widetilde{\tau}_q^{-1}$, which follow from (9.43) and (6.49), conclude the proof of (6.60) for $k = 1$.

In order to prove (6.60) for a general k, we proceed by induction. Assume the estimate holds for every $k' \leq k - 1$. Proving (6.60) at level k is done in the same way as we have established the induction step (in k) for (6.57). We let

$$\widetilde{P}_{k'} = \left(\prod_{i=1}^{k'}D^{\alpha_i}D_{t,q}^{\beta_i}\right)Dv_{\ell_q}$$

and decompose

$$\left(\prod_{i=1}^{k}D^{\alpha_i}D_{t,q}^{\beta_i}\right)Dv_{\ell_q} = (D^{\alpha_k+\alpha_{k-1}}D_{t,q}^{\beta_k+\beta_{k-1}})\widetilde{P}_{k-2}$$

$$+ D^{\alpha_k} \left[D_{t,q}^{\beta_k}, D^{\alpha_{k-1}} \right] D_{t,q}^{\beta_{k-1}} \widetilde{P}_{k-2}$$

and note that the first term is directly bounded using the induction assumption (at level $k-1$). To bound the commutator term, similarly to (6.64)–(6.66), we obtain from Lemmas A.12 and A.13 that

$$D^{\alpha_k} \left[D_{t,q}^{\beta_k}, D^{\alpha_{k-1}} \right] D_{t,q}^{\beta_{k-1}} \widetilde{P}_{k-2}$$

$$= D^{\alpha_k} \sum_{1 \leq |\gamma| \leq \beta_k} \frac{\beta_k!}{\gamma! (\beta_k - |\gamma|)!} \left(\sum_{j=1}^{\alpha_{k-1}} \widetilde{g}_j D^j \right) D_{t,q}^{\beta_k + \beta_{k-1} - |\gamma|} \widetilde{P}_{k-2} \,,$$

where one may use the previously established bound (6.60) with $k=1$ (instead of (6.67)) to estimate

$$\| D^a \widetilde{g}_j \|_{L^\infty (\text{supp}\, \psi_{i,q})}$$
$$\lesssim \mathcal{M} \left(a + \alpha_{k-1} - j, 2\mathsf{N}_{\text{ind},v}, \Gamma_q \lambda_q, \widetilde{\lambda}_q \right) \mathcal{M} \left(|\gamma|, \mathsf{N}_{\text{ind},t}, \Gamma_{q+1}^{i-c_0} \tau_q^{-1}, \Gamma_{q+1}^{-1} \widetilde{\tau}_q^{-1} \right).$$
$$(6.73)$$

Note that the above estimate is not merely (6.66) with q increased by 1. Rather, the above estimate is proven in the same way that (6.66) was proven, by first showing that the analogous version of (6.69) is

$$\| D^a f_{\gamma,q} \|_{L^\infty (\text{supp}\, \psi_{i,q})}$$
$$\lesssim \mathcal{M} \left(a, 2\mathsf{N}_{\text{ind},v}, \Gamma_q \lambda_q, \widetilde{\lambda}_q \right) \mathcal{M} \left(\gamma, \mathsf{N}_{\text{ind},t}, \Gamma_{q+1}^{i-c_0} \tau_q^{-1}, \Gamma_{q+1}^{-1} \widetilde{\tau}_q^{-1} \right) ,$$

from which the claimed estimate (6.73) on $D^a \widetilde{g}_j$ follows. The estimate

$$\left\| D^{\alpha_k} \left[D_{t,q}^{\beta_k}, D^{\alpha_{k-1}} \right] D_{t,q}^{\beta_{k-1}} \widetilde{P}_{k-2} \right\|_{L^\infty (\text{supp}\, \psi_{i,q})}$$
$$\lesssim (\Gamma_{q+1}^{i+1} \delta_q^{1/2}) \mathcal{M} \left(K+1, 2\mathsf{N}_{\text{ind},v}, \Gamma_q \lambda_q, \widetilde{\lambda}_q \right) \mathcal{M} \left(M, \mathsf{N}_{\text{ind},t}, \Gamma_{q+1}^{i-c_0} \tau_q^{-1}, \Gamma_{q+1}^{-1} \widetilde{\tau}_q^{-1} \right)$$
$$(6.74)$$

follows similarly to (6.70), from the estimate (6.73) for $\widetilde{g}_j$, and the bound (6.60) with $k-1$ terms in the product. This concludes the proof of estimate (6.60).

To conclude the proof of the lemma, we also need to establish the estimates for v_{ℓ_q} claimed in (6.61). The proof of this bound is nearly identical to that of (6.60), as is readily seen for $k=1$: we just need to replace Du_q estimates with u_q estimates, and $Dv_{\ell_{q-1}}$ bounds with $v_{\ell_{q-1}}$ bounds. For instance, instead of (6.71), we appeal to (6.59) and obtain a bound for $D^K D_{t,q}^M u_q$ which is better than (6.71) by a factor of $\widetilde{\lambda}_q$, and which holds for $K + M \leq {}^{3\mathsf{N}_{\text{fin}}}/2 + 1$. This estimate is sharper than required by (6.61). The estimate for $D^K D_{t,q}^M v_{\ell_{q-1}}$ is obtained similarly to (6.72), except that instead of appealing to the induction assumption

(3.23) at level $q' = q - 1$, we use (3.24) with $q' = q - 1$. The Sobolev loss λ_{q-1}^2 is then apparent from (3.24), and the estimates hold for $K + M \leq {}^{3N_{\text{fin}}}/_2 + 1$. These arguments establish (6.61) with $k = 1$. The case of general $k \geq 2$ is treated inductively exactly as before, because the commutator term is bounded in the same way as in (6.74), except that $K + 1$ is replaced by K. To avoid redundancy, we omit these details. $\square$

6.2.5   Material derivatives

The estimates in the previous sections, which have led up to Lemma 6.18, allow us to estimate mixed space, directional, and material derivatives of the velocity cutoff functions $\psi_{i,q}$, which in turn allow us to establish the inductive bounds (3.19) and (3.20) with $q' = q$.

In order to achieve this we crucially recall Remark 6.9. Note that if we were to directly differentiate (6.14), then we would need to consider all vectors $\vec{i} \in \mathbb{N}_0^{N_{\text{cut},t}+1}$ such that $\max_{0 \leq m \leq N_{\text{cut},t}} i_m = i$, and then for each one of these $\vec{i}$ consider the term $\mathbf{1}_{\text{supp}\,(\psi_{\vec{i},q})} D_{t,q-1}(\psi_{m,i_m,q}^2)$ for each $0 \leq m \leq N_{\text{cut},t}$; however, in this situation we encounter for instance a term with $i_0 = 0$ and $i_{m'} = i$ for all $1 \leq m' \leq N_{\text{cut},t}$; the bounds available on this term would be catastrophic due to the mismatch $i_0 < i_{m'}$ for all $m' > 0$. Identity (6.26) precisely permits us to avoid this situation, because it has essentially ordered the indices $\{i_m\}_{m=0}^{N_{\text{cut},t}}$ to be non-increasing in m. Indeed, inspecting (6.26) and using identity (6.25) and the definitions (6.15), (6.24), we see that

$$(x,t) \in \text{supp}\,(D_{t,q-1}\psi_{i,q}^2)$$

$$\Leftrightarrow \quad \exists \vec{i} \in \mathbb{N}_0^{N_{\text{cut},t}+1} \text{ and } \exists 0 \leq m \leq N_{\text{cut},t}$$

$$\text{with } i_m \in \{i-1, i\} \text{ and } \max_{0 \leq m' \leq N_{\text{cut},t}} i_{m'} = i$$

$$\text{such that } (x,t) \in \text{supp}\,(\psi_{\vec{i},q}) \cap \text{supp}\,(D_{t,q-1}\psi_{m,i_m,q})$$

$$\text{and } i_{m'} \leq i_m \text{ whenever } m < m' \leq N_{\text{cut},t}\,. \tag{6.75}$$

The generalization of characterization (6.75) to higher order material derivatives $D_{t,q-1}^M$ is direct: $(x,t) \in \text{supp}\,(D_{t,q-1}^M \psi_{i,q}^2)$ if and only if there exists $\vec{i} \in \mathbb{N}_0^{N_{\text{cut},t}+1}$ with maximal index equal to i, such that for every $0 \leq m \leq N_{\text{cut},t}$ for which $(x,t) \in \text{supp}\,(\psi_{\vec{i},q}) \cap \text{supp}\,(D_{t,q-1}\psi_{m,i_m,q})$ (there is potentially more than one such m if $M \geq 2$ due to the Leibniz rule), we have $i_{m'} \leq i_m \in \{i-1, i\}$ whenever $m < m'$. In light of this characterization, we have the following bounds:

Lemma 6.20. *Let $q \geq 1$ and $0 \leq i \leq i_{\max}(q)$, and fix $\vec{i} \in \mathbb{N}_0^{N_{\text{cut},t}+1}$ such that $\max_{0 \leq m \leq N_{\text{cut},t}} i_m = i$, as in the right side of (6.75). Fix $0 \leq m \leq N_{\text{cut},t}$ such that $i_m \in \{i-1, i\}$ and such that $i_{m'} \leq i_m$ for all $m \leq m' \leq N_{\text{cut},t}$. Lastly, fix j_m such that $i_*(j_m) \leq i_m$. For $N, K, M, k \geq 0$, $\alpha, \beta \in \mathbb{N}^k$ such that $|\alpha| = K$*

and $|\beta| = M$, we have

$$\frac{\mathbf{1}_{\text{supp}\,(\psi_{\widetilde{i},q})}\mathbf{1}_{\text{supp}\,(\psi_{j_m,q-1})}}{\psi_{m,i_m,j_m,q}^{1-(K+M)/N_{\text{fin}}}} \left| \left(\prod_{l=1}^{k} D^{\alpha_l} D_{t,q-1}^{\beta_l} \right) \psi_{m,i_m,j_m,q} \right|$$

$$\lesssim \mathcal{M}\left(K, N_{\text{ind},v}, \Gamma_q \lambda_q, \widetilde{\lambda}_q \Gamma_q \right) \mathcal{M}\left(M, N_{\text{ind},t} - N_{\text{cut},x}, \Gamma_{q+1}^{i+3} \tau_{q-1}^{-1}, \Gamma_{q+1}^{-1} \widetilde{\tau}_q^{-1} \right)$$

$$(6.76)$$

for all K such that $0 \le K + M \le N_{\text{fin}}$. Moreover,

$$\frac{\mathbf{1}_{\text{supp}\,(\psi_{\widetilde{i},q})}\mathbf{1}_{\text{supp}\,(\psi_{j_m,q-1})}}{\psi_{m,i_m,j_m,q}^{1-(N+K+M)/N_{\text{fin}}}} \left| D^N \left(\prod_{l=1}^{k} D_q^{\alpha_l} D_{t,q-1}^{\beta_l} \right) \psi_{m,i_m,j_m,q} \right|$$

$$\lesssim \mathcal{M}\left(N, N_{\text{ind},v}, \Gamma_q \lambda_q, \widetilde{\lambda}_q \Gamma_q \right) (\Gamma_{q+1}^{i-c_0} \tau_q^{-1})^K$$

$$\times \mathcal{M}\left(M, N_{\text{ind},t} - N_{\text{cut},x}, \Gamma_{q+1}^{3} \tau_{q-1}^{-1}, \Gamma_{q+1}^{-1} \widetilde{\tau}_q^{-1} \right) \quad (6.77)$$

holds whenever $0 \le N + K + M \le N_{\text{fin}}$.

Proof of Lemma 6.20. Note that for $M = 0$ estimate (6.76) was already established in (6.35). The bound (6.77) with $M = 0$, i.e., an estimate for the $D^N D_q^K \psi_{m,i_m,j_m,q}$, holds by appealing to the expansion (6.54)–(6.55) to the bound (6.56) (which is applicable since in the context of estimate (6.77) we work on the support of $\psi_{i,q}$), to the bound (6.76) with $M = 0$, and to the parameter inequality $\Gamma_{q+1}^{2+c_0} \delta_q^{1/2} \widetilde{\lambda}_q \le \tau_q^{-1}$ (which follows from (9.39)). The rest of the proof is dedicated to the case $M \ge 1$. The proofs are very similar to the proof of Lemma 6.13, but we additionally need to appeal to bounds and arguments from the proof of Lemma 6.18.

Proof of (6.76). As in the proof of Lemma 6.13, we start with the case $k = 1$, and estimate $D^K D_{t,q-1}^M \psi_{m,i_m,j_m,q}$ for $K + M \le N_{\text{fin}}$, with $M \ge 1$. We note that just like D, the operator $D_{t,q-1}$ is a scalar differential operator, and thus the Faà di Bruno argument which was used to bound (6.35) may be repeated. As was done there, we recall the definitions (6.7)–(6.8) and split the analysis in two cases, according to whether (6.37) or (6.42) holds.

Let us first consider the case (6.37). Our goal is to apply Lemma A.5 to the function $\psi = \psi_{m,q+1}$ or $\psi = \widetilde{\psi}_{m,q+1}$, with $\Gamma_\psi = \Gamma_{q+1}^{m+1}$, $\Gamma = \Gamma_{q+1}^{(m+1)(i_m - i_*(j_m))}$, $h(x,t) = h_{m,j_m,q}^2(x,t)$, and $D_t = D_{t,q-1}$. Estimate (A.24) holds by (6.3) and (6.5), so that it remains to obtain a bound on the material derivatives of $(h_{m,j_m,q}(x,t))^2$ and establish a bound which corresponds to (A.25) on the set $\text{supp}\,(\psi_{\widetilde{i},q}) \cap \text{supp}\,(\psi_{j_m,q-1}\psi_{m,i_m,j_m,q})$. Similarly to (6.38), for $K' + M' \le N_{\text{fin}}$ the Leibniz rule and definition (6.6) gives

$$\left| D^{K'} D_{t,q-1}^{M'} h_{m,j_m,q}^2 \right|$$

$$\lesssim (\lambda_q \Gamma_q)^{K'} (\tau_{q-1}^{-1} \Gamma_{q+1}^2)^{M'} \Gamma_{q+1}^{-2(m+1)i_*(j_m)}$$

$$\times \sum_{K''=0}^{K'} \sum_{M''=0}^{M'} \sum_{n=0}^{\mathsf{N}_{\mathrm{cut},x}} (\tau_{q-1}^{-1}\Gamma_{q+1}^2)^{-m-M''}(\lambda_q\Gamma_q)^{-n-K''}\delta_q^{-1/2}|D^{n+K''}D_{t,q-1}^{m+M''}u_q|$$

$$\times (\tau_{q-1}^{-1}\Gamma_{q+1}^2)^{-m-M'+M''}(\lambda_q\Gamma_q)^{-n-K'+K''}\delta_q^{-1/2}|D^{n+K'-K''}D_{t,q-1}^{m+M'-M''}u_q|. \tag{6.78}$$

By the characterization (6.75), for every (x,t) in the support described on the left side of (6.76) we have that for every $m \le R \le \mathsf{N}_{\mathrm{cut},t}$ there exists $i_R \le i_m$ and j_R with $i_*(j_R) \le i_R$, such that $(x,t) \in \mathrm{supp}\,\psi_{j_R,q-1}\psi_{R,i_R,j_R,q}$. As a consequence, for the terms in the sum (6.78) with $L \in \{n+K'', n+K'-K''\} \le \mathsf{N}_{\mathrm{cut},x}$ and $R \in \{m+M'', m+M'-M''\} \le \mathsf{N}_{\mathrm{cut},t}$, we may appeal to estimate (6.28) which gives a bound on $h_{R,j_R,q}$, and thus obtain

$$(\tau_{q-1}^{-1}\Gamma_{q+1}^2)^{-R}(\lambda_q\Gamma_q)^{-L}\delta_q^{-1/2}\left\|D^L D_{t,q-1}^R u_q\right\|_{L^\infty(\mathrm{supp}\,\psi_{R,i_R,j_R,q})}$$

$$\le \Gamma_{q+1}^{(R+1)i_*(j_R)}\Gamma_{q+1}^{(R+1)(i_R+1-i_*(j_R))}$$

$$\le \Gamma_{q+1}^{(R+1)(i_m+1)}.$$

On the other hand, if $L > \mathsf{N}_{\mathrm{cut},x}$, or if $R > \mathsf{N}_{\mathrm{cut},t}$, then by (5.6) and (5.9) we have that

$$(\tau_{q-1}^{-1}\Gamma_{q+1}^2)^{-R}(\lambda_q\Gamma_q)^{-L}\delta_q^{-1/2}\left\|D^L D_{t,q-1}^R u_q\right\|_{L^\infty(\mathrm{supp}\,\psi_{j_m,q-1})}$$

$$\le \widetilde{\lambda}_q^{3/2}\Gamma_q^{-L}\Gamma_{q+1}^{-2R}\mathcal{M}\left(L, 2\mathsf{N}_{\mathrm{ind},v}, 1, \lambda_q^{-1}\widetilde{\lambda}_q\right)\mathcal{M}\left(R, \mathsf{N}_{\mathrm{ind},t}, \Gamma_q^{j_m+1}, \tau_{q-1}\widetilde{\tau}_{q-1}^{-1}\right)$$

$$\le \mathcal{M}\left(L, 2\mathsf{N}_{\mathrm{ind},v}, 1, \lambda_q^{-1}\widetilde{\lambda}_q\right)\mathcal{M}\left(R, \mathsf{N}_{\mathrm{ind},t}, \Gamma_{q+1}^{i_m+1}, \tau_{q-1}\widetilde{\tau}_{q-1}^{-1}\right) \tag{6.79}$$

since $\mathsf{N}_{\mathrm{cut},x}$ and $\mathsf{N}_{\mathrm{cut},t}$ were taken sufficiently large to obey (9.51). Combining (6.78)–(6.79), we may derive that

$$\mathbf{1}_{\mathrm{supp}\,(\psi_{\widetilde{i},q})}\mathbf{1}_{\mathrm{supp}\,(\psi_{j_m,q-1})}\left|D^{K'}D_{t,q-1}^{M'}h_{m,j_m,q}^2\right|$$

$$\lesssim \Gamma_{q+1}^{2(m+1)(i_m-i_*(j_m)+1)}(\lambda_q\Gamma_q)^{K'}(\tau_{q-1}^{-1}\Gamma_{q+1}^2)^{M'}$$

$$\times \mathcal{M}\left(2\mathsf{N}_{\mathrm{cut},x}+K', 2\mathsf{N}_{\mathrm{ind},v}, 1, \lambda_q^{-1}\widetilde{\lambda}_q\right)\Gamma_{q+1}^{-2m(i_m+1)}$$

$$\times \sum_{M''=0}^{M'} \mathcal{M}\left(m+M'', \mathsf{N}_{\mathrm{ind},t}, \Gamma_{q+1}^{i_m+1}, \tau_{q-1}\widetilde{\tau}_{q-1}^{-1}\right)$$

$$\times \mathcal{M}\left(m+M'-M'', \mathsf{N}_{\mathrm{ind},t}, \Gamma_{q+1}^{i_m+1}, \tau_{q-1}\widetilde{\tau}_{q-1}^{-1}\right)$$

$$\lesssim \Gamma_{q+1}^{2(m+1)(i_m-i_*(j_m)+1)}(\lambda_q\Gamma_q)^{K'}(\tau_{q-1}^{-1}\Gamma_{q+1}^{i_m+3})^{M'}\mathcal{M}\left(K', \mathsf{N}_{\mathrm{ind},v}, 1, \lambda_q^{-1}\widetilde{\lambda}_q\right)$$

$$\times \mathcal{M}\left(M', \mathsf{N}_{\mathrm{ind},t}-\mathsf{N}_{\mathrm{cut},t}, 1, \tau_{q-1}\Gamma_{q+1}^{-(i_m+1)}\widetilde{\tau}_{q-1}^{-1}\right)$$

$$\lesssim \Gamma_{q+1}^{2(m+1)(i_m-i_*(j_m)+1)}\mathcal{M}\left(K', \mathsf{N}_{\mathrm{ind},v}, \Gamma_q\lambda_q, \Gamma_q\widetilde{\lambda}_q\right)$$

$$\times \mathcal{M}\left(M', \mathsf{N}_{\mathrm{ind},t} - \mathsf{N}_{\mathrm{cut},t}, \tau_{q-1}^{-1}\Gamma_{q+1}^{i+3}, \Gamma_{q+1}^{2}\tilde{\tau}_{q-1}^{-1}\right)$$

$$\lesssim \Gamma_{q+1}^{2(m+1)(i_m - i_*(j_m)+1)} \mathcal{M}\left(K', \mathsf{N}_{\mathrm{ind},v}, \Gamma_q\lambda_q, \Gamma_q\tilde{\lambda}_q\right)$$

$$\times \mathcal{M}\left(M', \mathsf{N}_{\mathrm{ind},t} - \mathsf{N}_{\mathrm{cut},t}, \tau_{q-1}^{-1}\Gamma_{q+1}^{i+3}, \Gamma_{q+1}^{-1}\tilde{\tau}_q^{-1}\right) \tag{6.80}$$

for all $K' + M' \leq \mathsf{N}_{\mathrm{fin}}$. Here we have used that $\mathsf{N}_{\mathrm{ind},v} \geq 2\mathsf{N}_{\mathrm{ind},t}$, that $m \leq \mathsf{N}_{\mathrm{cut},t}$, and that $i_m \leq i$. The upshot of (6.80) is that condition (A.25) in Lemma A.5 is now verified, with $\mathcal{C}_h = \Gamma_{q+1}^{2(m+1)(i_m - i_*(j_m)+1)}$, and $\lambda = \Gamma_q\lambda_q$, $\tilde{\lambda} = \Gamma_q\tilde{\lambda}_q$, $\mu = \tau_{q-1}^{-1}\Gamma_{q+1}^{i_m+3}$, $\tilde{\mu} = \Gamma_{q+1}^2\tilde{\tau}_{q-1}^{-1}$, $N_x = \mathsf{N}_{\mathrm{ind},v}$, and $N_t = \mathsf{N}_{\mathrm{ind},t} - \mathsf{N}_{\mathrm{cut},t}$. We obtain from (A.26) and the fact that $(\Gamma_\psi\Gamma)^{-2}\mathcal{C}_h = 1$ that (6.76) holds when $k = 1$ for those (x,t) such that $h_{m,j_m,q}(x,t)$ satisfies (6.37). The case when $h_{m,j_m,q}(x,t)$ satisfies the bound (6.42) is nearly identical, as was the case in the proof of Lemma 6.13. The only changes are that now $\Gamma_\psi = 1$ (according to (6.4)), and that the constant $\mathcal{C}_h$ which we read from the right side of (6.80) is now improved to $\Gamma_{q+1}^{2(m+1)(i_m - i_*(j_m))}$. These two changes offset each other, resulting in the same exact bound. Thus, we have shown that (6.76) holds when $k = 1$.

The general case $k \geq 1$ in (6.76) is obtained via induction on k, in precisely the same fashion as the proof of estimate (6.57) in Lemma 6.18. At the heart of the matter lies a commutator bound similar to (6.70), which is proven in precisely the same way by appealing to the fact that we work on $\mathrm{supp}\,(\psi_{\tilde{i},q}) \subset \mathrm{supp}\,(\psi_{i,q})$, and thus bound (6.66) is available; in turn, this bound provides sharper space and material estimates than required in (6.76), completing the proof. In order to avoid redundancy we omit further details.

Proof of (6.77). This estimate follows from Lemma A.10 with $v = u_q$, $B = D_{t,q-1}, f = \psi_{m,i_m,j_m,q}, \Omega = \mathrm{supp}\,(\psi_{\tilde{i},q}) \cap \mathrm{supp}\,(\psi_{j_m,q-1}) \cap \mathrm{supp}\,(\psi_{m,i_m,j_m,q})$, and $p = \infty$. Technically, the presence of the $\psi_{m,i_m,j_m,q}^{-1+(N+K+M)/\mathsf{N}_{\mathrm{fin}}}$ factor on the left side of (6.77) means that the bound doesn't follow from the statement of Lemma A.10, but instead, it follows from its proof; the changes to the argument are minor and we ignore this distinction. First, we note that since $\Omega \subset \mathrm{supp}\,(\psi_{i,q})$, estimate (6.57) allows us to verify condition (A.40) of Lemma A.10 with $N_* = {}^{3\mathsf{N}_{\mathrm{fin}}}/2 + 1$, $\mathcal{C}_v = \Gamma_{q+1}^{i+1}\delta_q^{1/2}$, $\lambda_v = \Gamma_q\lambda_q$, $\tilde{\lambda}_v = \tilde{\lambda}_q$, $N_x = 2\mathsf{N}_{\mathrm{ind},v} \geq \mathsf{N}_{\mathrm{ind},v}, \mu_v = \Gamma_{q+1}^{i+3}\tau_{q-1}^{-1}, \tilde{\mu}_v = \Gamma_{q+1}^{-1}\tilde{\tau}_q^{-1}$, and $N_t = \mathsf{N}_{\mathrm{ind},t} \geq \mathsf{N}_{\mathrm{ind},t} - \mathsf{N}_{\mathrm{cut},t}$. On the other hand, condition (A.41) of Lemma A.10 holds in view of (6.76) with $\mathcal{C}_f = 1$, $\lambda_f = \Gamma_q\lambda_q$, $\tilde{\lambda}_f = \Gamma_q\tilde{\lambda}_q$, $N_x = \mathsf{N}_{\mathrm{ind},v}, \mu_f = \Gamma_{q+1}^{i+3}\tau_{q-1}^{-1}, \tilde{\mu}_f = \Gamma_{q+1}^{-1}\tilde{\tau}_q^{-1}$, and $N_t = \mathsf{N}_{\mathrm{ind},t} - \mathsf{N}_{\mathrm{cut},t}$. As $|\alpha| = K$ and $|\beta| = M$, the bound (6.77) is now a direct consequence of (A.42) and the parameter inequality $\Gamma_{q+1}^{i+1}\delta_q^{1/2}\Gamma_q\tilde{\lambda}_q \leq \Gamma_{q+1}^{i-c_0}\tau_q^{-1} \Leftarrow \Gamma_{q+1}^{c_0+2}\delta_q^{1/2}\tilde{\lambda}_q \leq \tau_q^{-1}$; cf. (9.39).   □

A direct consequence of Lemma 6.20 and identity (6.75) is that the inductive bounds (3.19) and (3.20) hold for $q' = q$, as is shown by the following Lemma.

Lemma 6.21 (Mixed spatial and material derivatives for velocity cutoffs). *Let*

$q \geq 1$, $0 \leq i \leq i_{\max}(q)$, $N, K, M, k \geq 0$, and $\alpha, \beta \in \mathbb{N}^k$ be such that $|\alpha| = K$ and $|\beta| = M$. Then we have

$$\frac{1}{\psi_{i,q}^{1-(K+M)/\mathsf{N}_{\mathrm{fin}}}} \left| \left(\prod_{l=1}^{k} D^{\alpha_l} D_{t,q-1}^{\beta_l} \right) \psi_{i,q} \right|$$
$$\lesssim \mathcal{M} \left(K, \mathsf{N}_{\mathrm{ind},v}, \Gamma_q \lambda_q, \Gamma_q \widetilde{\lambda}_q \right) \mathcal{M} \left(M, \mathsf{N}_{\mathrm{ind},t} - \mathsf{N}_{\mathrm{cut},t}, \Gamma_{q+1}^{i+3} \tau_{q-1}^{-1}, \Gamma_{q+1}^{-1} \widetilde{\tau}_q^{-1} \right)$$
$$\text{(6.81)}$$

for $K + M \leq \mathsf{N}_{\mathrm{fin}}$, and

$$\frac{1}{\psi_{i,q}^{1-(N+K+M)/\mathsf{N}_{\mathrm{fin}}}} \left| D^N \left(\prod_{l=1}^{k} D_q^{\alpha_l} D_{t,q-1}^{\beta_l} \right) \psi_{i,q} \right|$$
$$\lesssim \mathcal{M} \left(N, \mathsf{N}_{\mathrm{ind},v}, \Gamma_q \lambda_q, \Gamma_q \widetilde{\lambda}_q \right) (\Gamma_{q+1}^{i-c_0} \tau_q^{-1})^K$$
$$\times \mathcal{M} \left(M, \mathsf{N}_{\mathrm{ind},t} - \mathsf{N}_{\mathrm{cut},t}, \Gamma_{q+1}^{i+3} \tau_{q-1}^{-1}, \Gamma_{q+1}^{-1} \widetilde{\tau}_q^{-1} \right) \qquad \text{(6.82)}$$

holds for $N + K + M \leq \mathsf{N}_{\mathrm{fin}}$.

Remark 6.22. As shown in Remark 3.4, the bound (6.82) and identity (A.39) imply that estimate (3.27) also holds with $q' = q$, namely that

$$\frac{1}{\psi_{i,q}^{1-(N+M)/\mathsf{N}_{\mathrm{fin}}}} \left| D^N D_{t,q}^M \psi_{i,q} \right|$$
$$\lesssim \mathcal{M} \left(N, \mathsf{N}_{\mathrm{ind},v}, \Gamma_q \lambda_q, \Gamma_q \widetilde{\lambda}_q \right) \mathcal{M} \left(M, \mathsf{N}_{\mathrm{ind},t} - \mathsf{N}_{\mathrm{cut},t}, \Gamma_{q+1}^{i-c_0} \tau_q^{-1}, \Gamma_{q+1}^{-1} \widetilde{\tau}_q^{-1} \right)$$
$$\text{(6.83)}$$

for $N + M \leq \mathsf{N}_{\mathrm{fin}}$. Note that for all $M \geq 0$ we have

$$\mathcal{M} \left(M, \mathsf{N}_{\mathrm{ind},t} - \mathsf{N}_{\mathrm{cut},t}, \Gamma_{q+1}^{i-c_0} \tau_q^{-1}, \Gamma_{q+1}^{-1} \widetilde{\tau}_q^{-1} \right)$$
$$\leq \Gamma_{q+1}^{-(\mathsf{N}_{\mathrm{ind},t} - \mathsf{N}_{\mathrm{cut},t})} \left(\tau_q \Gamma_{q+1}^{-1} \widetilde{\tau}_q^{-1} \right)^{\mathsf{N}_{\mathrm{cut}}} \mathcal{M} \left(M, \mathsf{N}_{\mathrm{ind},t}, \Gamma_{q+1}^{i-c_0+1} \tau_q^{-1}, \Gamma_{q+1}^{-1} \widetilde{\tau}_q^{-1} \right)$$
$$\leq \mathcal{M} \left(M, \mathsf{N}_{\mathrm{ind},t}, \Gamma_{q+1}^{i-c_0+1} \tau_q^{-1}, \Gamma_{q+1}^{-1} \widetilde{\tau}_q^{-1} \right)$$

once $\mathsf{N}_{\mathrm{ind},t}$ is taken to be sufficiently large when compared to $\mathsf{N}_{\mathrm{cut},t}$ to ensure that

$$\left(\tau_q \widetilde{\tau}_q^{-1} \right)^{\mathsf{N}_{\mathrm{cut}}} \leq \Gamma_{q+1}^{\mathsf{N}_{\mathrm{ind},t}}$$

for all $q \geq 1$. This condition holds in view of (9.52). In summary, we have thus obtained

$$\frac{1}{\psi_{i,q}^{1-(N+M)/\mathsf{N}_{\mathrm{fin}}}} \left| D^N D_{t,q}^M \psi_{i,q} \right|$$

$$\lesssim \mathcal{M}\left(N, \mathsf{N}_{\mathrm{ind},v}, \Gamma_q \lambda_q, \Gamma_q \widetilde{\lambda}_q\right) \mathcal{M}\left(M, \mathsf{N}_{\mathrm{ind},t}, \Gamma_{q+1}^{i-c_0+1} \tau_q^{-1}, \Gamma_{q+1}^{-1} \widetilde{\tau}_q^{-1}\right) \qquad (6.84)$$

for $N + M \leq \mathsf{N}_{\mathrm{fin}}$.

Proof of Lemma 6.21. Note that for $M = 0$ estimate (6.81) holds by (6.36). The bound (6.82) holds for $M = 0$, due to the expansion (6.54)–(6.55), to the bound (6.56) on the support of $\psi_{i,q}$, to the bound (6.82) with $M = 0$, and to the parameter inequality $\Gamma_{q+1}^{2+c_0} \delta_q^{1/2} \widetilde{\lambda}_q \leq \tau_q^{-1}$ (cf. (9.39)). The rest of the proof is dedicated to the case $M \geq 1$.

The argument is very similar to the proof of Lemma 6.13 and so we only emphasize the main differences. We start with the proof of (6.81). We claim that in the same way that (6.35) was shown to imply (6.45), one may show that estimate (6.76) implies that for any $\vec{i}$ and $0 \leq m \leq \mathsf{N}_{\mathrm{cut},t}$ as on the right side of (6.75) (in particular, as in Lemma 6.18), we have that

$$\frac{\mathbf{1}_{\mathrm{supp}\,(\psi_{\vec{i},q})}}{\psi_{m,i_m,q}^{1-(K+M)/\mathsf{N}_{\mathrm{fin}}}} \left| \left(\prod_{l=1}^{k} D^{\alpha_l} D_{t,q-1}^{\beta_l} \right) \psi_{m,i_m,q} \right|$$

$$\lesssim \mathcal{M}\left(K, \mathsf{N}_{\mathrm{ind},v}, \Gamma_q \lambda_q, \widetilde{\lambda}_q \Gamma_q\right) \mathcal{M}\left(M, \mathsf{N}_{\mathrm{ind},t} - \mathsf{N}_{\mathrm{cut},x}, \Gamma_{q+1}^{i+3} \tau_{q-1}^{-1}, \Gamma_{q+1}^{-1} \widetilde{\tau}_q^{-1}\right).$$
$$(6.85)$$

The proof of the above estimate is done by induction on k. For $k = 1$, the first step in establishing (6.85) is to use the Leibniz rule and induction on the number of material derivatives to reduce the problem to an estimate for $\psi_{m,i_m,q}^{-2+(K+M)/\mathsf{N}_{\mathrm{fin}}} D^K D_{t,q-1}^M (\psi_{m,i_m,q}^2)$; this is achieved in precisely the same way that (6.47) was proven. The derivatives of $\psi_{m,i_m,q}^2$ are now bounded via the Leibniz rule and the definition (6.11). Indeed, when $D^{K'} D_{t,q-1}^{M'}$ derivatives fall on $\psi_{m,i_m,j_m,q}^2$ the required bound is obtained from (6.76), which gives the same upper bound as the one required by (6.85). On the other hand, if $D^{K-K'} D_{t,q-1}^{M-M'}$ derivatives fall on $\psi_{j_m,q-1}^2$, the required estimate is provided by (3.27) with $q' = q - 1$ and i replaced by j_m; the resulting estimates are strictly better than what is required by (6.85). This shows that estimate (6.85) holds for $k = 1$. We then proceed inductively in $k \geq 1$, in the same fashion as the proof of estimate (6.57) in Lemma 6.18; the corresponding commutator bound is applicable because we work on $\mathrm{supp}\,(\psi_{m,i_m,q}) \cap \mathrm{supp}\,(\psi_{i,q})$. In order to avoid redundancy we omit these details, and conclude the proof of (6.85).

As in the proof of Lemma 6.13, we are now able to show that (6.81) is a consequence of (6.85). As before, by induction on the number of material derivatives and the Leibniz rule we reduce the problem to an estimate for $\psi_{i,q}^{-2+(K+M)/\mathsf{N}_{\mathrm{fin}}} \prod_{l=1}^{k} D^{\alpha_l} D_{t,q-1}^{\beta_l} (\psi_{i,q}^2)$; see the proof of (6.47) for details. In order to estimate derivatives of $\psi_{i,q}^2$, we use identities (6.25) and (6.26), which imply upon applying a differential operator, say $D_{t,q-1}$, that

$$D_{t,q-1}(\psi_{i,q}^2)$$

$$= D_{t,q-1}\left(\sum_{m=0}^{\mathsf{N}_{\mathrm{cut},t}} \prod_{m'=0}^{m-1} \Psi_{m',i,q}^2 \cdot \psi_{m,i,q}^2 \cdot \prod_{m''=m+1}^{\mathsf{N}_{\mathrm{cut},t}} \Psi_{m'',i-1,q}^2\right)$$

$$= \sum_{m=0}^{\mathsf{N}_{\mathrm{cut},t}} \sum_{\bar{m}'=0}^{m-1} D_{t,q-1}(\psi_{\bar{m}',i,q}^2) \prod_{\substack{0 \le m' \le m-1 \\ m' \ne \bar{m}'}} \Psi_{m',i,q}^2 \cdot \psi_{m,i,q}^2 \cdot \prod_{m''=m+1}^{\mathsf{N}_{\mathrm{cut},t}} \Psi_{m'',i-1,q}^2$$

$$+ \sum_{m=0}^{\mathsf{N}_{\mathrm{cut},t}} \sum_{\bar{m}''=m+1}^{\mathsf{N}_{\mathrm{cut},t}} \prod_{m'=0}^{m-1} \Psi_{m',i,q}^2 \cdot \psi_{m,i,q}^2 \cdot D_{t,q-1}(\Psi_{\bar{m}'',i-1,q}^2) \prod_{\substack{m+1 \le m'' \le \mathsf{N}_{\mathrm{cut},t} \\ m'' \ne \bar{m}''}} \Psi_{m'',i-1,q}^2$$

$$+ \sum_{m=0}^{\mathsf{N}_{\mathrm{cut},t}} \prod_{m'=0}^{m-1} \Psi_{m',i,q}^2 \cdot D_{t,q-1}(\psi_{m,i,q}^2) \cdot \prod_{m''=m+1}^{\mathsf{N}_{\mathrm{cut},t}} \Psi_{m'',i-1,q}^2 \cdot \tag{6.86}$$

Higher order material derivatives of $\psi_{i,q}^2$, and mixtures of space and material derivatives, are obtained similarly, by an application of the Leibniz rule. Equality (6.86) in particular justifies why we have only proven (6.85) for $\vec{i}$ and $0 \le m \le \mathsf{N}_{\mathrm{cut},t}$, as on the right side of (6.75)! With (6.85) and (6.86) in hand, we now repeat the argument from the proof of Lemma 6.13 (see the two displays below (6.47)) and conclude that (6.81) holds.

In order to conclude the proof of the lemma, it remains to establish (6.82). This bound follows now directly from (6.81) and an application of Lemma A.10 (to be more precise, we need to use the proof of this lemma), in precisely the same way that (6.76) was shown earlier to imply (6.77). As there are no changes to be made to this argument, we omit these details.   □

6.2.6 L^1 size of the velocity cutoffs

The purpose of this section is to show that the inductive estimate (3.21) holds with $q' = q$.

Lemma 6.23 (Support estimate). *For all $0 \le i \le i_{\max}(q)$ we have that*

$$\|\psi_{i,q}\|_{L^1} \lesssim \Gamma_{q+1}^{-2i+\mathsf{C}_b}, \tag{6.87}$$

where C_b is defined in (3.21) and thus depends only on b.

Proof of Lemma 6.23. If $i \le (\mathsf{C}_b - 1)/2$ then (6.87) trivially holds because $0 \le \psi_{i,q} \le 1$, and $|\mathbb{T}^3| \le \Gamma_{q+1}$ for all $q \ge 1$, once a is chosen to be sufficiently large. Thus, we only need to be concerned with i such that $(\mathsf{C}_b + 1)/2 \le i \le i_{\max}(q)$.

First, we note that Lemma 6.7 implies that the functions $\Psi_{m,i',q}$ defined in

(6.24) satisfy $0 \leq \Psi_{m,i',q}^2 \leq 1$, and thus (6.26) implies that

$$\|\psi_{i,q}\|_{L^1} \leq \sum_{m=0}^{N_{cut,t}} \|\psi_{m,i,q}\|_{L^1} . \tag{6.88}$$

Next, we let $j_*(i) = j_*(i,q)$ be the *maximal* index of j_m appearing in (6.11). In particular, recalling also (6.27), we have that

$$\Gamma_{q+1}^{i-1} < \Gamma_q^{j_*(i)} \leq \Gamma_{q+1}^i < \Gamma_q^{j_*(i)+1} . \tag{6.89}$$

Using (6.11), in which we simply write j instead of j_m, the fact that $0 \leq \psi_{j,q-1}^2, \psi_{m,i,j,q}^2 \leq 1$, and the inductive assumption (3.21) at level $q-1$, we may deduce that

$$\|\psi_{m,i,q}\|_{L^1} \leq \|\psi_{j_*(i),q-1}\|_{L^1} + \|\psi_{j_*(i)-1,q-1}\|_{L^1} + \sum_{j=0}^{j_*(i)-2} \|\psi_{j,q-1}\psi_{m,i,j,q}\|_{L^1}$$

$$\leq \Gamma_q^{-2j_*(i)+C_b} + \Gamma_q^{-2j_*(i)+2+C_b} + \sum_{j=0}^{j_*(i)-2} |supp(\psi_{j,q-1}\psi_{m,i,j,q})| . \tag{6.90}$$

The second term on the right side of (6.90) is estimated using the last inequality in (6.89) as

$$\Gamma_q^{-2j_*(i)+2+C_b} \leq \Gamma_{q+1}^{-2i}\Gamma_q^{4+C_b} \leq \Gamma_{q+1}^{-2i+C_b-1}\Gamma_q^{4+C_b-b(C_b-1)} = \Gamma_{q+1}^{-2i+C_b-1}, \tag{6.91}$$

where in the last equality we have used the definition of C_b in (3.21). Clearly, the first term on the right side of (6.90) is also bounded by the right side of (6.91). We are left to estimate the terms appearing in the sum on the right side of (6.90). The key fact is that for any $j \leq j_*(i) - 2$ we have that $i \geq i_*(j) + 1$; this can be seen to hold because $b < 2$. Recalling the definition (6.7) and item 2 of Lemma 6.2, we obtain that for $j \leq j_*(i) - 2$ we have

$$supp(\psi_{j,q-1}\psi_{m,i,j,q}) \subseteq \left\{ (x,t) \in supp(\psi_{j,q-1}): h_{m,j,q}^2 \geq \frac{1}{4}\Gamma_{q+1}^{2(m+1)(i-i_*(j))} \right\}$$

$$\subseteq \left\{ (x,t): \psi_{j\pm,q-1}^2 h_{m,j,q}^2 \geq \frac{1}{4}\Gamma_{q+1}^{2(m+1)(i-i_*(j))} \right\} . \tag{6.92}$$

In the second inclusion of (6.92) we have appealed to (6.23) at level $q-1$. By Chebyshev's inequality and the definition of $h_{m,j,q}$ in (6.6) we deduce that

$$|supp(\psi_{j,q-1}\psi_{m,i,j,q})|$$

$$\leq 4\Gamma_{q+1}^{-2(m+1)(i-i_*(j))} \sum_{n=0}^{N_{\text{cut},x}} \Gamma_{q+1}^{-2i_*(j)} \delta_q^{-1} (\lambda_q \Gamma_q)^{-2n} \left(\tau_{q-1}^{-1} \Gamma_{q+1}^{i_*(j)+2}\right)^{-2m}$$

$$\times \left\| \psi_{j\pm,q-1} D^n D_{t,q-1}^m u_q \right\|_{L^2}^2 .$$

Since in the above display we have that $n \leq N_{\text{cut},x} \leq 2N_{\text{ind},v}$ and $m \leq N_{\text{cut},t} \leq N_{\text{ind},t}$, we may combine the above estimate with (5.5) and deduce that

$$\left| \text{supp} \left(\psi_{j,q-1} \psi_{m,i,j,q} \right) \right|$$

$$\leq 4\Gamma_{q+1}^{-2(m+1)(i-i_*(j))} \Gamma_{q+1}^{-2i_*(j)} \left(\Gamma_q^{j+1} \Gamma_{q+1}^{-i_*(j)-2}\right)^{2m} \sum_{n=0}^{N_{\text{cut},x}} \Gamma_q^{-2n}$$

$$\leq 8\Gamma_{q+1}^{-2i} \left(\Gamma_q^{j+1} \Gamma_{q+1}^{-i-2}\right)^{2m}$$

$$\leq \Gamma_{q+1}^{-2i+C_b-1} . \tag{6.93}$$

In the last inequality we have used that $\Gamma_q^j \leq \Gamma_{q+1}^i$, that $m \geq 0$, and that $C_b \geq 2$ (since $b \leq 6$).

Combining (6.88), (6.90), (6.91), and (6.93) we deduce that

$$\|\psi_{i,q}\|_{L^1} \leq N_{\text{cut},t} \, j_*(i) \, \Gamma_{q+1}^{-2i+C_b-1} .$$

In order to conclude the proof of the lemma, we use the fact that $N_{\text{cut},t}$ is a constant independent of q, and that by (6.90) and (3.17) we have

$$j_*(i) \leq i \frac{\log \Gamma_{q+1}}{\log \Gamma_q} \leq i_{\max}(q) b \leq \frac{4b}{\varepsilon_\Gamma(b-1)} .$$

Thus $j_*(i)$ is also bounded from above by a constant independent of q and upon taking a sufficiently large we have

$$N_{\text{cut},t} \, j_*(i) \, \Gamma_{q+1}^{-1} \leq \frac{4N_{\text{cut},t}b}{\varepsilon_\Gamma(b-1)} \Gamma_{q+1}^{-1} \leq 1,$$

which concludes the proof. $\qquad\square$

6.3 DEFINITION OF THE TEMPORAL CUTOFF FUNCTIONS

Let $\chi : (-1,1) \to [0,1]$ be a C^∞ function which induces a partition of unity according to

$$\sum_{k \in \mathbb{Z}} \chi^2(\cdot - k) \equiv 1. \tag{6.94}$$

Consider the translated and rescaled function

$$\chi \left(t\tau_q^{-1} \Gamma_{q+1}^{i-c_0+2} - k \right) ,$$

which is supported in the set of times t satisfying

$$\left| t - \tau_q \Gamma_{q+1}^{-i+c_0-2} k \right| \leq \tau_q \Gamma_{q+1}^{-i+c_0-2}$$
$$\Longleftrightarrow t \in \left[(k-1)\tau_q \Gamma_{q+1}^{-i+c_0-2}, (k+1)\tau_q \Gamma_{q+1}^{-i+c_0-2} \right] . \tag{6.95}$$

We then define temporal cutoff functions

$$\chi_{i,k,q}(t) = \chi_{(i)}(t) = \chi \left(t\tau_q^{-1} \Gamma_{q+1}^{i-c_0+2} - k \right) . \tag{6.96}$$

It is then clear that

$$\left| \partial_t^m \chi_{i,k,q} \right| \lesssim (\Gamma_{q+1}^{i-c_0+2} \tau_q^{-1})^m \tag{6.97}$$

for $m \geq 0$ and

$$\chi_{i,k_1,q}(t) \chi_{i,k_2,q}(t) = 0 \tag{6.98}$$

for all $t \in \mathbb{R}$ unless $|k_1 - k_2| \leq 1$. In analogy with $\psi_{i\pm,q}$, we define

$$\chi_{(i,k\pm,q)}(t) := \left(\chi_{(i,k-1,q)}^2(t) + \chi_{(i,k,q)}^2(t) + \chi_{(i,k+1,q)}^2(t) \right)^{\frac{1}{2}}, \tag{6.99}$$

which are cutoffs with the property that

$$\chi_{(i,k\pm,q)} \equiv 1 \text{ on supp} \left(\chi_{(i,k,q)} \right). \tag{6.100}$$

Next, we define the cutoffs $\tilde{\chi}_{i,k,q}$ by

$$\tilde{\chi}_{i,k,q}(t) = \tilde{\chi}_{(i)}(t) = \chi \left(t\tau_q^{-1} \Gamma_{q+1}^{i-c_0} - \Gamma_{q+1}^{-c_0} k \right). \tag{6.101}$$

For comparison with (6.95), we have that $\tilde{\chi}_{i,k,q}$ is supported in the set of times t satisfying

$$\left| t - \tau_q \Gamma_{q+1}^{-i+c_0} k \right| \leq \tau_q \Gamma_{q+1}^{-i+c_0}. \tag{6.102}$$

As a consequence of these definitions and a sufficiently large choice of λ_0, let (i, k) and (i^*, k^*) be such that $\text{supp} \chi_{i,k,q} \cap \text{supp} \chi_{i^*,k^*,q} \neq \emptyset$ and $i^* \in \{i-1, i, i+1\}$; then

$$\text{supp} \chi_{i,k,q} \subset \text{supp} \tilde{\chi}_{i^*,k^*,q}. \tag{6.103}$$

Finally, we shall require cutoffs $\overline{\chi}_{q,n,p}$ which satisfy the following three properties:

1. $\overline{\chi}_{q,n,p}(t) \equiv 1$ on $\text{supp}_t \mathring{R}_{q,n,p}$.

2. $\overline{\chi}_{q,n,p}(t) = 0$ if $\left\| \mathring{R}_{q,n,p}(\cdot, t') \right\|_{L^\infty(\mathbb{T}^3)} = 0$ for all $|t - t'| \leq \left(\delta_q^{1/2} \lambda_q \Gamma_{q+1}^2 \right)^{-1}$.

3. $\partial_t^m \overline{\chi}_{q,n,p} \lesssim \left(\delta_q^{1/2} \lambda_q \Gamma_{q+1}^2 \right)^m$.

For the sake of specificity, recalling (9.63), we may set

$$\overline{\chi}_{q,n,p} = \phi^{(t)}_{\left(\delta_q^{1/2}\lambda_q\Gamma_{q+1}^2\right)} * 1_{\left\{ t : \left\| \mathring{R}_{q,n,p} \right\|_{L^\infty} \left(\left[t - \left(\delta_q^{1/2}\lambda_q\Gamma_{q+1}^2\right)^{-1}, t + \left(\delta_q^{1/2}\lambda_q\Gamma_{q+1}^2\right)^{-1} \right] \times \mathbb{T}^3 \right) > 0 \right\}}.$$
(6.104)

It is then clear that $\overline{\chi}_{q,n,p}$ slightly expands and then mollifies the characteristic function of the time support of $\mathring{R}_{q,n,p}$ so that the inductive assumptions (7.12), (7.19), and (7.26) regarding the time support of $w_{q+1,n,p}$ may be verified.

6.4   ESTIMATES ON FLOW MAPS

We can now make estimates regarding the flows of the vector field v_{ℓ_q} on the support of a cutoff function.

Lemma 6.24 (Lagrangian paths don't jump many supports). *Let $q \geq 0$ and (x_0, t_0) be given. Assume that the index i is such that $\psi_{i,q}^2(x_0, t_0) \geq \kappa^2$, where $\kappa \in \left[\frac{1}{16}, 1 \right]$. Then the forward flow $(X(t), t) := (X(x_0, t_0; t), t)$ of the velocity field v_{ℓ_q} originating at (x_0, t_0) has the property that $\psi_{i,q}^2(X(t), t) \geq \kappa^2/2$ for all t be such that $|t - t_0| \leq \left(\delta_q^{1/2} \lambda_q \Gamma_{q+1}^{i+3} \right)^{-1}$, which by (9.39) and (9.19) is satisfied for $|t - t_0| \leq \tau_q \Gamma_{q+1}^{-i+5+c_0}$.*

Proof of Lemma 6.24. By the mean value theorem in time along the Lagrangian flow $(X(t), t)$ and (6.83), we have that

$$|\psi_{i,q}(X(t), t) - \psi_{i,q}(x_0, t_0)| \leq |t - t_0| \, \|D_{t,q}\psi_{i,q}\|_{L^\infty}$$
$$\leq |t - t_0| \, \|D_{t,q-1}\psi_{i,q}\|_{L^\infty} + |t - t_0| \, \|u_q \cdot \nabla\psi_{i,q}\|_{L^\infty}.$$

From Lemma 6.21, Lemma 6.13, Lemma 6.11, and (9.41), we have that

$$\|D_{t,q-1}\psi_{i,q}\|_{L^\infty} + \|u_q \cdot \nabla\psi_{i,q}\|_{L^\infty} \lesssim \Gamma_{q+1}^{i+3} \tau_{q-1}^{-1} + \delta_q^{1/2} \Gamma_{q+1}^{i+1} \lambda_q \Gamma_q$$
$$\lesssim \delta_q^{1/2} \lambda_q \Gamma_{q+1}^{i+2},$$

and hence, under the working assumption on $|t - t_0|$ we obtain

$$|\psi_{i,q}(X(x_0, t_0; t), t) - \psi_{i,q}(x_0, t_0)| \lesssim \Gamma_{q+1}^{-1}, \tag{6.105}$$

for some implicit constant $C > 0$ which is independent of $q \geq 0$. From the assumption of the lemma and (6.105) it follows that

$$\psi_{i,q}(X(t), t) \geq \kappa - C\Gamma_{q+1}^{-1} \geq \kappa/\sqrt{2}$$

for all $q \geq 0$, since we have that $\kappa \geq 1/16$ and $C\Gamma_{q+1}^{-1} \leq 1/100$, which holds independently of q once λ_0 is chosen sufficiently large. $\qquad\qquad\square$

Corollary 6.25. *Suppose (x,t) is such that $\psi_{i,q}^2(x,t) \geq \kappa^2$, where $\kappa \in [1/16, 1]$. For t_0 such that $|t - t_0| \leq \left(\delta_q^{1/2} \lambda_q \Gamma_{q+1}^{i+4}\right)^{-1}$, which is in particular satisfied for $|t - t_0| \leq \tau_q \Gamma_{q+1}^{-i+4+c_0}$, define x_0 to satisfy*

$$x = X(x_0, t_0; t).$$

That is, the forward flow X of the velocity field v_{ℓ_q}, originating at x_0 at time t_0, reaches the point x at time t. Then we have

$$\psi_{i,q}(x_0, t_0) \neq 0.$$

Proof of Corollary 6.25. By contradiction, suppose that $\psi_{i,q}(x_0, t_0) = 0$. Without loss of generality we can assume $t < t_0$. By continuity, there exists a minimal time $t' \in (t, t_0]$ such that for $x' = x'(t')$ defined by

$$x = X(x', t'; t),$$

we have

$$\psi_{i,q}(x', t') = 0.$$

By minimality and (6.19), there exists an $i' \in \{i-1, i+1\}$ such that

$$\psi_{i',q}(x', t') = 1.$$

Applying Lemma 6.24, estimate (6.105), we obtain

$$|\psi_{i',q}\left(X(x', t'; t), t\right) - \psi_{i',q}(x', t')| = |\psi_{i',q}(x, t) - \psi_{i',q}(x', t')| \lesssim \Gamma_{q+1}^{-1}. \quad (6.106)$$

Here we have used that $|t' - t| \leq |t_0 - t| \leq \left(\delta_q^{1/2} \lambda_q \Gamma_{q+1}^{i+4}\right)^{-1} \leq \left(\delta_q^{1/2} \lambda_q \Gamma_{q+1}^{i'+3}\right)^{-1}$, so that Lemma 6.24 is applicable. Since $\psi_{i',q}(x', t') = 1$, from (6.106) we see that $\psi_{i',q}(x, t) > 0$, and so $\psi_{i,q}^2(x, t) = 1 - \psi_{i',q}^2(x, t)$. Then we obtain

$$\begin{aligned}
\psi_{i,q}^2(x, t) &= 1 - \psi_{i',q}^2(x, t) \\
&= (1 + \psi_{i',q}(x, t))(1 - \psi_{i',q}(x, t)) \\
&= (1 + \psi_{i',q}(x, t))(\psi_{i',q}(x', t') - \psi_{i',q}(x, t)) \\
&\lesssim \Gamma_{q+1}^{-1},
\end{aligned}$$

which is a contradiction once λ_0 is chosen sufficiently large, since we assumed that $\psi_{i,q}^2(x, t) \geq \kappa^2$ and $\kappa \geq 1/16$. $\qquad\qquad\square$

Definition 6.26. *We define* $\Phi_{i,k,q}(x,t) := \Phi_{(i,k)}(x,t)$ *to be the flows induced by* v_{ℓ_q} *with initial datum at time* $k\tau_q\Gamma_{q+1}^{-i}$ *given by the identity, i.e.,*

$$\begin{cases} (\partial_t + v_{\ell_q} \cdot \nabla)\Phi_{i,k,q} = 0 \\ \Phi_{i,k,q}(x, k\tau_q\Gamma_{q+1}^{-i}) = x. \end{cases} \tag{6.107}$$

We will use $D\Phi_{(i,k)}$ to denote the gradient of $\Phi_{(i,k)}$ (which is thus a matrix-valued function). The inverse of the matrix $D\Phi_{(i,k)}$ is denoted by $\left(D\Phi_{(i,k)}\right)^{-1}$, in contrast to $D\Phi_{(i,k)}^{-1}$, which is the gradient of the inverse map $\Phi_{(i,k)}^{-1}$.

Corollary 6.27 (Deformation bounds). *For* $k \in \mathbb{Z}$, $0 \le i \le i_{max}$, $q \ge 0$, *and* $2 \le N \le {}^{3N_{fin}}/2 + 1$, *we have the following bounds on the support of* $\psi_{i,q}(x,t)\widetilde{\chi}_{i,k,q}(t)$.

$$\left\| D\Phi_{(i,k)} - \mathrm{Id} \right\|_{L^\infty(\mathrm{supp}\,(\psi_{i,q}\widetilde{\chi}_{i,k,q}))} \lesssim \Gamma_{q+1}^{-1} \tag{6.108}$$

$$\left\| D^N\Phi_{(i,k)} \right\|_{L^\infty(\mathrm{supp}\,(\psi_{i,q}\widetilde{\chi}_{i,k,q}))} \lesssim \Gamma_{q+1}^{-1}\mathcal{M}\left(N-1, 2N_{ind,v}, \Gamma_q\lambda_q, \widetilde{\lambda}_q\right) \tag{6.109}$$

$$\left\| (D\Phi_{(i,k)})^{-1} - \mathrm{Id} \right\|_{L^\infty(\mathrm{supp}\,(\psi_{i,q}\widetilde{\chi}_{i,k,q}))} \lesssim \Gamma_{q+1}^{-1} \tag{6.110}$$

$$\left\| D^{N-1}\left((D\Phi_{(i,k)})^{-1}\right) \right\|_{L^\infty(\mathrm{supp}\,(\psi_{i,q}\widetilde{\chi}_{i,k,q}))} \lesssim \Gamma_{q+1}^{-1}\mathcal{M}\left(N-1, 2N_{ind,v}, \Gamma_q\lambda_q, \widetilde{\lambda}_q\right) \tag{6.111}$$

$$\left\| D^N\Phi_{(i,k)}^{-1} \right\|_{L^\infty(\mathrm{supp}\,(\psi_{i,q}\widetilde{\chi}_{i,k,q}))} \lesssim \Gamma_{q+1}^{-1}\mathcal{M}\left(N-1, 2N_{ind,v}, \Gamma_q\lambda_q, \widetilde{\lambda}_q\right) \tag{6.112}$$

Furthermore, we have the following bounds for $1 \le N + M \le {}^{3N_{fin}}/2$:

$$\left\| D^{N-N'}D_{t,q}^M D^{N'+1}\Phi_{(i,k)} \right\|_{L^\infty(\mathrm{supp}\,(\psi_{i,q}\widetilde{\chi}_{i,k,q}))}$$
$$\le \widetilde{\lambda}_q^N\mathcal{M}\left(M, N_{ind,t}, \Gamma_{q+1}^{i-c_0}\tau_q^{-1}, \widetilde{\tau}_q^{-1}\Gamma_{q+1}^{-1}\right) \tag{6.113}$$

$$\left\| D^{N-N'}D_{t,q}^M D^{N'}(D\Phi_{(i,k)})^{-1} \right\|_{L^\infty(\mathrm{supp}\,(\psi_{i,q}\widetilde{\chi}_{i,k,q}))}$$
$$\le \widetilde{\lambda}_q^N\mathcal{M}\left(M, N_{ind,t}, \Gamma_{q+1}^{i-c_0}\tau_q^{-1}, \widetilde{\tau}_q^{-1}\Gamma_{q+1}^{-1}\right) \tag{6.114}$$

for all $0 \le N' \le N$.

Proof of Corollary 6.27. Let $t_k := \tau_q\Gamma_{q+1}^{-i}k$. For t is on the support of $\widetilde{\chi}_{i,k,q}$, we may assume from (6.102) that $|t - t_k| \le \tau_q\Gamma_{q+1}^{-i+c_0}$. Moreover, since the $\{\psi_{i',q}\}_{i'\ge0}$ form a partition of unity, we know that there exists i' such that $\psi_{i',q}^2(x,t) \ge {}^1/2$ and $i' \in \{i-1, i, i+1\}$. Thus, we have that $|t - t_k| \le \tau_q\Gamma_{q+1}^{-i'+1+c_0}$, and Corollary 6.25 is applicable. For this purpose, let x_0 be defined by $X(x_0, t_k; t) = x$, where X is the forward flow of the velocity field v_{ℓ_q}, which equals the identity at time t_k. Corollary 6.25 guarantees that $(x_0, t_k) \in \mathrm{supp}\,(\psi_{i',q})$.

The above argument shows that the flow $(X(x_0, t_k; t), t)$ remains in the support of $\psi_{i',q}$ for all t such that $|t - t_k| \leq \tau_q \Gamma_{q+1}^{-i+c_0}$, where $i' \in \{i-1, i, i+1\}$. In turn, using estimate (6.60), this shows that

$$\sup_{|t-t_k| \leq \tau_q \Gamma_{q+1}^{-i+c_0}} |Dv_{\ell_q}(X(x_0, t_k; t), t)| \lesssim \|Dv_{\ell_q}\|_{L^\infty(\mathrm{supp}\,(\psi_{i\pm,q}))} \lesssim \Gamma_{q+1}^{i+2} \delta_q^{1/2} \widetilde{\lambda}_q.$$

To conclude, using (4) from Lemma A.1 and (9.39), we obtain

$$\left\| D\Phi_{(i,k)} - \mathrm{Id} \right\|_{L^\infty(\mathrm{supp}\,(\psi_{i,q}\widetilde{\chi}_{i,k,q}))} \lesssim \tau_q \Gamma_{q+1}^{-i+c_0} \Gamma_{q+1}^{i+2} \delta_q^{1/2} \widetilde{\lambda}_q \lesssim \Gamma_{q+1}^{-1},$$

which implies the desired estimate in (6.108).

Similarly, since the flow $(X(x_0, t_k; t), t)$ remains in the support of $\psi_{i',q}$ for all t such that $|t - t_k| \leq \tau_q \Gamma_{q+1}^{-i+c_0}$, for $N \geq 2$ the estimates in (3) from Lemma A.1 give that

$$\left\| D^N \Phi_{(i,k)} \right\|_{L^\infty(\mathrm{supp}\,(\psi_{i,q}\widetilde{\chi}_{i,k,q}))}$$
$$\lesssim \tau_q \Gamma_{q+1}^{-i+c_0} \left\| D^N v_{\ell_q} \right\|_{L^\infty(\mathrm{supp}\,(\psi_{i\pm,q}))}$$
$$\lesssim \tau_q \Gamma_{q+1}^{-i+c_0} (\Gamma_{q+1}^{i+2} \delta_q^{1/2}) \widetilde{\lambda}_q \mathcal{M}\left(N-1, 2\mathsf{N}_{\mathrm{ind},v}, \Gamma_q \lambda_q, \widetilde{\lambda}_q\right)$$
$$\lesssim \Gamma_{q+1}^{-1} \mathcal{M}\left(N-1, 2\mathsf{N}_{\mathrm{ind},v}, \Gamma_q \lambda_q, \widetilde{\lambda}_q\right).$$

Here we have used the bound (6.60) with $M = 0$ and $K = N - 1$ up to $N = 3\mathsf{N}_{\mathrm{fin}}/2 + 1$.

The first bound on the inverse matrix follows from the fact that matrix inversion is a smooth function in a neighborhood of the identity and fixes the identity. The second bound on the inverse matrix follows from the fact that $\det D\Phi_{(i,k)} = 1$, so that we have the formula

$$\mathrm{cof}\, D\Phi_{(i,k)}^T = (D\Phi_{(i,k)})^{-1}.$$

Then since the cofactor matrix is a C^∞ function of the entries of $D\Phi$, we can apply Lemma A.4 and the bound on $D^N \Phi_{(i,k)}$. Note that in the application of Lemma A.4, we set $h = D\Phi_{(i,k)} - \mathrm{Id}$, $\Gamma = \Gamma_\psi = 1$, $\mathcal{C}_h = \Gamma_{q+1}^{-1}$, and the cost of the spatial derivatives to be that given in (6.109). The final bound on the inverse flow $\Phi_{(i,k)}^{-1}$ follows from the identity

$$D^N \left(\Phi_{(i,k)}^{-1} \right)(x) = D^{N-1} \left(\left(D\Phi_{(i,k)} \right)^{-1} \left(\Phi^{-1}(x) \right) \right), \tag{6.115}$$

the Faà di Bruno formula in Lemma A.4, induction on N, and the previously demonstrated bounds.

The bound in (6.113) will be achieved by bounding

$$D^{N-N'} \left[D_{t,q}^M, D^{N'+1} \right] \Phi_{(i,k)},$$

which after using that $D_{t,q}\Phi_{(i,k)} = 0$ will conclude the proof. Towards this end, we apply Lemma A.14, specifically Remark A.16 and Remark A.15, with $v = v_{\ell_q}$ and $f = \Phi_{(i,k)}$. The assumption (A.50) (adjusted to fit Remark A.15) follows from (6.60) with $N_0 = {}^{3N_{fin}}/2$, $\mathcal{C}_v = \Gamma_{q+1}^{i+1}\delta_q^{1/2}$, $\lambda_v = \widetilde{\lambda}_v = \widetilde{\lambda}_q$, $\mu_v = \Gamma_{q+1}^{i-c_0}\tau_q^{-1}$, $\widetilde{\mu}_v = \Gamma_{q+1}^{-1}\widetilde{\tau}_q^{-1}$, and $N_t = N_{ind,t}$. The assumption (A.51) follows with $\mathcal{C}_f = \Gamma_{q+1}^{-1}$ from (6.109) and the fact that $D_{t,q}\Phi_{(i,k)} = 0$. The desired bound then follows from the conclusion (A.56) from Remark A.16 after using Γ_{q+1}^{-1} to absorb implicit constants. The bound in (6.114) will follow again from Lemma A.5 after using the fact that $\left(D\Phi_{(i,k)}\right)^{-1}$ is a smooth function of $D\Phi_{(i,k)}$ in a neighborhood of the identity, which is guaranteed from (6.108). As before, we set $\Gamma = \Gamma_\psi = 1$ and $\mathcal{C}_h = \Gamma_{q+1}^{-1}$ in the application of Lemma A.5. The derivative costs are precisely those in (6.113). $\qquad\square$

6.5   STRESS ESTIMATES ON THE SUPPORT OF THE NEW VELOCITY CUTOFF FUNCTIONS

Before giving the definition of the stress cutoffs, we first note that we can upgrade the L^1 bounds for $\psi_{i,q-1}D^n D_{t,q-1}^m \mathring{R}_{\ell_q}$ available in (5.7), to L^1 bounds for $\psi_{i,q}D^n D_{t,q}^m \mathring{R}_{\ell_q}$. We claim that:

Lemma 6.28 (L^1 **estimates for zeroth order stress**). *Let* $\mathring{R}_{\ell_q}$ *be as defined in* (5.1)*. For* $q \geq 1$ *and* $0 \leq i \leq i_{max}(q)$ *we have the estimate*

$$\left\| D^k D_{t,q}^m \mathring{R}_{\ell_q} \right\|_{L^1(\mathrm{supp}\,(\psi_{i,q}))} \leq \Gamma_q^{-C_R}\delta_{q+1}\mathcal{M}\left(k, 2N_{ind,v}, \lambda_q\Gamma_q, \widetilde{\lambda}_q\right)$$
$$\times \mathcal{M}\left(m, N_{ind,t}, \Gamma_{q+1}^{i-c_0}\tau_q^{-1}, \Gamma_{q+1}^{-1}\widetilde{\tau}_q^{-1}\right) \quad (6.116)$$

for all $k + m \leq {}^{3N_{fin}}/2$.

Proof of Lemma 6.28. The first step is to apply Remark A.15, to the functions $v = v_{\ell_{q-1}}$, $f = \mathring{R}_{\ell_q}$, with $p = 1$, and on the domain $\Omega = \mathrm{supp}\,(\psi_{i,q-1})$. The bound (A.50) holds in view of the inductive assumption (3.23) with $q' = q-1$, for the parameters $\mathcal{C}_v = \Gamma_q^{i+1}\delta_{q-1}^{1/2}$, $\lambda_v = \widetilde{\lambda}_v = \widetilde{\lambda}_{q-1}$, $\mu_v = \Gamma_q^{i-c_0}\tau_{q-1}^{-1}$, $\widetilde{\mu}_v = \Gamma_q^{-1}\widetilde{\tau}_{q-1}^{-1}$, $N_x = 2N_{ind,v}$, $N_t = N_{ind,t}$, and $N_\circ = {}^{3N_{fin}}/2$. On the other hand, the assumption (A.51) holds due to (5.7) and the fact that $\psi_{i\pm,q-1} \equiv 1$ on $\mathrm{supp}\,(\psi_{i,q-1})$, with the parameters $\mathcal{C}_f = \Gamma_q^{-C_R}\delta_{q+1}$, $\lambda_f = \lambda_q$, $\widetilde{\lambda}_f = \widetilde{\lambda}_q$, $N_x = 2N_{ind,v}$, $\mu_f = \Gamma_q^{i+3}\tau_{q-1}^{-1}$, $\widetilde{\mu}_f = \widetilde{\tau}_{q-1}^{-1}$, $N_t = N_{ind,t}$, and $N_\circ = 2N_{fin}$. We thus conclude from (A.54) that

$$\left\| \left(\prod_{i=1}^k D^{\alpha_i} D_{t,q-1}^{\beta_i}\right) \mathring{R}_{\ell_q} \right\|_{L^1(\mathrm{supp}\,(\psi_{i,q-1}))}$$

$$\lesssim \Gamma_q^{-\mathsf{C_R}}\delta_{q+1}\mathcal{M}\left(|\alpha|, 2\mathsf{N}_{\mathrm{ind},v}, \lambda_q, \widetilde{\lambda}_q\right)\mathcal{M}\left(|\beta|, \mathsf{N}_{\mathrm{ind},t}, \Gamma_q^{i+3}\tau_{q-1}^{-1}, \widetilde{\tau}_{q-1}^{-1}\right)$$

whenever $|\alpha| + |\beta| \leq {}^{3\mathsf{N}_{\mathrm{fin}}}/2$. Here we have used that $\widetilde{\lambda}_{q-1} \leq \lambda_q$ and that $\Gamma_q^{i+1}\delta_{q-1}^{1/2}\widetilde{\lambda}_{q-1} \leq \Gamma_q^{i+3}\tau_{q-1}^{-1} \leq \widetilde{\tau}_{q-1}^{-1}$ (in view of (9.39), (9.43), and (3.18)). In particular, the definitions of $\psi_{i,q}$ in (6.14) and of $\psi_{m,i_m,q}$ in (6.11) imply that

$$\left\|\left(\prod_{i=1}^{k}D^{\alpha_i}D_{t,q-1}^{\beta_i}\right)\mathring{R}_{\ell_q}\right\|_{L^1(\mathrm{supp}\,(\psi_{i,q}))}$$
$$\lesssim \Gamma_q^{-\mathsf{C_R}}\delta_{q+1}\mathcal{M}\left(|\alpha|, 2\mathsf{N}_{\mathrm{ind},v}, \lambda_q, \widetilde{\lambda}_q\right)\mathcal{M}\left(|\beta|, \mathsf{N}_{\mathrm{ind},t}, \Gamma_{q+1}^{i+3}\tau_{q-1}^{-1}, \widetilde{\tau}_{q-1}^{-1}\right) \quad (6.117)$$

for all $|\alpha| + |\beta| \leq {}^{3\mathsf{N}_{\mathrm{fin}}}/2$.

The second step is to apply Lemma A.10 with $B = D_{t,q-1}$, $A = u_q \cdot \nabla$, $v = u_q$, $f = \mathring{R}_{\ell_q}$, $p = 1$, and $\Omega = \mathrm{supp}\,(\psi_{i,q})$. In this case $D^k(A + B)^m f = D^k D_{t,q}^m \mathring{R}_{\ell_q}$, which is exactly the object that we need to estimate in (6.116). The assumption (A.40) holds due to (6.57) with $\mathcal{C}_v = \Gamma_{q+1}^{i+1}\delta_q^{1/2}$, $\lambda_v = \Gamma_q \lambda_q$, $\widetilde{\lambda}_v = \widetilde{\lambda}_q$, $N_x = 2\mathsf{N}_{\mathrm{ind},v}$, $\mu_v = \Gamma_{q+1}^{i+3}\tau_{q-1}^{-1}$, $\widetilde{\mu}_v = \Gamma_{q+1}^{-1}\widetilde{\tau}_q^{-1}$, $N_t = \mathsf{N}_{\mathrm{ind},t}$, and $N_* = {}^{3\mathsf{N}_{\mathrm{fin}}}/2+1$. The assumption (A.41) holds due to (6.117) with the parameters $\mathcal{C}_f = \Gamma_q^{-\mathsf{C_R}}\delta_{q+1}$, $\lambda_f = \lambda_q$, $\widetilde{\lambda}_f = \widetilde{\lambda}_q$, $N_x = 2\mathsf{N}_{\mathrm{ind},v}$, $\mu_f = \Gamma_{q+1}^{i+3}\tau_{q-1}^{-1}$, $\widetilde{\mu}_f = \widetilde{\tau}_{q-1}^{-1}$, $N_t = \mathsf{N}_{\mathrm{ind},t}$, and $N_* = {}^{3\mathsf{N}_{\mathrm{fin}}}/2$. The bound (A.44) and the parameter inequalities $\Gamma_{q+1}^{i+1}\delta_q^{1/2}\widetilde{\lambda}_q \leq \Gamma_{q+1}^{i-c_0-2}\tau_q^{-1} \leq \Gamma_{q+1}^{-1}\widetilde{\tau}_q^{-1}$ and $\Gamma_{q+1}^{i+3}\tau_{q-1}^{-1} \leq \Gamma_{q+1}^{i-c_0}\tau_q^{-1}$ (which hold due to (9.40), (9.39), (9.43), and (3.18)) then directly imply (6.116), concluding the proof. $\qquad\square$

Remark 6.29. As discussed in Sections 2.4 and 2.7, in order to verify at level $q + 1$ the inductive assumptions in (3.13) for the new stress $\mathring{R}_{q+1}$, it will be necessary to consider a sequence of intermediate (in terms of the cost of a spatial derivative) objects $\mathring{R}_{q,n,p}$ indexed by n for $1 \leq n \leq n_{\max}$ and $1 \leq p \leq p_{\max}$. For notational convenience, when $n = 0$ and $p = 1$, we define $\mathring{R}_{q,0,1} := \mathring{R}_{\ell_q}$, and estimates on $\mathring{R}_{q,0}$ are already provided by Lemma 6.28. When $n = 0$ and $p \geq 2$, $\mathring{R}_{q,0,p} = 0$. For $1 \leq n \leq n_{\max}$ and $1 \leq p \leq p_{\max}$, the higher order stresses $\mathring{R}_{q,n,p}$ are defined in Section 8.1, specifically in (8.7). Note that the definition of $\mathring{R}_{q,n,p}$ is given as a finite sum of sub-objects $\mathring{H}_{q,n,p}^{n'}$ for $n' \leq n - 1$ and thus requires induction on n. The definition of $\mathring{H}_{q,n,p}^{n'}$ is contained in Section 8.3, specifically in (8.36) and (8.53). Estimates on $\mathring{H}_{q,n,p}^{n'}$ on the support of $\psi_{i,q}$ are stated in (7.15), (7.22), and (7.29) and proven in Section 8.6. For the time being, we *assume* that $\mathring{R}_{q,n,p}$ is *well-defined* and satisfies L^1 estimates similar to those alluded to in (2.19); more precisely, we assume that

$$\left\|D^k D_{t,q}^m \mathring{R}_{q,n,p}\right\|_{L^1(\mathrm{supp}\,\psi_{i,q})} \lesssim \delta_{q+1,n,p}\lambda_{q,n,p}^k\mathcal{M}\left(m, \mathsf{N}_{\mathrm{ind},t}, \Gamma_{q+1}^{i-\mathsf{c_n}}\tau_q^{-1}, \Gamma_{q+1}^{-1}\widetilde{\tau}_q^{-1}\right)$$
$$(6.118)$$

for all $0 \leq k+m \leq N_{\text{fin},n}$. For the purpose of defining the stress cutoff functions, the precise definitions of the n- and p-dependent parameters $\delta_{q+1,n,p}, \lambda_{q,n,p}$, $N_{\text{fin},n}$, and c_n present in (6.118) are not relevant. Note, however, that definitions for $\lambda_{q,n,p}$ for $n = 0$ are given in (9.26), while for $1 \leq n \leq n_{\max}$ and $1 \leq p \leq p_{\max}$, the definitions are given in (9.29). Similarly, when $n = 0$, we let $\delta_{q+1,0,p} = \Gamma_q^{-c_R} \delta_{q+1}$ as is consistent with (9.32), and when $1 \leq n \leq n_{\max}$ and $1 \leq p \leq p_{\max}$, $\delta_{q+1,n,p}$ is defined in (9.34). Finally, note that there are losses in the sharpness and order of the available derivative estimates in (6.118) relative to (6.116). Specifically, the higher order estimates will only be proven up to $N_{\text{fin},n}$, which is a parameter that is decreasing with respect to n and defined in (9.37). For the moment it is only important to note that $N_{\text{fin},n} \gg 14 N_{\text{ind},v}$ for all $0 \leq n \leq n_{\max}$, which is necessary in order to establish (3.13) and (3.15) at level $q + 1$. Similarly, there is a loss in the cost of sharp material derivatives in (6.118), as c_n will be a parameter which is decreasing with respect to n. When $n = 0$, we set $c_n = c_0$ so that (6.116) is consistent with (6.118). For $1 \leq n \leq n_{\max}$, c_n is defined in (9.35).

6.6   DEFINITION OF THE STRESS CUTOFF FUNCTIONS

For $q \geq 1$, $0 \leq i \leq i_{\max}$, $0 \leq n \leq n_{\max}$, and $1 \leq p \leq p_{\max}$, in analogy with the functions $h_{m,j_m,q}$ in (6.6), and keeping in mind the bound (6.118), we define

$$g_{i,q,n,p}^2(x,t) = 1 + \sum_{k=0}^{N_{\text{cut},x}} \sum_{m=0}^{N_{\text{cut},t}} \delta_{q+1,n,p}^{-2} (\Gamma_{q+1} \lambda_{q,n,p})^{-2k} (\Gamma_{q+1}^{i-c_n+2} \tau_q^{-1})^{-2m}$$
$$\times |D^k D_{t,q}^m \mathring{R}_{q,n,p}(x,t)|^2. \tag{6.119}$$

With this notation, for $j \geq 1$ the stress cutoff functions are defined by

$$\omega_{i,j,q,n,p}(x,t) = \psi_{0,q+1}\left(\Gamma_{q+1}^{-2j} g_{i,q,n,p}(x,t)\right), \tag{6.120}$$

while for $j = 0$ we let

$$\omega_{i,0,q,n,p}(x,t) = \widetilde{\psi}_{0,q+1}\left(g_{i,q,n,p}(x,t)\right), \tag{6.121}$$

where $\psi_{0,q+1}$ and $\widetilde{\psi}_{0,q+1}$ are as in Lemma 6.2. The above defined cutoff functions $\omega_{i,j,q,n,p}$ will be shown to obey good estimates on the support of the velocity cutoffs $\psi_{i,q}$ defined earlier.

6.7 PROPERTIES OF THE STRESS CUTOFF FUNCTIONS

6.7.1 Partition of unity

An immediate consequence of (6.1) with $m = 0$ is that for every fixed i, n, we have

$$\sum_{j \geq 0} \omega_{i,j,q,n,p}^2 = 1 \tag{6.122}$$

on $\mathbb{T}^3 \times \mathbb{R}$. Thus, $\{\omega_{i,j,q,n,p}^2\}_{j \geq 0}$ is a partition of unity.

6.7.2 L^∞ estimates for the higher order stresses

We recall from (6.4) and (6.5) that the cutoff function $\psi_{0,q+1}$ appearing in the definition (6.120) satisfies different derivative bounds according to the size of its argument. Accordingly, we introduce the following notation.

Definition 6.30 (Left side of the cutoff function $\omega_{i,j,q,n,p}$). *For $j \geq 1$ we say that*

$$(x, t) \in \operatorname{supp}(\omega_{i,j,q,n,p}^{\mathsf{L}}) \qquad if \qquad 1/4 \leq \Gamma_{q+1}^{-2j} g_{i,q,n,p}(x,t) \leq 1. \tag{6.123}$$

When $j = 0$ we do not define the left side of the cutoff function $\omega_{i,0,q,n,p}$.

Directly from the definition (6.119)–(6.121) and the support properties of the functions $\psi_{0,q+1}$ and $\widetilde{\psi}_{0,q+1}$ stated in Lemma 6.2, and using Definition 6.30, it follows that:

Lemma 6.31. *For all $0 \leq m \leq \mathsf{N}_{\mathrm{cut},t}$, $0 \leq k \leq \mathsf{N}_{\mathrm{cut},x}$, and $j \geq 0$, we have that*

$$\mathbf{1}_{\operatorname{supp}(\omega_{i,j,q,n,p})} |D^k D_{t,q}^m \mathring{R}_{q,n,p}(x,t)|$$
$$\leq \Gamma_{q+1}^{2(j+1)} \delta_{q+1,n,p} (\Gamma_{q+1} \lambda_{q,n,p})^k (\Gamma_{q+1}^{i-\mathsf{c_n}+2} \tau_q^{-1})^m.$$

In the above estimate, if we replace $\mathbf{1}_{\operatorname{supp}(\omega_{i,j,q,n,p})}$ with $\mathbf{1}_{\operatorname{supp}(\omega_{i,j,q,n,p}^{\mathsf{L}})}$ (cf. Definition 6.30), then the factor $\Gamma_{q+1}^{2(j+1)}$ may be sharpened to Γ_{q+1}^{2j}. Moreover, if $j \geq 1$, then $g_{i,q,n,p}(x,t) \geq (1/4)\Gamma_{q+1}^{2j}$.

Lemma 6.31 provides sharp L^∞ bounds for the space and material derivatives of $\mathring{R}_{q,n,p}$, at least when the number of space derivatives is less than $\mathsf{N}_{\mathrm{cut},x}$, and the number of material derivatives is less than $\mathsf{N}_{\mathrm{cut},t}$. If we are willing to pay a Sobolev-embedding loss, then (6.118) implies lossy L^∞ bounds for large numbers of space and material derivatives.

Lemma 6.32 (Derivative bounds with Sobolev loss). *For $q \geq 1$, $n \geq 0$,* *and $0 \le i \le i_{\max}$, we have that:*

$$\left\| D^k D^m_{t,q} \mathring{R}_{q,n,p} \right\|_{L^\infty(\operatorname{supp} \psi_{i,q})} \lesssim \delta_{q+1,n,p} \lambda^{k+3}_{q,n,p} \mathcal{M}\left(m, \mathsf{N}_{\mathrm{ind},t}, \Gamma^{i-c_n+1}_{q+1} \tau^{-1}_q, \Gamma^{-1}_{q+1} \widetilde{\tau}^{-1}_q \right)$$
(6.124)

for all $k + m \le \mathsf{N}_{\mathrm{fin},n} - 4$.

Proof of Lemma 6.32. We apply Lemma A.3 to $f = \mathring{R}_{q,n,p}$, with $\psi_i = \psi_{i,q}$, and with $p = 1$. Assumption (A.16) holds in view of (6.36), with the parameter choice $\rho = \Gamma_q \widetilde{\lambda}_q < \Gamma_{q+1} \widetilde{\lambda}_q = \lambda_{q,0,1} \le \lambda_{q,n,p}$, where the inequalities follow immediately from (9.26)–(9.29). The assumption (A.17) holds due to (6.118), with the parameter choices $\mathcal{C}_f = \delta_{q+1,n,p}$, $\lambda = \widetilde{\lambda} = \lambda_{q,n,p}$, $\mu_i = \Gamma^{i-c_n+1}_{q+1} \tau^{-1}_q$, $\widetilde{\mu}_i = \Gamma^{-1}_{q+1} \widetilde{\tau}^{-1}_q$, $N_t = \mathsf{N}_{\mathrm{ind},t}$, and $N_\circ = \mathsf{N}_{\mathrm{fin},n}$. The lemma now directly follows from (A.18b) with $p = 1$. $\qquad\square$

We note that Lemmas 6.31 and 6.32 imply the following estimate:

Corollary 6.33 (L^∞ **bounds for the stress**). *For $q \ge 0$, $0 \le i \le i_{\max}$, $0 \le n \le n_{\max}$, and $1 \le p \le p_{\max}$ we have*

$$\left\| D^k D^m_{t,q} \mathring{R}_{q,n,p} \right\|_{L^\infty(\operatorname{supp} \psi_{i,q} \cap \operatorname{supp} \omega_{i,j,q,n,p})}$$
$$\lesssim \Gamma^{2(j+1)}_{q+1} \delta_{q+1,n,p} (\Gamma_{q+1} \lambda_{q,n,p})^k \mathcal{M}\left(m, \mathsf{N}_{\mathrm{ind},t}, \Gamma^{i-c_n+2}_{q+1} \tau^{-1}_q, \Gamma^{-1}_{q+1} \widetilde{\tau}^{-1}_q \right)$$
(6.125)

for all $k + m \le \mathsf{N}_{\mathrm{fin},n} - 4$. In the above estimate, if we replace $\operatorname{supp}(\omega_{i,j,q,n,p})$ with $\operatorname{supp}(\omega^{\mathsf{L}}_{i,j,q,n,p})$ (cf. Definition 6.30), then the factor $\Gamma^{2(j+1)}_{q+1}$ may be sharpened to Γ^{2j}_{q+1}.

Proof of Corollary 6.33. For $m \le \mathsf{N}_{\mathrm{cut},t}$ and $k \le \mathsf{N}_{\mathrm{cut},x}$, the bound (6.125) is already contained in Lemma 6.31 (for both $\operatorname{supp}(\omega_{i,j,q,n,p})$ and the improved bound for $\operatorname{supp}(\omega^{\mathsf{L}}_{i,j,q,n,p})$). When either $k > \mathsf{N}_{\mathrm{cut},x}$ or $m > \mathsf{N}_{\mathrm{cut},t}$, we appeal to estimate (6.124) and the parameter bound

$$\delta_{q+1,n,p} \lambda^{k+3}_{q,n,p} \mathcal{M}\left(m, \mathsf{N}_{\mathrm{ind},t}, \Gamma^{i-c_n+1}_{q+1} \tau^{-1}_q, \Gamma^{-1}_{q+1} \widetilde{\tau}^{-1}_q \right)$$
$$\le \left(\Gamma^{-k-\min\{m, \mathsf{N}_{\mathrm{ind},t}\}}_{q+1} \lambda^3_{q,n,p} \right) \delta_{q+1,n,p} (\Gamma_{q+1} \lambda_{q,n,p})^k$$
$$\times \mathcal{M}\left(m, \mathsf{N}_{\mathrm{ind},t}, \Gamma^{i-c_n+2}_{q+1} \tau^{-1}_q, \Gamma^{-1}_{q+1} \widetilde{\tau}^{-1}_q \right)$$
$$\le \delta_{q+1,n,p} (\Gamma_{q+1} \lambda_{q,n,p})^k \mathcal{M}\left(m, \mathsf{N}_{\mathrm{ind},t}, \Gamma^{i-c_n+2}_{q+1} \tau^{-1}_q, \Gamma^{-1}_{q+1} \widetilde{\tau}^{-1}_q \right).$$

The second estimate in the above display is a consequence of the fact that when either $k > \mathsf{N}_{\mathrm{cut},x}$ or $m > \mathsf{N}_{\mathrm{cut},t}$, since $\mathsf{N}_{\mathrm{cut},x} \ge \mathsf{N}_{\mathrm{cut},t}$, we have

$$\Gamma^{-k-\min\{m, \mathsf{N}_{\mathrm{ind},t}\}}_{q+1} \lambda^3_{q,n,p} \le \Gamma^{-\mathsf{N}_{\mathrm{cut},t}}_{q+1} \lambda^3_{q+1} \le 1$$
(6.126)

once $\mathsf{N}_{\text{cut},t}$ (and hence $\mathsf{N}_{\text{cut},x}$) are chosen large enough, as in (9.51).   □

In the proof of Lemma 6.36 below, we shall require one more L^∞ bound for $\mathring{R}_{q,n,p}$, which is for iterates of space and material derivatives. It is convenient to record this bound now, as it follows directly from Corollary 6.33.

Corollary 6.34. *For $q \geq 0$, $0 \leq i \leq i_{\max}$, $0 \leq n \leq n_{\max}$, $1 \leq p \leq p_{\max}$, and $\alpha, \beta \in \mathbb{N}_0^k$ we have*

$$\left\| \left(\prod_{\ell=1}^{k} D^{\alpha_\ell} D_{t,q}^{\beta_\ell} \right) \mathring{R}_{q,n,p} \right\|_{L^\infty(\text{supp}\,\psi_{i,q}\cap\text{supp}\,\omega_{i,j,q,n,p})}$$
$$\lesssim \Gamma_{q+1}^{2(j+1)} \delta_{q+1,n,p} (\Gamma_{q+1}\lambda_{q,n,p})^{|\alpha|} \mathcal{M}\left(|\beta|, \mathsf{N}_{\text{ind},t}, \Gamma_{q+1}^{i-\mathsf{c}_n+2}\tau_q^{-1}, \Gamma_{q+1}^{-1}\tilde{\tau}_q^{-1} \right) \quad (6.127)$$

for all $|\alpha| + |\beta| \leq \mathsf{N}_{\text{fin},n} - 4$. In the above estimate, if we replace $\text{supp}\,(\omega_{i,j,q,n,p})$ with $\text{supp}\,(\omega_{i,j,q,n,p}^{\mathsf{L}})$ (cf. Definition 6.30), then the factor $\Gamma_{q+1}^{2(j+1)}$ may be sharpened to Γ_{q+1}^{2j}.

Proof of Corollary 6.34. The proof follows from Corollary 6.33 and Lemma A.14. The bounds corresponding to $\text{supp}\,\omega_{i,j,q,n,p}$ and $\text{supp}\,\omega_{i,j,q,n,p}^{\mathsf{L}}$ are identical (except for the improvement $\Gamma_{q+1}^{2(j+1)} \mapsto \Gamma_{q+1}^{2j}$ in the later case), so we only give details for the former. Since $D_{t,q} = \partial_t + v_{\ell_q} \cdot \nabla$, Lemma A.14 is applied with $v = v_{\ell_q}$, $f = \mathring{R}_{q,n,p}$, $\Omega = \text{supp}\,\psi_{i,q} \cap \text{supp}\,\omega_{i,j,q,n,p}$, and $p = \infty$. In view of estimate (6.60) and the fact that $3\mathsf{N}_{\text{fin}}/2 \geq \mathsf{N}_{\text{fin},n}$, the assumption (A.50) holds with $\mathcal{C}_v = \Gamma_{q+1}^{i+1}\delta_q^{1/2}$, $\lambda_v = \Gamma_q\lambda_q$, $\tilde{\lambda}_v = \tilde{\lambda}_q$, $\mathsf{N}_x = 2\mathsf{N}_{\text{ind},v}$, $\mu_v = \Gamma_{q+1}^{i-\mathsf{c}_n}\tau_q^{-1}$, $\tilde{\mu}_v = \Gamma_{q+1}^{-1}\tilde{\tau}_q^{-1}$, and $\mathsf{N}_t = \mathsf{N}_{\text{ind},t}$. On the other hand, the bound (6.127) implies assumption (A.51) with $\mathcal{C}_f = \Gamma_{q+1}^{2(j+1)}\delta_{q+1,n,p}$, $\lambda_f = \tilde{\lambda}_f = \Gamma_{q+1}\lambda_{q,n,p}$, $\mu_f = \Gamma_{q+1}^{i-\mathsf{c}_n+2}\tau_q^{-1}$, $\tilde{\mu}_f = \Gamma_{q+1}^{-1}\tilde{\tau}_q^{-1}$, and $\mathsf{N}_t = \mathsf{N}_{\text{ind},t}$. Since $\lambda_v \leq \lambda_f$, $\tilde{\lambda}_v \leq \tilde{\lambda}_f$, $\mu_v \leq \mu_f$, and $\tilde{\mu}_v = \tilde{\mu}_f$, we deduce from the bound (A.54) (in fact, its version mentioned in Remark A.15) that (6.127) holds, thereby concluding the proof. Here we are also implicitly using the parameter estimate $\mathcal{C}_v\tilde{\lambda}_v \leq \mu_f$, which holds due to (9.39).   □

6.7.3   Maximal j index in the stress cutoffs

Lemma 6.35 (Maximal j index in the stress cutoffs). *Fix $q \geq 0$, $0 \leq n \leq n_{\max}$, and $1 \leq p \leq p_{\max}$. There exists a $j_{\max} = j_{\max}(q, n, p) \geq 1$, determined by (6.128) below, which is bounded independently of q, n, and p as in (6.129), such that for any $0 \leq i \leq i_{\max}(q)$, we have*

$$\psi_{i,q}\,\omega_{i,j,q,n,p} \equiv 0 \qquad \text{for all} \qquad j > j_{\max}.$$

Moreover, the bound

$$\Gamma_{q+1}^{2(j_{\max}-1)} \lesssim \lambda_{q,n,p}^3$$

holds, with an implicit constant that is independent of q and n.

Proof of Lemma 6.35. We define $j_{\max}$ by

$$j_{\max} = j_{\max}(q,n,p) = \frac{1}{2}\left\lceil \frac{\log(M_b\sqrt{8N_{\mathrm{cut},x}N_{\mathrm{cut},t}}\lambda_{q,n,p}^3)}{\log(\Gamma_{q+1})} \right\rceil, \qquad (6.128)$$

where M_b is the implicit q-, n-, p-, and i-independent constant in (6.124); that is, we take the largest such constant among all values of k and m with $k+m \leq N_{\mathrm{fin},n} - 4$. To see that $j_{\max}$ may be bounded independently of q, n, and p, we note that $\lambda_{q,n,p} \leq \lambda_{q+1}$, and thus

$$2j_{\max} \leq 1 + \frac{\log(M_b\sqrt{8N_{\mathrm{cut},x}N_{\mathrm{cut},t}}) + 3\log(\lambda_{q+1})}{\log(\Gamma_{q+1})} \to 1 + \frac{3b}{\varepsilon_\Gamma(b-1)}$$

as $q \to \infty$. Thus, assuming that $a = \lambda_0$ is sufficiently large, we obtain that

$$2j_{\max}(q,n,p) \leq \frac{4b}{\varepsilon_\Gamma(b-1)} \qquad (6.129)$$

for all $q \geq 0$, $0 \leq n \leq n_{\max}$, and $1 \leq p \leq p_{\max}$.

To conclude the proof of the lemma, let $j > j_{\max}$, as defined in (6.128), and assume by contradiction that there exists a point $(x,t) \in \mathrm{supp}\,(\psi_{i,q}\omega_{i,j,q,n,p}) \neq \emptyset$. In particular, $j \geq 1$. Then, by (6.119)–(6.120) and the pigeonhole principle, we see that there exist $0 \leq k \leq N_{\mathrm{cut},x}$ and $0 \leq m \leq N_{\mathrm{cut},t}$ such that

$$|D^k D_{t,q}^m \mathring{R}_{q,n,p}(x,t)| \geq \frac{\Gamma_{q+1}^{2j}}{\sqrt{8N_{\mathrm{cut},x}N_{\mathrm{cut},t}}}\delta_{q+1,n,p}(\Gamma_{q+1}\lambda_{q,n,p})^k(\Gamma_{q+1}^{i-\mathsf{c}_n+2}\tau_q^{-1})^m.$$

On the other hand, from (6.124), we have that

$$|D^k D_{t,q}^m \mathring{R}_{q,n,p}(x,t)| \leq M_b\lambda_{q,n,p}^3\delta_{q+1,n,p}\lambda_{q,n,p}^k(\Gamma_{q+1}^{i-\mathsf{c}_n+1}\tau_q^{-1})^m.$$

The above two estimates imply that

$$\Gamma_{q+1}^{2(j_{\max}+1)} \leq \Gamma_{q+1}^{2j} \leq M_b\sqrt{8N_{\mathrm{cut},x}N_{\mathrm{cut},t}}\Gamma_{q+1}^{-k-m}\lambda_{q,n,p}^3, \leq M_b\sqrt{8N_{\mathrm{cut},x}N_{\mathrm{cut},t}}\lambda_{q,n,p}^3,$$

which contradicts the fact that $j > j_{\max}$, as defined in (6.128). $\qquad\square$

6.7.4 Bounds for space and material derivatives of the stress cutoffs

Lemma 6.36 (Derivative bounds for the stress cutoffs). *For $q \geq 0$, $0 \leq n \leq n_{\max}$, $1 \leq p \leq p_{\max}$, $0 \leq i \leq i_{\max}$, and $0 \leq j \leq j_{\max}$, we have that*

$$\frac{\mathbf{1}_{\operatorname{supp} \psi_{i,q}} |D^N D^M_{t,q} \omega_{i,j,q,n,p}|}{\omega_{i,j,q,n,p}^{1-(N+M)/N_{\mathrm{fin}}}}$$
$$\lesssim (\Gamma_{q+1} \lambda_{q,n,p})^N \mathcal{M} \left(M, N_{\mathrm{ind},t} - N_{\mathrm{cut},t}, \Gamma_{q+1}^{i-c_n+2} \tau_q^{-1}, \Gamma_{q+1}^{-1} \widetilde{\tau}_q^{-1} \right) \qquad (6.130)$$

for all $N + M \leq N_{\mathrm{fin},n} - N_{\mathrm{cut},x} - N_{\mathrm{cut},t} - 4$.

Remark 6.37. Notice that the sharp derivative bounds in (6.130) are only up to $N_{\mathrm{ind},t} - N_{\mathrm{cut},t}$. In order to obtain bounds up to $N_{\mathrm{ind},t}$, we may argue exactly as in the string of inequalities which converted (6.83) into (6.84), resulting in the bound

$$\frac{\mathbf{1}_{\operatorname{supp} \psi_{i,q}} |D^N D^M_{t,q} \omega_{i,j,q,n,p}|}{\omega_{i,j,q,n,p}^{1-(N+M)/N_{\mathrm{fin}}}}$$
$$\lesssim (\Gamma_{q+1} \lambda_{q,n,p})^N \mathcal{M} \left(M, N_{\mathrm{ind},t}, \Gamma_{q+1}^{i-c_n+3} \tau_q^{-1}, \Gamma_{q+1}^{-1} \widetilde{\tau}_q^{-1} \right) . \qquad (6.131)$$

Proof of Lemma 6.36. For simplicity, we only treat here the case $j \geq 1$. Indeed, for $j = 0$ we simply replace $\psi_{0,q+1}$ with $\widetilde{\psi}_{0,q+1}$, which by Lemma 6.2 has similar properties to $\psi_{0,q+1}$.

The goal is to apply the Faà di Bruno Lemma A.5 with $\psi = \psi_{0,q+1}$, $\Gamma = \Gamma_{q+1}^{-j}$, $D_t = D_{t,q}$, and $h(x,t) = g_{i,q,n,p}(x,t)$, so that $g = \omega_{i,j,q,n,p}$.

Because the cutoff function $\psi = \psi_{0,q+1}$ satisfies slightly different estimates depending on whether we are in the case (6.4) or (6.5), assumption (A.24) holds with $\Gamma_\psi = 1$, and respectively $\Gamma_\psi = \Gamma_{q+1}^{-1}$, depending on whether we work on the set $\operatorname{supp}(\omega_{i,j,q,n,p}^{\mathsf{L}})$ or on the set $\operatorname{supp}(\omega_{i,j,q,n,p}) \setminus \operatorname{supp}(\omega_{i,j,q,n,p}^{\mathsf{L}})$ (cf. Definition 6.30). We have in fact encountered this same issue in the proof of Lemmas 6.13 and 6.20. The slightly worse value of Γ_ψ for $(x,t) \in \operatorname{supp}(\omega_{i,j,q,n,p}^{\mathsf{L}})$ is, however, precisely balanced out by the fact that in Corollary 6.34 the bound (6.127) is improved by a factor for Γ_{q+1}^2 on $\operatorname{supp}(\omega_{i,j,q,n,p}^{\mathsf{L}})$. Since these two factors of Γ_{q+1}^2 cancel out, as they did in Lemmas 6.13 and 6.20, we only give the proof of the bound (6.130) for $(x,t) \in \operatorname{supp}(\omega_{i,j,q,n,p}) \setminus \operatorname{supp}(\omega_{i,j,q,n,p}^{\mathsf{L}})$, which is equivalent to the condition that $1 < \Gamma_{q+1}^{-2j} g_{i,q,n,p}(x,t) \leq \Gamma_{q+1}^2$. Note moreover that we do not perform any estimates for (x,t) such that $1 < \Gamma_{q+1}^{-2j} g_{i,q,n,p}(x,t) < (1/4)\Gamma_{q+1}^2$ since in this region $\psi_{0,q+1} \equiv 1$ (see item 2(b) in Lemma 6.2) and so its derivative is equal to 0. Therefore, for the remainder of the proof we work with the subset of $\operatorname{supp} \omega_{i,j,q,n,p}$, on which we have

$$(1/4)\Gamma_{q+1}^2 \leq \Gamma_{q+1}^{-2j} g_{i,q,n,p}(x,t) \leq \Gamma_{q+1}^2 . \qquad (6.132)$$

This ensures that assumption (A.24) of Lemma A.5 holds with $\Gamma_\psi = \Gamma_{q+1}^{-1}$.

In order to verify condition (A.25), the main requirement is a supremum bound for $D^N D^M_{t,q} g_{i,q,n,p}$ in L^∞ on the support of $\psi_{i,q} \omega_{i,j,q,n,p}$. In this direction, we claim that for all (x,t) as in (6.132), we have

$$
\mathbf{1}_{\mathrm{supp}\,\psi_{i,q}} \left| D^N D^M_{t,q} g_{i,q,n,p}(x,t) \right|
$$
$$
\lesssim \Gamma_{q+1}^{2j+2} (\Gamma_{q+1} \lambda_{q,n,p})^N \mathcal{M}\left(M, \mathsf{N}_{\mathrm{ind},t} - \mathsf{N}_{\mathrm{cut},t}, \Gamma_{q+1}^{i-\mathsf{c}_{\mathrm{n}}+2} \tau_q^{-1}, \Gamma_{q+1}^{-1} \widetilde{\tau}_q^{-1}\right) \quad (6.133)
$$

for all $N + M \leq \mathsf{N}_{\mathrm{fin},n} - \mathsf{N}_{\mathrm{cut},x} - \mathsf{N}_{\mathrm{cut},t} - 4$. Thus, assumption (A.25) of Lemma A.5 holds with $\mathcal{C}_h = \Gamma_{q+1}^{2j+2}$, $\lambda = \widetilde{\lambda} = \Gamma_{q+1} \lambda_{q,n,p}$, $\mu = \Gamma_{q+1}^{i-\mathsf{c}_{\mathrm{n}}+2} \tau_q^{-1}$, $\widetilde{\mu} = \Gamma_{q+1}^{-1} \widetilde{\tau}_q^{-1}$, and $N_t = \mathsf{N}_{\mathrm{ind},t} - \mathsf{N}_{\mathrm{cut},t}$. In particular, we note that $(\Gamma_\psi \Gamma)^{-2} \mathcal{C}_h = 1$, and estimate (A.26) of Lemma A.5 directly implies (6.130).

Thus, in order to complete the proof of the lemma it remains to establish estimate (6.133). As in the proof of Lemma 6.13, it is more convenient to first estimate $D^N D^M_{t,q}(g_{i,q,n,p}(x,t)^2)$, as its definition (cf. (6.119)) makes it more amenable to the use of the Leibniz rule. Indeed, for all $N + M \leq \mathsf{N}_{\mathrm{fin},n} - \mathsf{N}_{\mathrm{cut},x} - \mathsf{N}_{\mathrm{cut},t} - 4$ we have that

$$
D^N D^M_{t,q} g^2_{i,q,n,p} = \sum_{N'=0}^{N} \sum_{M'=0}^{M} \binom{N}{N'} \binom{M}{M'} \sum_{k=0}^{\mathsf{N}_{\mathrm{cut},x}} \sum_{m=0}^{\mathsf{N}_{\mathrm{cut},t}}
$$
$$
\times \frac{D^{N'} D^{M'}_{t,q} D^k D^m_{t,q} \mathring{R}_{q,n,p}\, D^{N-N'} D^{M-M'} D^k D^m_{t,q} \mathring{R}_{q,n,p}}{\delta^2_{q+1,n,p} (\Gamma_{q+1} \lambda_{q,n,p})^{2k} (\Gamma_{q+1}^{i-\mathsf{c}_{\mathrm{n}}+2} \tau_q^{-1})^{2m}}.
$$

Combining the above display with estimate (6.127) and the fact that $k + m + N + M \leq \mathsf{N}_{\mathrm{fin},n} - 4$, we deduce

$$
\mathbf{1}_{\mathrm{supp}\,\psi_{i,q} \cap \mathrm{supp}\,\omega_{i,j,q,n,p}} \left| D^N D^M_{t,q} g^2_{i,q,n,p} \right|
$$
$$
\lesssim \sum_{N'=0}^{N} \sum_{M'=0}^{M} \sum_{k=0}^{\mathsf{N}_{\mathrm{cut},x}} \sum_{m=0}^{\mathsf{N}_{\mathrm{cut},t}} \frac{1}{\delta^2_{q+1,n,p} (\Gamma_{q+1} \lambda_{q,n,p})^{2k} (\Gamma_{q+1}^{i-\mathsf{c}_{\mathrm{n}}+2} \tau_q^{-1})^{2m}}
$$
$$
\times \Gamma_{q+1}^{2(j+1)} \delta_{q+1,n,p} (\Gamma_{q+1} \lambda_{q,n,p})^{N'+k}
$$
$$
\times \mathcal{M}\left(M'+m, \mathsf{N}_{\mathrm{ind},t}, \Gamma_{q+1}^{i-\mathsf{c}_{\mathrm{n}}+2} \tau_q^{-1}, \Gamma_{q+1}^{-1} \widetilde{\tau}_q^{-1}\right)
$$
$$
\times \Gamma_{q+1}^{2(j+1)} \delta_{q+1,n,p} (\Gamma_{q+1} \lambda_{q,n,p})^{N-N'+k}
$$
$$
\times \mathcal{M}\left(M-M'+m, \mathsf{N}_{\mathrm{ind},t}, \Gamma_{q+1}^{i-\mathsf{c}_{\mathrm{n}}+2} \tau_q^{-1}, \Gamma_{q+1}^{-1} \widetilde{\tau}_q^{-1}\right)
$$
$$
\lesssim \Gamma_{q+1}^{4(j+1)} (\Gamma_{q+1} \lambda_{q,n,p})^N \mathcal{M}\left(M, \mathsf{N}_{\mathrm{ind},t} - \mathsf{N}_{\mathrm{cut},t}, \Gamma_{q+1}^{i-\mathsf{c}_{\mathrm{n}}+2} \tau_q^{-1}, \Gamma_{q+1}^{-1} \widetilde{\tau}_q^{-1}\right). \quad (6.134)
$$

Lastly, we show that the bound (6.134), and the fact that we work with (x,t) such that (6.132) holds, implies (6.133). This argument is the same as the one found earlier in (6.45)–(6.47). We establish (6.133) inductively in K for $N+M \leq K$. We know from (6.132) that (6.133) holds for $K = 0$, i.e., for $N = M = 0$. So let us assume by induction that (6.133) was previously established for any pair $N' + M' \leq K - 1$, and fix a new pair with $N + M = K$. Similarly to (6.46), the Leibniz rule gives

$$
D^N D_{t,q}^M (g_{i,q,n,p}^2) - 2g_{i,q,n,p} D^N D_{t,q}^M g_{i,q,n,p}
$$

$$
= \sum_{\substack{0 \le N' \le N \\ 0 \le M' \le M \\ 0 < N'+M' < N+M}} \binom{N}{N'} \binom{M}{M'} D^{N'} D_{t,q}^{M'} g_{i,q,n,p} \, D^{N-N'} D_{t,q}^{M-M'} g_{i,q,n,p} \, .
$$

Since every term in the sum on the right side of the above display satisfies
$1 \le N' + M' \le K - 1$, these terms are bounded by our inductive assumption,
and we deduce that

$$
\mathbf{1}_{\mathrm{supp}\,\psi_{i,q}} \left| D^N D_{t,q}^M g_{i,q,n,p} \right|
$$

$$
\lesssim \frac{\left| D^N D_{t,q}^M (g_{i,q,n,p}^2) \right|}{g_{i,q,n,p}}
$$

$$
+ \frac{\Gamma_{q+1}^{2(2j+2)} (\Gamma_{q+1} \lambda_{q,n,p})^N \mathcal{M}\left(M, \mathsf{N}_{\mathrm{ind},t} - \mathsf{N}_{\mathrm{cut},t}, \Gamma_{q+1}^{i-c_n+2} \tau_q^{-1}, \Gamma_{q+1}^{-1} \tilde{\tau}_q^{-1} \right)}{g_{i,q,n,p}} \, .
$$

Thus, (6.133) also holds for $N + M = K$ by combining the above display with
(6.132) (which implies $g_{i,q,n,p} \ge \Gamma_{q+1}^{2j+2}$), and with estimate (6.134) (which gives
the bounds for the derivatives of $g_{i,q,n,p}^2$). This concludes the proof of (6.133)
and thus of the lemma.   □

6.7.5 L^r norm of the stress cutoffs

Lemma 6.38. *Let $q \ge 0$. For $r \ge 1$ we have that*

$$
\| \omega_{i,j,q,n,p} \|_{L^r(\mathrm{supp}\,\psi_{i\pm,q})} \lesssim \Gamma_{q+1}^{-2j/r} \tag{6.135}
$$

*holds for all $0 \le i \le i_{\max}$, $0 \le j \le j_{\max}$, $0 \le n \le n_{\max}$, and $1 \le p \le p_{\max}$. The
implicit constant is independent of $i, j, q, n,$ and p.*

Proof of Lemma 6.38. The argument is similar to the proof of (6.87). We begin
with the case $r = 1$. The other cases $r \in (1, \infty]$ follow from the fact that
$\omega_{i,j,q,n,p} \le 1$ and Lebesgue interpolation.

For $j = 0$ we are done, since, by definition, $0 \le \omega_{i,j,q,n,p} \le 1$; thus we
consider only $j \ge 1$. Since $\psi_{i\pm2,q} \equiv 1$ on $\mathrm{supp}\,(\psi_{i\pm,q})$, and using Lemma 6.31,
we see that for any $(x,t) \in \mathrm{supp}\,(\psi_{i\pm,q}\omega_{i,j,q,n,p})$ we have

$$
\psi_{i\pm2,q}^2 g_{i,q,n,p}^2 = \psi_{i\pm2,q}^2 + \sum_{k=0}^{\mathsf{N}_{\mathrm{cut},x}} \sum_{m=0}^{\mathsf{N}_{\mathrm{cut},t}} \frac{|\psi_{i\pm2,q} D^k D_{t,q}^m \mathring{R}_{q,n,p}(x,t)|^2}{\delta_{q+1,n,p}^2 (\Gamma_{q+1} \lambda_{q,n,p})^{2k} (\Gamma_{q+1}^{i-c_n+2} \tau_q^{-1})^{2m}}
$$

$$
\ge \frac{1}{16} \Gamma_{q+1}^{4j} \, .
$$

Using that $a + b \geq \sqrt{a^2 + b^2}$ for $a, b \geq 0$, and using $\Gamma_{q+1}^{4j} \geq 64$ for $j \geq 1$, we conclude that

$$\sum_{k=0}^{\mathsf{N}_{\text{cut},x}} \sum_{m=0}^{\mathsf{N}_{\text{cut},t}} \frac{|\psi_{i\pm2,q} D^k D_{t,q}^m \mathring{R}_{q,n,p}(x,t)|}{\delta_{q+1,n,p}(\Gamma_{q+1}\lambda_{q,n,p})^k (\Gamma_{q+1}^{i-\mathsf{c}_n+2}\tau_q^{-1})^m} \geq \frac{1}{16}\Gamma_{q+1}^{2j}.$$

Therefore, using Chebyshev's inequality and the inductive assumption (6.118), we obtain

$$|\text{supp}(\psi_{i\pm,q}\omega_{i,j,q,n,p})|$$

$$\leq \left|\left\{(x,t): \psi_{i\pm2,q}g_{i,q,n,p} \geq (1/16)\Gamma_{q+1}^{2j}\right\}\right|$$

$$\leq \left|\left\{(x,t): \sum_{k=0}^{\mathsf{N}_{\text{cut},x}} \sum_{m=0}^{\mathsf{N}_{\text{cut},t}} \frac{|\psi_{i\pm2,q} D^k D_{t,q}^m \mathring{R}_{q,n,p}(x,t)|}{\delta_{q+1,n,p}(\Gamma_{q+1}\lambda_{q,n,p})^k (\Gamma_{q+1}^{i-\mathsf{c}_n+2}\tau_q^{-1})^m} \geq (1/16)\Gamma_{q+1}^{2j}\right\}\right|$$

$$\leq 16\Gamma_{q+1}^{-2j} \sum_{k=0}^{\mathsf{N}_{\text{cut},x}} \sum_{m=0}^{\mathsf{N}_{\text{cut},t}} \delta_{q+1,n,p}^{-1}(\Gamma_{q+1}\lambda_{q,n,p})^{-k} (\Gamma_{q+1}^{i-\mathsf{c}_n+2}\tau_q^{-1})^{-m}$$

$$\times \left\|\psi_{i\pm2,q} D^k D_{t,q}^m \mathring{R}_{q,n,p}\right\|_{L^1}$$

$$\lesssim 16\Gamma_{q+1}^{-2j} \sum_{k=0}^{\mathsf{N}_{\text{cut},x}} \sum_{m=0}^{\mathsf{N}_{\text{cut},t}} \Gamma_{q+1}^{-k}$$

$$\lesssim \Gamma_{q+1}^{-2j},$$

where the implicit constant depends only on $\mathsf{N}_{\text{cut},t}$. The proof is concluded since the L^1 norm of a function with range in $[0,1]$ is bounded by the measure of its support.   □

6.8   DEFINITION AND PROPERTIES OF THE CHECKERBOARD CUTOFF FUNCTIONS

For $0 \leq n \leq n_{\text{max}}$, consider all the $\frac{\mathbb{T}^3}{\lambda_{q,n,0}}$-periodic cells contained in $\mathbb{T}^3$, of which there are $\lambda_{q,n,0}^3$. Index these cells by integer triples $\vec{l} = (l, w, h)$ for $l, w, h \in \{0, ..., \lambda_{q,n,0} - 1\}$. Let $\mathcal{X}_{q,n,\vec{l}}$ be a partition of unity adapted to this checkerboard of periodic cells which satisfies, for any q and n,

$$\sum_{\vec{l}=(l,w,h)} \left(\mathcal{X}_{q,n,\vec{l}}\right)^2 = 1. \tag{6.136}$$

Furthermore, for $\vec{l} = (l, w, h), \vec{l}^* = (l^*, w^*, h^*) \in \{0, ..., \lambda_{q,n,0} - 1\}^3$ such that

$$|l - l^*| \geq 2, \qquad |w - w^*| \geq 2, \qquad |h - h^*| \geq 2,$$

we impose that

$$\mathcal{X}_{q,n,\vec{l}} \mathcal{X}_{q,n,\vec{l}^*} = 0. \tag{6.137}$$

Definition 6.39 (Checkerboard cutoff function). *Given* q, $0 \leq n \leq n_{\max}$, $i \leq i_{\max}$, *and* $k \in \mathbb{Z}$, *we define*

$$\zeta_{q,i,k,n,\vec{l}}(x, t) = \mathcal{X}_{q,n,\vec{l}}(\Phi_{i,k,q}(x, t)). \tag{6.138}$$

Lemma 6.40. *The cutoff functions* $\left\{ \zeta_{q,i,k,n,\vec{l}} \right\}_{\vec{l}}$ *satisfy the following properties:*

1. *The material derivative* $D_{t,q} \left(\zeta_{q,i,k,n,\vec{l}} \right)$ *vanishes.*
2. *For each* $t \in \mathbb{R}$ *and all* $x \in \mathbb{T}^3$,

$$\sum_{\vec{l}=(l,w,h)} \left(\zeta_{q,i,k,n,\vec{l}}(x, t) \right)^2 = 1. \tag{6.139}$$

3. *We have the spatial derivative estimate for all* $m \leq 3\mathsf{N}_{\mathrm{fin}}/2 + 1$,

$$\left\| D^m \zeta_{q,i,k,n,\vec{l}} \right\|_{L^\infty(\mathrm{supp}\, \psi_{i,q} \widetilde{\chi}_{i,k,q})} \lesssim \lambda_{q,n,0}^m. \tag{6.140}$$

4. *There exists an implicit dimensional constant independent of* q, n, k, i, *and* $\vec{l}$ *such that for all* $(x, t) \in \mathrm{supp}\, \psi_{i,q} \widetilde{\chi}_{i,k,q}$,

$$\mathrm{diam} \left(\mathrm{supp} \left(\zeta_{q,i,k,n,\vec{l}}(\cdot, t) \right) \right) \lesssim (\lambda_{q,n,0})^{-1}. \tag{6.141}$$

Proof of Lemma 6.40. The proof of (1) is immediate given that $\zeta_{q,i,k,n,\vec{l}}$ is precomposed with the flow map $\Phi_{i,k,q}$. (6.139) follows from (1), (6.136), and the fact that for each $t \in \mathbb{R}$, $\Phi_{i,k,q}(t, \cdot)$ is a diffeomorphism of $\mathbb{T}^3$. The spatial derivative estimate in (6.140) follows from Lemma A.4, (6.109), and the parameter definitions in (9.19), (9.26), and (9.29). The property in (6.141) follows from the construction of the $\mathcal{X}_{q,n,\vec{l}}$ functions (which can be taken simply as a dilation by a factor of $\lambda_{q,n,1}$ of a q-independent partition of unity on $\mathbb{R}^3$) and (6.108). $\qquad \square$

6.9   DEFINITION OF THE CUMULATIVE CUTOFF FUNCTION

Finally, combining the cutoff functions defined in Definition 6.6, (6.120)–(6.121), and (6.96), we define the cumulative cutoff function by

$$\eta_{i,j,k,q,n,p,\vec{l}}(x,t) = \psi_{i,q}(x,t)\omega_{i,j,q,n,p}(x,t)\chi_{i,k,q}(t)\overline{\chi}_{q,n,p}(t)\zeta_{q,i,k,n,\vec{l}}(x,t).$$

Since the values of q and n are clear from the context, the values in $\vec{l}$ are irrelevant in many arguments, and the time cutoffs $\overline{\chi}_{q,n,p}$ are only used in Section 8.9, we may abbreviate the above using any of

$$\eta_{i,j,k,q,n,p,\vec{l}}(x,t) = \eta_{i,j,k,q,n,p}(x,t) = \eta_{(i,j,k)}(x,t)$$
$$= \psi_{(i)}(x,t)\omega_{(i,j)}(x,t)\chi_{(i,k)}(t)\zeta_{(i,k)}(x,t).$$

It follows from Lemma 6.8, (6.122), (6.94), and (6.139) that for every (q,n,p) fixed, we have a partition of unity

$$\sum_{i,j\geq0}\sum_{k\in\mathbb{Z}}\sum_{\vec{l}}\eta^2_{i,j,k,q,n,p,\vec{l}}(x,t) = 1. \tag{6.142}$$

The sum in i goes up to $i_{\max}$ (defined in (6.53)), while the sum in j goes up to $j_{\max}$ (defined in (6.128)). In analogy with $\psi_{i\pm,q}$, we define

$$\omega_{(i,j\pm)}(x,t) := \left(\omega^2_{(i,j-1)}(x,t) + \omega^2_{(i,j)}(x,t) + \omega^2_{(i,j+1)}(x,t)\right)^{\frac{1}{2}}, \tag{6.143}$$

which are cutoffs with the property that

$$\omega_{(i,j\pm)} \equiv 1 \text{ on } \mathrm{supp}\,(\omega_{(i,j)}). \tag{6.144}$$

We then define

$$\eta_{(i\pm,j\pm,k,\pm)}(x,t) := \psi_{i\pm,q}(x,t)\omega_{(i,j\pm)}(x,t)\widetilde{\chi}_{i,k,q}(t)\zeta_{q,i,k,n,\vec{l}}(x,t), \tag{6.145}$$

which are cutoffs with the property that

$$\eta_{(i,\pm,j\pm,k\pm)} \equiv \zeta_{q,i,k,n,\vec{l}} \quad \text{on } \mathrm{supp}\,(\psi_{(i)}\omega_{(i,j)}\chi_{(i,k)}). \tag{6.146}$$

We conclude this section with estimates on the supports of the cumulative cutoff function $\eta_{(i,j,k)}$.

Lemma 6.41. *For $r_1,r_2 \in [1,\infty]$ with $\frac{1}{r_1} + \frac{1}{r_2} = 1$ we have*

$$\sum_{\vec{l}}\left|\mathrm{supp}\,(\eta_{i,j,k,q,n,p,\vec{l}})\right| \lesssim \Gamma_{q+1}^{-2\left(\frac{i}{r_1}+\frac{j}{r_2}\right)+\frac{c_b}{r_1}+2}. \tag{6.147}$$

Proof of Lemma 6.41. Applying Lemma 6.23, Lemma 6.38, Hölder's inequality, and interpolating, we obtain

$$
\begin{aligned}
|\operatorname{supp}(\psi_{i,q}) \cap \operatorname{supp}(\omega_{i,j,q,n,p})| &\leq \left\| \psi_{i\pm,q} \omega_{(i,j\pm)} \right\|_{L^1} \\
&\leq \left\| \psi_{i\pm,q} \right\|_{L^{r_1}} \left\| \omega_{(i,j\pm)} \right\|_{L^{r_2}} \\
&\lesssim \Gamma_{q+1}^{-\frac{2(i-1)-c_b}{r_1} - \frac{2(j-1)}{r_2}}.
\end{aligned}
$$

Using $\frac{1}{r_1} + \frac{1}{r_2} = 1$ and (6.139), which give that the $\zeta_{q,i,k,n,\vec{l}}$ form a partition of unity, yields (6.147). $\qquad\square$

Chapter Seven

From q to $q+1$: breaking down the main inductive estimates

The overarching goal of this section is to state several propositions which decompose the verification of the main inductive assumptions (3.13) and (3.14) for the perturbation w_{q+1} and assumption (3.15) for the stress $\mathring{R}_{q+1}$ into digestible components. We remind the reader—cf. Remark 6.1—that the rest of the inductive estimates stated in Section 3.2.3 are proven in Chapter 6. We begin in Section 7.1 with Proposition 7.1, which simply translates the main inductive assumptions into statements phrased at level $q+1$. At this point, we then introduce in Section 7.2 a handful of notations which will be necessary in order to state the propositions which form the constituent parts of the proof of Proposition 7.1. The next three propositions (7.3, 7.4, and 7.5) are described and presented in Section 7.3. They are significantly more detailed than Proposition 7.1, as they contain the precise estimates that will be propagated throughout the construction and cancellation of the higher order stresses $\mathring{R}_{q,\tilde{n}}$. These three propositions will be verified in Chapter 8.

7.1 INDUCTION ON Q

The main claim of this section is an induction on q.

Proposition 7.1 (Inductive Step on q). *Given v_{ℓ_q}, $\mathring{R}_{\ell_q}$, and $\mathring{R}_q^{\mathrm{comm}}$ satisfying the Euler-Reynolds system*

$$\partial_t v_{\ell_q} + \mathrm{div}\,(v_{\ell_q} \otimes v_{\ell_q}) + \nabla p_{\ell_q} = \mathrm{div}\,\mathring{R}_{\ell_q} + \mathrm{div}\,\mathring{R}_q^{\mathrm{comm}} \qquad (7.1\mathrm{a})$$

$$\mathrm{div}\,v_{\ell_q} = 0, \qquad (7.1\mathrm{b})$$

with v_{ℓ_q}, $\mathring{R}_{\ell_q}$, and $\mathring{R}_q^{\mathrm{comm}}$ satisfying the conclusions of Lemma 5.1, in addition to (3.12)–(3.25b), there exist $v_{q+1} = v_{\ell_q} + w_{q+1}$ and $\mathring{R}_{q+1}$ which satisfy the following:

1. v_{q+1} and $\mathring{R}_{q+1}$ solve the Euler-Reynolds system

$$\partial_t v_{q+1} + \mathrm{div}\,(v_{q+1} \otimes v_{q+1}) + \nabla p_{q+1} = \mathring{R}_{q+1} \qquad (7.2\mathrm{a})$$

$$\mathrm{div}\,v_{q+1} = 0. \qquad (7.2\mathrm{b})$$

2. *For all* $k, m \leq 7N_{\mathrm{ind},v}$,

$$\left\| \psi_{i,q} D^k D^m_{t,q} w_{q+1} \right\|_{L^2} \leq \Gamma^{-1}_{q+1} \delta^{\frac{1}{2}}_{q+1} \lambda^k_{q+1} \mathcal{M} \left(m, N_{\mathrm{ind},t}, \tau^{-1}_q \Gamma^{i-1}_{q+1}, \tilde{\tau}^{-1}_q \Gamma^{-1}_{q+1} \right). \tag{7.3}$$

Furthermore, we have that

$$\mathrm{supp}_t (\mathring{R}_q) \subset [T_1, T_2]$$

$$\Rightarrow \quad \mathrm{supp}_t (w_{q+1}) \subset \left[T_1 - (\lambda_q \delta^{1/2}_q)^{-1}, T_2 + (\lambda_q \delta^{1/2}_q)^{-1} \right]. \tag{7.4}$$

3. *For all* $k, m \leq 3N_{\mathrm{ind},v}$,

$$\left\| \psi_{i,q} D^k D^m_{t,q} \mathring{R}_{q+1} \right\|_{L^1} \leq \Gamma^{-C_R}_{q+1} \delta_{q+2} \lambda^k_{q+1} \mathcal{M} \left(m, N_{\mathrm{ind},t}, \Gamma^{i+1}_{q+1} \tau^{-1}_q, \Gamma^{-1}_{q+1} \tilde{\tau}^{-1}_q \right). \tag{7.5}$$

Remark 7.2. In achieving the conclusions (7.2), (7.3), and (7.5), we have verified the inductive assumptions (3.13)–(3.15) at level $q+1$. The inductive assumption (3.12) at levels $q' < q+1$ follows from Lemma (5.1). The proof of Proposition 7.1 will entail many estimates which are much more detailed than (7.3) and (7.5), but for the time being we record only the basic estimates, which are direct translations of (3.13)–(3.15) at level $q + 1$.

7.2 NOTATIONS

The proof of Proposition 7.1 will be achieved through an induction with respect to $\tilde{n}$, where $0 \leq \tilde{n} \leq n_{\mathrm{max}}$ corresponds to the addition of the perturbation $w_{q+1,\tilde{n}} = \sum_{\tilde{p}=1}^{p_{\mathrm{max}}} w_{q+1,\tilde{n},\tilde{p}}$. The addition of each perturbation $w_{q+1,\tilde{n}}$ will move the minimum effective frequency present in the stress terms to $\lambda_{q,\tilde{n}+1,0}$. This induction on $\tilde{n}$ requires three subpropositions; the base case $\tilde{n} = 0$, the inductive step from $\tilde{n} - 1$ to $\tilde{n}$ for $\tilde{n} \leq n_{\mathrm{max}} - 1$, and the final step from $n_{\mathrm{max}} - 1$ to n_{max}. Throughout these propositions, we shall employ the following notations.

1. $\tilde{n}$: An integer taking values $0 \leq \tilde{n} \leq n_{\mathrm{max}}$ over which induction is performed. At every step in the induction, we add another component $w_{q+1,\tilde{n}}$ of the final perturbation

$$w_{q+1} = \sum_{\tilde{n}=0}^{n_{\mathrm{max}}} \sum_{\tilde{p}=1}^{p_{\mathrm{max}}} w_{q+1,\tilde{n},\tilde{p}}.$$

We emphasize that the use of $\tilde{n}$ at various points in statements and estimates means that we are *currently* working on the inductive step at level $\tilde{n}$.

2. $\boldsymbol{n}$: An integer taking values $1 \leq n \leq n_{\max}$ which correspond to the higher order stresses $\mathring{R}_{q,n}$. Occasionally, we shall use the notation $\mathring{R}_{q,0} = \mathring{R}_{\ell_q}$ to streamline an argument. We emphasize that n will be used at various points in statements and estimates to reference *higher order* objects in addition to those at level $\tilde{n}$, and so will satisfy the inequality $\tilde{n} \leq n$.

3. $\boldsymbol{\mathring{H}^{n'}_{q,n,p}}$: The component of $\mathring{R}_{q,n,p}$ originating from an error term produced by the addition of $w_{q+1,n'}$. The parameter n' will always be a *subsidiary* parameter used to reference objects created at or *below* the level $\tilde{n}$ that we are currently working on, and so will satisfy $n' \leq \tilde{n}$.

4. $\mathbb{P}_{[\boldsymbol{q,n,p}]}$: We use the spatial Littlewood-Paley projectors $\mathbb{P}_{[q,n,p]}$ defined by

$$\mathbb{P}_{[q,n,p]} = \begin{cases} \mathbb{P}_{\geq \lambda_{q,n_{\max},p_{\max}}} & \text{if } n = n_{\max}, p = p_{\max} + 1 \\ \mathbb{P}_{[\lambda_{q,n,p-1}, \lambda_{q,n,p})} & \text{if } 1 \leq n \leq n_{\max}, 1 \leq p \leq p_{\max} \end{cases}, \quad (7.6)$$

where $\mathbb{P}_{[\lambda_1, \lambda_2)}$ is defined in Section 9.4 as $\mathbb{P}_{\geq \lambda_1} \mathbb{P}_{< \lambda_2}$. Note that for $n = n_{\max}$ and $p = p_{\max} + 1$, $\mathbb{P}_{[q,n_{\max},p_{\max}+1]}$ projects onto *all* frequencies larger than $\lambda_{q,n_{\max},p_{\max}} = \lambda_{q,n_{\max}+1,0}$. Errors which include the frequency projector $\mathbb{P}_{[q,n_{\max},p_{\max}+1]}$ will be small enough to be absorbed into $\mathring{R}_{q+1}$. We shall frequently utilize sums of Littlewood-Paley projectors $\mathbb{P}_{[q,n,p]}$ to decompose products of intermittent pipe flows periodized to scale $\lambda_{q,\tilde{n}}^{-1}$. These sums will be written in terms of three parameters—n, p, and $\tilde{n}$. As a consequence of (7.6), (9.29), (9.23), and (9.22), we have that $\lambda_{q,\tilde{n}+1,0} \leq \lambda_{q,\tilde{n}}$ for $0 \leq \tilde{n} \leq n_{\max}$, so that

$$\left(\sum_{n=\tilde{n}+1}^{n_{\max}} \sum_{p=1}^{p_{\max}} \mathbb{P}_{[q,n,p]} + \mathbb{P}_{[q,n_{\max},p_{\max}+1]} \right) \mathbb{P}_{\geq \lambda_{q,\tilde{n}}} = \mathbb{P}_{\geq \lambda_{q,\tilde{n}+1,0}} \mathbb{P}_{\geq \lambda_{q,\tilde{n}}}$$

$$= \mathbb{P}_{\geq \lambda_{q,\tilde{n}}}. \quad (7.7)$$

A consequence of (7.7) is that for $\frac{\mathbb{T}^3}{\lambda_{q,\tilde{n}}}$-periodic functions[1] where $0 \leq \tilde{n} \leq n_{\max}$,

$$f = \fint_{\mathbb{T}^3} f + \mathbb{P}_{\geq \lambda_{q,\tilde{n}}} f$$

$$= \fint_{\mathbb{T}^3} f + \mathbb{P}_{\geq \lambda_{q,\tilde{n}}} \left(\sum_{n=\tilde{n}+1}^{n_{\max}} \sum_{p=1}^{p_{\max}} \mathbb{P}_{[q,n,p]} + \mathbb{P}_{[q,n_{\max},p_{\max}+1]} \right) f. \quad (7.8)$$

These equalities will be useful in the calculations in Section 8.3, and we will recall their significance when we estimate the Type 1 errors in Section 8.6.

---

[1] We note that in the second equality in (7.8), such functions do not have active frequencies between $\lambda_{q,\tilde{n}+1,0}$ and $\lambda_{q,\tilde{n}}$.

5. $\mathring{R}^{\tilde{n}}_{q+1}$: Any stress term which satisfies the estimates required of $\mathring{R}_{q+1}$ and which has already been estimated at the $\tilde{n}^{\text{th}}$ stage of the induction; that is, error terms arising from the addition of $w_{q+1,n'}$ for $n' \leq \tilde{n}$. We *exclude* $\mathring{R}^{\text{comm}}_q$ from $\mathring{R}^{\tilde{n}}_{q+1}$, absorbing it only at the very end when we define $\mathring{R}_{q+1}$. Thus

$$\mathring{R}^{\tilde{n}+1}_{q+1} = \mathring{R}^{\tilde{n}}_{q+1} + \left(\text{errors coming from } w_{q+1,\tilde{n}+1} \text{ that also go into } \mathring{R}_{q+1} \right). \tag{7.9}$$

7.3 INDUCTION ON $\tilde{N}$

The first proposition asserts that there exists a perturbation $w_{q+1,0}$ which we add to v_{ℓ_q} so that $v_{q,0} := v_{\ell_q} + w_{q+1,0}$ satisfies the following. First, $v_{q,0}$ solves the Euler-Reynolds system with a right-hand side consisting of stresses $\mathring{R}^0_{q+1}$ and $\mathring{H}^0_{q,n,p}$ which belong respectively to $\mathring{R}_{q+1}$ and $\mathring{R}_{q,n,p}$ for $1 \leq n \leq n_{\max}$ and $1 \leq p \leq p_{\max}$. Secondly, $w_{q+1,0}$ satisfies estimates which in particular imply the inductive assumptions required of the velocity perturbation w_{q+1} in (7.3).[2] Thirdly, $\mathring{R}^0_{q+1}$ satisfies the estimates required of $\mathring{R}_{q+1}$ in the inductive assumption (6.118) (with an extra factor of smallness). Finally, each $\mathring{H}^0_{q,n,p}$ satisfies the inductive assumptions required of $\mathring{R}_{q,n,p}$ in (6.118).

Proposition 7.3 (Induction on $\tilde{n}$: The base case $\tilde{n} = 0$). *Under the assumptions of Proposition 7.1 (equivalently the conclusions of Lemma 5.1), there exist* $w_{q+1,0} = \displaystyle\sum_{\tilde{p}=1}^{p_{\max}} w_{q+1,0,p} = w_{q+1,0,1}$, $\mathring{R}^0_{q+1}$, *and* $\mathring{H}^0_{q,n,p}$ *for* $1 \leq n \leq n_{\max}$ *and* $1 \leq p \leq p_{\max}$ *such that the following hold.*

1. $v_{q,0} := v_{\ell_q} + w_{q+1,0}$ *solves*

$$\partial_t v_{q,0} + \text{div} \left(v_{q,0} \otimes v_{q,0} \right) + \nabla p_{q,0}$$

$$= \text{div} \left(\mathring{R}^0_{q+1} \right) + \text{div} \left(\sum_{n=1}^{n_{\max}} \sum_{p=1}^{p_{\max}} \mathring{H}^0_{q,n,p} \right) + \text{div} \, \mathring{R}^{\text{comm}}_q \tag{7.10a}$$

$$\text{div} \, v_{q,0} = 0. \tag{7.10b}$$

2. *For all* $k + m \leq N_{\text{fin},0} - N_{\text{cut},t} - N_{\text{cut},x} - 2N_{\text{dec}} - 9$ *and* $1 \leq \tilde{p} \leq p_{\max}$ *(although only $w_{q+1,0,1}$ is non-zero)*

$$\left\| D^k D^m_{t,q} w_{q+1,0,\tilde{p}} \right\|_{L^2 (\text{supp} \, \psi_{i,q})}$$

---

[2]This is checked in Remark 8.3.

$$\lesssim \delta_{q+1,0,\tilde{p}}^{\frac{1}{2}} \Gamma_{q+1}^{3+\frac{c_b}{2}} \lambda_{q+1}^k \mathcal{M}\left(m, \mathsf{N}_{\mathrm{ind},t}, \tau_q^{-1}\Gamma_{q+1}^{i-c_0+4}, \tilde{\tau}_q^{-1}\Gamma_{q+1}^{-1}\right). \qquad (7.11)$$

Furthermore, we have that

$$\mathrm{supp}_t(\mathring{R}_q) \subset [T_1, T_2]$$
$$\Rightarrow \mathrm{supp}_t(w_{q+1,0,\tilde{p}}) \subset \left[T_1 - (\lambda_q \delta_q^{1/2}\Gamma_{q+1})^{-1}, T_2 + (\lambda_q \delta_q^{1/2}\Gamma_{q+1})^{-1}\right]. \qquad (7.12)$$

3. *For all* $k, m \leq 3\mathsf{N}_{\mathrm{ind},v}$,

$$\left\|\psi_{i,q} D^k D_{t,q}^m \mathring{R}_{q+1}^0\right\|_{L^1}$$
$$\lesssim \Gamma_{q+1}^{-C_R} \Gamma_{q+1}^{-1} \delta_{q+2} \lambda_{q+1}^k \mathcal{M}\left(m, \mathsf{N}_{\mathrm{ind},t}, \tau_q^{-1}\Gamma_{q+1}^{i+1}, \tilde{\tau}_q^{-1}\Gamma_{q+1}^{-1}\right). \qquad (7.13)$$

Furthermore, we have that

$$\mathrm{supp}_t \mathring{R}_{q+1}^0 \subseteq \mathrm{supp}_t w_{q+1,0}. \qquad (7.14)$$

4. *For all* $k + m \leq \mathsf{N}_{\mathrm{fin},n}$ *and* $1 \leq n \leq n_{\max}, 1 \leq p \leq p_{\max}$,

$$\left\|D^k D_{t,q}^m \mathring{H}_{q,n,p}^0\right\|_{L^1(\mathrm{supp}\,\psi_{i,q})}$$
$$\lesssim \delta_{q+1,n,p} \lambda_{q,n,p}^k \mathcal{M}\left(m, \mathsf{N}_{\mathrm{ind},t}, \tau_q^{-1}\Gamma_{q+1}^{i-c_n}, \tilde{\tau}_q^{-1}\Gamma_{q+1}^{-1}\right). \qquad (7.15)$$

Furthermore, we have that

$$\mathrm{supp}_t \mathring{H}_{q,n,p}^0 \subseteq \mathrm{supp}_t w_{q+1,0}. \qquad (7.16)$$

The second proposition assumes that perturbations $w_{q+1,n'}$ have been added for $n' \leq \tilde{n}-1$ while satisfying four criteria. Firstly, $v_{q,\tilde{n}-1} = v_{\ell_q} + \sum_{n' \leq \tilde{n}-1} w_{q+1,n'}$ solves an Euler-Reynolds system with stresses $\mathring{R}_{q+1}^{\tilde{n}-1}$ and $\mathring{H}_{q,n,p}^{n'}$. Secondly, the perturbations $w_{q+1,n'}$ satisfy the inductive assumptions required of w_{q+1} in (7.3) for $n' \leq \tilde{n}-1$. Thirdly, $\mathring{R}_{q+1}^{\tilde{n}-1}$ satisfies the inductive assumption (7.5) at level $q+1$. Finally, $\mathring{H}_{q,n,p}^{n'}$ satisfies the assumption (6.118) in the parameter regime $\tilde{n} \leq n \leq n_{\max}, n' \leq \tilde{n}-1, 1 \leq p \leq p_{\max}$. The conclusion of the proposition replaces each $\tilde{n}-1$ in the assumptions with $\tilde{n}$.

Proposition 7.4 (Induction on $\tilde{n}$: From $\tilde{n}-1$ to $\tilde{n}$ for $1 \leq \tilde{n} \leq n_{\max}-1$). *Let* $1 \leq \tilde{n} \leq n_{\max} - 1$ *be given, and let*

$$v_{q,\tilde{n}-1} := v_{\ell_q} + \sum_{n'=0}^{\tilde{n}-1} w_{q+1,n'} = v_{\ell_q} + \sum_{n'=0}^{\tilde{n}-1} \sum_{p'=1}^{p_{\max}} w_{q+1,n',p'},$$

$\mathring{R}_{q+1}^{\tilde{n}-1}$, *and* $\mathring{H}_{q,n,p}^{n'}$ *be given for* $n' \leq \tilde{n}-1$, $\tilde{n} \leq n \leq n_{\max}$ *and* $1 \leq p, p' \leq p_{\max}$

such that the following are satisfied.

1. $v_{q,\tilde{n}-1}$ *solves:*

$$\partial_t v_{q,\tilde{n}-1} + \mathrm{div}\left(v_{q,\tilde{n}-1} \otimes v_{q,\tilde{n}-1}\right) + \nabla p_{q,\tilde{n}-1}$$

$$= \mathrm{div}\left(\mathring{R}_{q+1}^{\tilde{n}-1}\right) + \mathrm{div}\left(\sum_{n=\tilde{n}}^{n_{\max}} \sum_{p=1}^{p_{\max}} \sum_{n'=0}^{\tilde{n}-1} \mathring{H}_{q,n,p}^{n'}\right) + \mathrm{div}\,\mathring{R}_q^{\mathrm{comm}}$$

$$\tag{7.17a}$$

$$\mathrm{div}\,v_{q,\tilde{n}-1} = 0. \tag{7.17b}$$

2. *For all* $k + m \leq N_{\mathrm{fin},n'} - N_{\mathrm{cut},t} - N_{\mathrm{cut},x} - 2N_{\mathrm{dec}} - 9$, $n' \leq \tilde{n} - 1$, *and* $1 \leq p' \leq p_{\max}$,

$$\left\| D^k D_{t,q}^m w_{q+1,n',p'} \right\|_{L^2(\mathrm{supp}\,\psi_{i,q})}$$

$$\lesssim \delta_{q+1,n',p'}^{\frac{1}{2}} \Gamma_{q+1}^{3+\frac{c_b}{2}} \lambda_{q+1}^k \mathcal{M}\left(m, N_{\mathrm{ind},t}, \tau_q^{-1}\Gamma_{q+1}^{i-c_{n'}+4}, \tilde{\tau}_q^{-1}\Gamma_{q+1}^{-1}\right). \tag{7.18}$$

Furthermore, we have that

$$\mathrm{supp}_t(\mathring{R}_{q,n',p'}) \subset [T_{1,n',p'}, T_{2,n',p'}]$$

$$\Rightarrow \mathrm{supp}_t(w_{q+1,n',p'})$$

$$\subset \left[T_{1,n',p'} - (\lambda_q \delta_q^{1/2}\Gamma_{q+1})^{-1}, T_{2,n',p'} + (\lambda_q \delta_q^{1/2}\Gamma_{q+1})^{-1}\right]. \tag{7.19}$$

3. *For all* $k, m \leq 3N_{\mathrm{ind},v}$,

$$\left\| \psi_{i,q} D^k D_{t,q}^m \mathring{R}_{q+1}^{\tilde{n}-1} \right\|_{L^1}$$

$$\lesssim \Gamma_{q+1}^{-C_R}\Gamma_{q+1}^{-1}\delta_{q+2}\lambda_{q+1}^k \mathcal{M}\left(m, N_{\mathrm{ind},t}, \Gamma_{q+1}^{i+1}\tau_q^{-1}, \Gamma_{q+1}^{-1}\tilde{\tau}_q^{-1}\right). \tag{7.20}$$

Furthermore, we have that

$$\mathrm{supp}_t \mathring{R}_{q+1}^{\tilde{n}-1} \subseteq \bigcup_{n' \leq \tilde{n}-1} \mathrm{supp}_t w_{q+1,n'}. \tag{7.21}$$

4. *For all* $k + m \leq N_{\mathrm{fin},n}$, $\tilde{n} \leq n \leq n_{\max}$, $n' \leq \tilde{n} - 1$, *and* $1 \leq p \leq p_{\max}$,

$$\left\| D^k D_{t,q}^m \mathring{H}_{q,n,p}^{n'} \right\|_{L^1(\mathrm{supp}\,\psi_{i,q})}$$

$$\lesssim \delta_{q+1,n,p}\lambda_{q,n,p}^k \mathcal{M}\left(m, N_{\mathrm{ind},t}, \tau_q^{-1}\Gamma_{q+1}^{i-c_n}, \tilde{\tau}_q^{-1}\Gamma_{q+1}^{-1}\right). \tag{7.22}$$

Furthermore, we have that

$$\mathrm{supp}_t \mathring{H}_{q,n,p}^{n'} \subseteq \mathrm{supp}_t w_{q+1,n'}. \tag{7.23}$$

Then there exists $w_{q+1,\tilde{n}}$ *such that (1)-(4) are satisfied with* $\tilde{n} - 1$ *replaced by*

$\widetilde{n}$.

The final proposition considers the case $\widetilde{n} = n_{\max}$ and shows that, under assumptions analogous to those in Proposition 7.4, there exists $w_{q+1,n_{\max}}$ such that all remaining errors after the addition of $w_{q+1,n_{\max}}$ can be absorbed into $\mathring{R}_{q+1}$, thus verifying the conclusions of Proposition 7.1.

Proposition 7.5 (Induction on $\widetilde{n}$: The final case $\widetilde{n} = n_{\max}$). *Let*

$$v_{q,n_{\max}-1} := v_{\ell_q} + \sum_{n'=0}^{n_{\max}-1} w_{q+1,n'} = v_{\ell_q} + \sum_{n'=0}^{n_{\max}-1} \sum_{p'=1}^{p_{\max}} w_{q+1,n',p'},$$

$\mathring{R}_{q+1}^{n_{\max}-1}$, *and* $\mathring{H}_{q,n_{\max},p}^{n'}$ *be given for* $n' \leq n_{\max} - 1$ *and* $1 \leq p, p' \leq p_{\max}$ *such that the following are satisfied.*

1. *$v_{q,n_{\max}-1}$ solves:*

$$\partial_t v_{q,n_{\max}-1} + \mathrm{div}\left(v_{q,n_{\max}-1} \otimes v_{q,n_{\max}-1}\right) + \nabla p_{q,n_{\max}-1}$$
$$= \mathrm{div}\left(\mathring{R}_{q+1}^{n_{\max}-1}\right) + \mathrm{div}\left(\sum_{n'=0}^{n_{\max}-1} \sum_{p=1}^{p_{\max}} \mathring{H}_{q,n_{\max},p}^{n'}\right) + \mathrm{div}\,\mathring{R}_q^{\mathrm{comm}}$$

$$\tag{7.24a}$$

$$\mathrm{div}\,v_{q,n_{\max}-1} = 0. \tag{7.24b}$$

2. *For all* $k + m \leq \mathsf{N}_{\mathrm{fin},n'} - \mathsf{N}_{\mathrm{cut},t} - \mathsf{N}_{\mathrm{cut},x} - 2\mathsf{N}_{\mathrm{dec}} - 9$, $n' \leq n_{\max} - 1$, *and* $1 \leq p' \leq p_{\max}$,

$$\left\| D^k D_{t,q}^m w_{q+1,n',p'} \right\|_{L^2(\mathrm{supp}\,\psi_{i,q})}$$
$$\lesssim \delta_{q+1,n',p'}^{\frac{1}{2}} \Gamma_{q+1}^{3+\frac{\mathsf{C}_b}{2}} \lambda_{q+1}^k \mathcal{M}\left(m, \mathsf{N}_{\mathrm{ind},t}, \tau_q^{-1}\Gamma_{q+1}^{i-\mathsf{c}_{n'}+4}, \widetilde{\tau}_q^{-1}\Gamma_{q+1}^{-1}\right). \tag{7.25}$$

Furthermore, we have that

$$\mathrm{supp}_t(\mathring{R}_{q,n',p'}) \subset [T_{1,n',p'}, T_{2,n',p'}]$$
$$\Rightarrow \mathrm{supp}_t(w_{q+1,n',p'})$$
$$\subset \left[T_{1,n',p'} - (\lambda_q \delta_q^{1/2}\Gamma_{q+1})^{-1}, T_{2,n',p'} + (\lambda_q \delta_q^{1/2}\Gamma_{q+1})^{-1}\right]. \tag{7.26}$$

3. *For all* $k, m \leq 3\mathsf{N}_{\mathrm{ind},v}$,

$$\left\| \psi_{i,q} D^k D_{t,q}^m \mathring{R}_{q+1}^{n_{\max}-1} \right\|_{L^1}$$
$$\lesssim \Gamma_{q+1}^{-\mathsf{C}_{\mathsf{R}}} \Gamma_{q+1}^{-1} \delta_{q+2} \lambda_{q+1}^k \mathcal{M}\left(m, \mathsf{N}_{\mathrm{ind},t}, \Gamma_{q+1}^{i+1}\tau_q^{-1}, \Gamma_{q+1}^{-1}\widetilde{\tau}_q^{-1}\right). \tag{7.27}$$

Furthermore, we have that

$$\operatorname{supp}_t \mathring{R}_{q+1}^{n_{\max}-1} \subseteq \bigcup_{n' \leq n_{\max}-1} \operatorname{supp}_t w_{q+1,n'} . \tag{7.28}$$

4. *For all $k + m \leq \mathsf{N}_{\mathrm{fin},n_{\max}}$, $n' \leq n_{\max} - 1$, and $1 \leq p \leq p_{\max}$*

$$\left\| D^k D_{t,q}^m \mathring{H}_{q,n_{\max},p}^{n'} \right\|_{L^1(\operatorname{supp} \psi_{i,q})}$$
$$\lesssim \delta_{q+1,n_{\max},p} \lambda_{q,n_{\max},p} \mathcal{M}\left(m, \mathsf{N}_{\mathrm{ind},t}, \tau_q^{-1} \Gamma_{q+1}^{i-\mathsf{c}_{n_{\max}}}, \widetilde{\tau}_q^{-1} \Gamma_{q+1}^{-1} \right) . \tag{7.29}$$

Furthermore, we have that

$$\operatorname{supp}_t \mathring{H}_{q,n,p}^{n'} \subseteq \operatorname{supp}_t w_{q+1,n'} . \tag{7.30}$$

Then there exist $w_{q+1,n_{\max}}$ and $\mathring{R}_{q+1}$ such that $v_{q+1} := v_{q,n_{\max}-1} + w_{q+1,n_{\max}}$ and $\mathring{R}_{q+1}$ satisfy conclusions (7.2), (7.3), (7.4), and (7.5) from Proposition 7.1.

Chapter Eight

Proving the main inductive estimates

Because the proofs of Propositions 7.3, 7.4, and 7.5 will be comprised of multiple arguments with many similarities, we divide up the proofs of the propositions into sections corresponding to these arguments.[1] First, we define $\mathring{R}_{q,\tilde{n},\tilde{p}}$ and $w_{q+1,\tilde{n},\tilde{p}}$ in Section 8.1 for each $0 \leq \tilde{n} \leq n_{\max}$ and $1 \leq \tilde{p} \leq p_{\max}$. Then, Section 8.2 collects estimates on $w_{q+1,\tilde{n},\tilde{p}}$, thus verifying (7.11) and (7.12), (7.18) and (7.19), and (7.25) and (7.26) at levels $\tilde{n} = 0$, $1 \leq \tilde{n} \leq n_{\max} - 1$, and $\tilde{n} = n_{\max}$, respectively. Next, in Section 8.3 we separate out the different types of error terms and write down the Euler-Reynolds system satisfied by $v_{q,\tilde{n}}$, which verifies (7.10), (7.17), and (7.24), again at the respective values of $\tilde{n}$.

The error estimates are then divided into five sections. We first estimate the transport and Nash errors in Sections 8.4 and 8.5. The next section estimates the Type 1 oscillation errors (notated with $\mathring{H}_{q,n,p}^{\tilde{n}}$), which are obtained via Littlewood-Paley projectors $\mathbb{P}_{[q,n,p]}$. In the parameter regime $1 \leq n \leq n_{\max}$ and $1 \leq p \leq p_{\max}$, Type 1 oscillation errors will satisfy the estimates (7.15), (7.22), and (7.29) at respective parameter values $\tilde{n} = 0$, $1 \leq \tilde{n} \leq n_{\max} - 1$, and $\tilde{n} = n_{\max}$. Type 1 oscillation errors obtained from $\mathbb{P}_{[q,n_{\max},p_{\max}+1]}$ have a sufficiently high minimum frequency (from (7.6), specifically $\lambda_{q,n_{\max}+1,0}$, which by a large choice of $n_{\max}$ is very close to λ_{q+1}) to be absorbed into $\mathring{R}_{q+1}$. Then in Section 8.7, we use Proposition 4.8 to show that on the support of a checkerboard cutoff function, Type 2 oscillation errors vanish. The divergence corrector errors are estimated in Sections 8.8. The divergence corrector, Nash, and transport errors will always be absorbed into $\mathring{R}_{q+1}$ and thus must again satisfy one of (7.13), (7.20), and (7.27). Finally, the conclusions (7.12), (7.14), (7.16), (7.19), (7.21), (7.23), (7.26), (7.28), and (7.30) concerning the time support will be verified in Section 8.9.

8.1 DEFINITION OF $\mathring{R}_{q,\tilde{n},\tilde{p}}$ AND $W_{q+1,\tilde{n},\tilde{p}}$

In this section we construct the perturbations $w_{q+1,\tilde{n}}$. Before doing so, we recall the significance of each parameter used to define the perturbations.

---

[1]This organization of proof avoids having to alternate between the definitions of $w_{q+1,\tilde{n},\tilde{p}}$ and $\mathring{R}_{q,\tilde{n},\tilde{p}}$ for all $1 \leq \tilde{n} \leq n_{\max}$ and $1 \leq \tilde{p} \leq p_{\max}$. We judge that it is wiser to define all the perturbations simultaneously under the assumptions of Propositions 7.3, 7.4, and 7.5. Namely, we assume that each $\mathring{R}_{q,\tilde{n},\tilde{p}}$ exists and satisfies the enumerated properties, some of which may not be verified until later.

1. ξ is the vector direction of the axis of the pipe.
2. i quantifies the amplitude of the velocity field v_{ℓ_q} along which the pipe will flow.
3. j quantifies the amplitude of the Reynolds stress.
4. k describes which time cutoff $\chi_{i,k,q}$ is active.
5. $q+1$ is the stage of the overall convex integration scheme.
6. $\widetilde{n}$ and $\widetilde{p}$ signify which higher order stress $\mathring{R}_{q,\widetilde{n},\widetilde{p}}$ is being corrected, and $\widetilde{n}$ also denotes the intermittency parameter $r_{q+1,\widetilde{n}}$.
7. $\vec{l} = (l, w, h)$ is used to index the checkerboard cutoff functions. Recall that the admissible values of l, w, and h range from 0 to $\lambda_{q,\widetilde{n},0} - 1$ and thus depend on $\widetilde{n}$.

8.1.1   The case $\widetilde{n} = 0$

To define $w_{q+1,0} = \displaystyle\sum_{\widetilde{p}=1}^{p_{\max}} w_{q+1,0,p} = w_{q+1,0,1}$, we recall the notation $\mathring{R}_{\ell_q} = \mathring{R}_{q,0}$
and set

$$R_{q,0,1,j,i,k} = \nabla\Phi_{(i,k)} \left(\delta_{q+1,0,1}\Gamma_{q+1}^{2j+4}\mathrm{Id} - \mathring{R}_{q,0} \right) \nabla\Phi_{(i,k)}^{T}. \tag{8.1}$$

For $\widetilde{p} \geq 2$, we set $R_{q,0,\widetilde{p},j,i,k} = 0$. Fix values of i, j, and k. Let $\xi \in \Xi$ be a vector from Proposition 4.1. For all $\xi \in \Xi$, we define the coefficient function $a_{\xi,i,j,k,q,0,\widetilde{p},\vec{l}}$ by

$$
\begin{aligned}
a_{\xi,i,j,k,q,0,\widetilde{p},\vec{l}} &:= a_{\xi,i,j,k,q,0,\widetilde{p}} \\
&:= a_{(\xi)} \\
&= \delta_{q+1,0,\widetilde{p}}^{1/2}\Gamma_{q+1}^{j+2}\eta_{i,j,k,q,0,\widetilde{p},\vec{l}}\gamma_{\xi}\left(\frac{R_{q,0,\widetilde{p},j,i,k}}{\delta_{q+1,0,\widetilde{p}}\Gamma_{q+1}^{2j+4}} \right).
\end{aligned} \tag{8.2}
$$

From Lemma 6.31, we see that on the support of $\eta_{(i,j,k)}$ we have $|\mathring{R}_{q,0,\widetilde{p}}| \leq \Gamma_{q+1}^{2j+2}\delta_{q+1,0,\widetilde{p}}$, and thus by estimate (6.108) from Corollary 6.27, for $\widetilde{p} = 1$ we have that

$$\left| \frac{R_{q,0,\widetilde{p},j,i,k}}{\delta_{q+1,0,\widetilde{p}}\Gamma_{q+1}^{2j+4}} - \mathrm{Id} \right| \leq \Gamma_{q+1}^{-1} < \frac{1}{2}$$

once λ_0 is sufficiently large. Thus we may apply Proposition 4.1.

The coefficient function $a_{(\xi)}$ is then multiplied by an intermittent pipe flow

$$\nabla\Phi_{(i,k)}^{-1}\mathbb{W}_{\xi,q+1,0} \circ \Phi_{(i,k)},$$

where we have used the objects defined in Proposition 4.4 and the shorthand notation

$$\mathbb{W}_{\xi,q+1,0} = \mathbb{W}_{\xi,q+1,0}^{(i,j,k,0,\vec{l})} = \mathbb{W}_{\xi,q+1,0}^{s} = \mathbb{W}_{\xi,\lambda_{q+1},r_{q+1,0}}^{s}. \tag{8.3}$$

The superscript $s = (i, j, k, 0, \vec{l})$ indicates the placement of the intermittent pipe flow $\mathbb{W}_{\xi,q+1,0}^{i,j,k,0,p,\vec{l}}$ (cf. (2) from Proposition 4.4), which depends on i, j, k, $\tilde{n} = 0$, and $\vec{l}$ and is only relevant in Section 8.7.[2] To ease notation, we will suppress the superscript except in Section 8.7. Furthermore, item 1 from Proposition 4.4 gives that

$$\nabla \Phi_{(i,k)}^{-1} \mathbb{W}_{\xi,q+1,0} \circ \Phi_{(i,k)} = \mathrm{curl}\left(\nabla \Phi_{(i,k)}^{T} \mathbb{U}_{\xi,q+1,0} \circ \Phi_{(i,k)} \right).$$

We can now write the principal part of the first term of the perturbation as

$$w_{q+1,0}^{(p)} = \sum_{i,j,k,\tilde{p}} \sum_{\vec{l}} \sum_{\xi} a_{(\xi)} \mathrm{curl}\left(\nabla \Phi_{(i,k)}^{T} \mathbb{U}_{\xi,q+1,0} \circ \Phi_{(i,k)} \right) := \sum_{i,j,k,\tilde{p}} \sum_{\vec{l}} \sum_{\xi} w_{(\xi)}.$$
(8.4)

The notation $w_{(\xi)}$ implicitly encodes all indices and thus will be a useful shorthand for the principal part of the perturbation. To make the perturbation divergence-free, we add

$$w_{q+1,0}^{(c)} = \sum_{i,j,k,\tilde{p}} \sum_{\vec{l}} \sum_{\xi} \nabla a_{(\xi)} \times \left(\nabla \Phi_{(i,k)}^{T} \mathbb{U}_{\xi,q+1,0} \circ \Phi_{(i,k)} \right) = \sum_{i,j,k,\tilde{p}} \sum_{\vec{l}} \sum_{\xi} w_{(\xi)}^{(c)}$$
(8.5)

so that

$$w_{q+1,0} = w_{q+1,0}^{(p)} + w_{q+1,0}^{(c)} = \sum_{i,j,k,\tilde{p}} \sum_{\vec{l}} \sum_{\xi} \mathrm{curl}\left(a_{(\xi)} \nabla \Phi_{(i,k)}^{T} \mathbb{U}_{\xi,q+1,0} \circ \Phi_{(i,k)} \right)$$
(8.6)

is divergence-free and mean-zero.

8.1.2   The case $1 \leq \tilde{n} \leq n_{\max}$

With $w_{q+1,0}$ constructed, we construct $w_{q+1,\tilde{n}} = \sum_{\tilde{p}=1}^{p_{\max}} w_{q+1,\tilde{n},\tilde{p}}$ for $1 \leq \tilde{n} \leq n_{\max}$.

For $1 \leq \tilde{p} \leq p_{\max}$, we define

$$\mathring{R}_{q,\tilde{n},\tilde{p}} = \sum_{n' \leq \tilde{n}-1} \mathring{H}_{q,\tilde{n},\tilde{p}}^{n'}.$$
(8.7)

With this definition in hand, we set

$$R_{q,\tilde{n},\tilde{p},j,i,k} = \nabla \Phi_{(i,k)} \left(\delta_{q+1,\tilde{n},\tilde{p}} \Gamma_{q+1}^{2j+4} \mathrm{Id} - \mathring{R}_{q,\tilde{n},\tilde{p}} \right) \nabla \Phi_{(i,k)}^{T},$$
(8.8)

---

[2]Note that for $\tilde{p} \geq 2$, $\delta_{q+1,0,\tilde{p}} = 0$, so there is no need for the placement to depend on $\tilde{p}$ in this case, as $w_{q+1,0,\tilde{p}}$ will uniformly vanish.

and define the coefficient function $a_{\xi,i,j,k,q,\tilde{n},\tilde{p},\vec{l}}$ by

$$a_{\xi,i,j,k,q,\tilde{n},\tilde{p},\vec{l}} = a_{\xi,i,j,k,q,\tilde{n},\tilde{p}}$$

$$= a_{(\xi)} = \delta_{q+1,\tilde{n},\tilde{p}}^{1/2}\Gamma_{q+1}^{j+2}\eta_{i,j,k,q,\tilde{n},\tilde{p},\vec{l}}\gamma_{\xi}\left(\frac{R_{q,\tilde{n},\tilde{p},j,i,k}}{\delta_{q+1,\tilde{n},\tilde{p}}\Gamma_{q+1}^{2j+4}}\right). \quad (8.9)$$

By Lemma 6.31 and Corollary 6.27 as before, $R_{q,\tilde{n},\tilde{p},j,i,k}/(\delta_{q+1,\tilde{n},\tilde{p}}\Gamma_{q+1}^{2j+4})$ lies in the domain of γ_{ξ}, as soon as λ_0 is sufficiently large (similarly to the display below (8.2)). The coefficient function is multiplied by an intermittent pipe flow

$$\nabla\Phi_{(i,k)}^{-1}\mathbb{W}_{\xi,q+1,\tilde{n}} \circ \Phi_{(i,k)} = \mathrm{curl}\left(\nabla\Phi_{(i,k)}^{T}\mathbb{U}_{\xi,q+1,\tilde{n}} \circ \Phi_{(i,k)}\right),$$

where we have used the shorthand notation

$$\mathbb{W}_{\xi,q+1,\tilde{n}} = \mathbb{W}_{\xi,q+1,\tilde{n}}^{i,j,k,\tilde{n},\tilde{p},\vec{l}} = \mathbb{W}_{\xi,q+1,\tilde{n}}^{s} = \mathbb{W}_{\xi,\lambda_{q+1},r_{q+1,\tilde{n}}}^{s}. \quad (8.10)$$

As before, the superscript $s = (i,j,k,\tilde{n},\tilde{p},\vec{l})$ refers to the placement of the pipe, depends on i, j, k, $\tilde{n}$, $\tilde{p}$, and $\vec{l}$, and will be chosen in Section 8.7. Thus the principal part of the perturbation is defined by

$$w_{q+1,\tilde{n},\tilde{p}}^{(p)} = \sum_{i,j,k}\sum_{\vec{l}}\sum_{\xi} a_{(\xi)}\mathrm{curl}\left(\nabla\Phi_{(i,k)}^{T}\mathbb{U}_{\xi,q+1,\tilde{n}} \circ \Phi_{(i,k)}\right)$$

$$= \sum_{i,j,k}\sum_{\vec{l}}\sum_{\xi} w_{(\xi)}. \quad (8.11)$$

As before, we add a corrector

$$w_{q+1,\tilde{n},\tilde{p}}^{(c)} = \sum_{i,j,k}\sum_{\vec{l}}\sum_{\xi} \nabla a_{(\xi)} \times \left(\nabla\Phi_{(i,k)}^{T}\mathbb{U}_{\xi,q+1,\tilde{n}} \circ \Phi_{(i,k)}\right)$$

$$= \sum_{i,j,k}\sum_{\vec{l}}\sum_{\xi} w_{(\xi)}^{(c)}, \quad (8.12)$$

producing the divergence-free perturbation

$$w_{q+1,\tilde{n}} = \sum_{\tilde{p}=1}^{p_{\max}} w_{q+1,\tilde{n},\tilde{p}} = \sum_{\tilde{p}=1}^{p_{\max}}\left(w_{q+1,\tilde{n},\tilde{p}}^{(p)} + w_{q+1,\tilde{n},\tilde{p}}^{(c)}\right)$$

$$= \sum_{i,j,k,\tilde{p}}\sum_{\vec{l}}\sum_{\xi}\mathrm{curl}\left(a_{(\xi)}\nabla\Phi_{(i,k)}^{T}\mathbb{U}_{\xi,q+1,\tilde{n}} \circ \Phi_{(i,k)}\right). \quad (8.13)$$

8.2 ESTIMATES FOR $W_{q+1,\widetilde{n},\widetilde{p}}$

In this section, we verify (7.11), (7.18), and (7.25). We first estimate the L^r norms of the coefficient functions $a_{(\xi)}$. We have consolidated the proofs for each value of $\widetilde{n}$ into the following lemma.

Lemma 8.1. *For $N + M \leq N_{\mathrm{fin},\widetilde{n}} - N_{\mathrm{cut},t} - N_{\mathrm{cut},x} - 4$, $r \geq 1$, and $r_1, r_2 \in [1, \infty]$ with $\frac{1}{r_1} + \frac{1}{r_2} = 1$, we have*

$$\left\| D^N D_{t,q}^M a_{\xi,i,j,k,q,\widetilde{n},\widetilde{p},\vec{l}} \right\|_{L^r}$$

$$\lesssim \left| \mathrm{supp}\,(\eta_{i,j,k,q,\widetilde{n},\widetilde{p},\vec{l}}) \right|^{\frac{1}{r}} \delta_{q+1,\widetilde{n},\widetilde{p}}^{1/2} \Gamma_{q+1}^{j+2}$$

$$\times (\Gamma_{q+1} \lambda_{q,\widetilde{n},\widetilde{p}})^N \mathcal{M}\left(M, N_{\mathrm{ind},t}, \tau_q^{-1} \Gamma_{q+1}^{i-c_{\widetilde{n}}+3}, \widetilde{\tau}_q^{-1} \Gamma_{q+1}^{-1} \right). \quad (8.14)$$

Proof of Lemma 8.1. We begin by considering the case $r = \infty$. The general case $r \geq 1$ will then follow from the size of the support of $a_{(\xi)}$. Recalling estimate (6.125), we have that for all $N + M \leq N_{\mathrm{fin},\widetilde{n}} - 4$,

$$\left\| D^N D_{t,q}^M \mathring{R}_{q,\widetilde{n},\widetilde{p}} \right\|_{L^\infty(\mathrm{supp}\,\eta_{(i,j,k)})}$$

$$\lesssim \delta_{q+1,\widetilde{n},\widetilde{p}} \Gamma_{q+1}^{2j+2} (\Gamma_{q+1} \lambda_{q,\widetilde{n},\widetilde{p}})^N \mathcal{M}\left(M, N_{\mathrm{ind},t}, \tau_q^{-1} \Gamma_{q+1}^{i-c_{\widetilde{n}}+2}, \widetilde{\tau}_q^{-1} \Gamma_{q+1}^{-1} \right).$$

From Corollary 6.27, we have that for all $N + M \leq 3N_{\mathrm{fin}}/2$,

$$\left\| D^N D_{t,q}^M D\Phi_{(i,k)} \right\|_{L^\infty(\mathrm{supp}\,(\psi_{i,q}\chi_{i,k,q}))} \leq \widetilde{\lambda}_q^N \mathcal{M}\left(M, N_{\mathrm{ind},t}, \Gamma_{q+1}^{i-c_0} \tau_q^{-1}, \widetilde{\tau}_q^{-1} \Gamma_{q+1}^{-1} \right).$$

Thus from the Leibniz rule and the definitions (8.8), (8.1), for $N + M \leq N_{\mathrm{fin},\widetilde{n}} - 4$,

$$\left\| D^N D_{t,q}^M R_{q,\widetilde{n},\widetilde{p},j,i,k} \right\|_{L^\infty(\mathrm{supp}\,\eta_{(i,j,k)})}$$

$$\lesssim \delta_{q+1,\widetilde{n},\widetilde{p}} \Gamma_{q+1}^{2j+4} (\Gamma_{q+1} \lambda_{q,\widetilde{n},\widetilde{p}})^N \mathcal{M}\left(M, N_{\mathrm{ind},t}, \tau_q^{-1} \Gamma_{q+1}^{i-c_{\widetilde{n}}+2}, \widetilde{\tau}_q^{-1} \Gamma_{q+1}^{-1} \right). \quad (8.15)$$

The above estimates allow us to apply Lemma A.5 with $N = N'$, $M = M'$ so that $N + M \leq N_{\mathrm{fin},\widetilde{n}} - 4$, $\psi = \gamma_\xi$, (which is allowable since by Proposition 4.1 we have that $D^B \gamma_\xi$ is bounded uniformly with respect to q, and we have checked in Section 8.1 that the argument of γ_ξ remains strictly within a ball of radius ε of the identity), $\Gamma_\psi = 1$, $v = v_{\ell_q}$, $D_t = D_{t,q}$, $h(x,t) = R_{q,\widetilde{n},\widetilde{p},j,i,k}(x,t)$, $C_h = \delta_{q+1,\widetilde{n},\widetilde{p}} \Gamma_{q+1}^{2j+4} = \Gamma^2$, $\lambda = \widetilde{\lambda} = \lambda_{q,\widetilde{n},\widetilde{p}} \Gamma_{q+1}$, $\mu = \tau_q^{-1} \Gamma_{q+1}^{i-c_{\widetilde{n}}+2}$, $\widetilde{\mu} = \widetilde{\tau}_q^{-1} \Gamma_{q+1}^{-1}$, and $N_t = N_{\mathrm{ind},t}$. We obtain that for all $N + M \leq N_{\mathrm{fin},\widetilde{n}} - 4$,

$$\left\| D^N D_{t,q}^M \gamma_\xi \left(\frac{R_{q,\widetilde{n},\widetilde{p},j,i,k}}{\delta_{q+1,\widetilde{n},\widetilde{p}} \Gamma_{q+1}^{2j+4}} \right) \right\|_{L^\infty(\mathrm{supp}\,\eta_{(i,j,k)})}$$

$$\lesssim (\Gamma_{q+1}\lambda_{q,\tilde{n},\tilde{p}})^N \, \mathcal{M}\left(M, \mathsf{N}_{\mathrm{ind},t}, \tau_q^{-1}\Gamma_{q+1}^{i-\mathsf{c}_{\tilde{n}}+2}, \widetilde{\tau}_q^{-1}\Gamma_{q+1}^{-1}\right).$$

From the above bound, definitions (8.2) and (8.9), the Leibniz rule, estimates (6.84), (6.97), and (6.131), and Lemma 6.40, we obtain that for $N + M \leq \mathsf{N}_{\mathrm{fin},\tilde{n}} - \mathsf{N}_{\mathrm{cut},t} - \mathsf{N}_{\mathrm{cut},x} - 4$,[3]

$$\left\| D^N D_{t,q}^M a_{(\xi)} \right\|_{L^\infty(\mathrm{supp}\,\eta_{(i,j,k)})}$$

$$\lesssim \delta_{q+1,\tilde{n},\tilde{p}}^{1/2}\Gamma_{q+1}^{j+2} \sum_{\substack{N'+N''=N,\\M'+M''=M}} \left\| D^{N'} D_{t,q}^{M'} \eta_{(i,j,k)} \right\|_{L^\infty}$$

$$\times \left\| D^{N''} D_{t,q}^{M''} \gamma_\xi \left(\frac{R_{q,\tilde{n},\tilde{p},j,i,k}}{\delta_{q+1,\tilde{n},\tilde{p}}\Gamma_{q+1}^{2j+4}} \right) \right\|_{L^\infty(\mathrm{supp}\,\eta_{(i,j,k)})}$$

$$\lesssim \delta_{q+1,\tilde{n},\tilde{p}}^{1/2}\Gamma_{q+1}^{j+2} \sum_{\substack{N'+N''=N,\\M'+M''=M}} (\Gamma_{q+1}\lambda_{q,\tilde{n},\tilde{p}})^{N'} \, \mathcal{M}\left(M', \mathsf{N}_{\mathrm{ind},t}, \tau_q^{-1}\Gamma_{q+1}^{i-\mathsf{c}_{\tilde{n}}+3}, \widetilde{\tau}_q^{-1}\Gamma_{q+1}^{-1}\right)$$

$$\times (\Gamma_{q+1}\lambda_{q,\tilde{n},\tilde{p}})^{N''} \, \mathcal{M}\left(M'', \mathsf{N}_{\mathrm{ind},t}, \tau_q^{-1}\Gamma_{q+1}^{i-\mathsf{c}_{\tilde{n}}+2}, \widetilde{\tau}_q^{-1}\Gamma_{q+1}^{-1}\right)$$

$$\lesssim \delta_{q+1,\tilde{n},\tilde{p}}^{1/2}\Gamma_{q+1}^{j+2} (\Gamma_{q+1}\lambda_{q,\tilde{n},\tilde{p}})^N \, \mathcal{M}\left(M, \mathsf{N}_{\mathrm{ind},t}, \tau_q^{-1}\Gamma_{q+1}^{i-\mathsf{c}_{\tilde{n}}+3}, \widetilde{\tau}_q^{-1}\Gamma_{q+1}^{-1}\right).$$

This concludes the proof of (8.14) when $r = \infty$. Recall from Lemma 6.41 that

$$\left|\mathrm{supp}\,(\eta_{(i,j,k)})\right| \lesssim \Gamma_{q+1}^{-2\left(\frac{i}{r_1}+\frac{j}{r_2}\right)+\mathsf{C}_b+2}. \tag{8.16}$$

The general result then follows. $\square$

An immediate consequence of Lemma 8.1 is that we have estimates for the velocity increments themselves. These are summarized in the following corollary.

Corollary 8.2. *For $N + M \leq \mathsf{N}_{\mathrm{fin},\tilde{n}} - \mathsf{N}_{\mathrm{cut},t} - \mathsf{N}_{\mathrm{cut},x} - 2\mathsf{N}_{\mathrm{dec}} - 8$ we have the following estimate:*

$$\left\| D^N D_{t,q}^M w_{(\xi)} \right\|_{L^r} \lesssim \left|\mathrm{supp}\,(\eta_{i,j,k,q,\tilde{n},\tilde{p},l})\right|^{\frac{1}{r}} \delta_{q+1,\tilde{n},\tilde{p}}^{1/2}\Gamma_{q+1}^{j+2} (r_{q+1,\tilde{n}})^{2/r-1}$$

$$\times \lambda_{q+1}^N \mathcal{M}\left(M, \mathsf{N}_{\mathrm{ind},t}, \tau_q^{-1}\Gamma_{q+1}^{i-\mathsf{c}_{\tilde{n}}+3}, \widetilde{\tau}_q^{-1}\Gamma_{q+1}^{-1}\right). \tag{8.17}$$

For $N+M \leq \mathsf{N}_{\mathrm{fin},\tilde{n}} - \mathsf{N}_{\mathrm{cut},t} - \mathsf{N}_{\mathrm{cut},x} - 2\mathsf{N}_{\mathrm{dec}} - 9$ and $(r, r_1, r_2) \in \{(1,2,2),(2,\infty,1)\}$,

---

[3]The limit on the number of derivatives comes from (6.131) and (8.15). The sharp cost of a material derivative comes from (6.131).

we have the following estimates:

$$\left\| D^N D_{t,q}^M w_{(\xi)}^{(c)} \right\|_{L^r} \lesssim \frac{\Gamma_{q+1} \lambda_{q,\tilde{n},\tilde{p}}}{\lambda_{q+1}} \left| \text{supp} \left(\eta_{i,j,k,q,\tilde{n},\tilde{p},l} \right) \right|^{\frac{1}{r}} \delta_{q+1,\tilde{n},\tilde{p}}^{1/2} \Gamma_{q+1}^{j+2} (r_{q+1,\tilde{n}})^{2/r-1}$$
$$\times \lambda_{q+1}^N \mathcal{M} \left(M, \mathsf{N}_{\text{ind},t}, \tau_q^{-1} \Gamma_{q+1}^{i-c_{\tilde{n}}+3}, \tilde{\tau}_q^{-1} \Gamma_{q+1}^{-1} \right), \tag{8.18}$$

$$\left\| D^N D_{t,q}^M w_{q+1,\tilde{n},\tilde{p}} \right\|_{L^r(\text{supp}\,\psi_{i,q})} \lesssim \delta_{q+1,\tilde{n},\tilde{p}}^{1/2} \Gamma_{q+1}^{\frac{-2i+\mathsf{C}_b}{r_1}+2+\frac{2}{r}} (r_{q+1,\tilde{n}})^{2/r-1} \lambda_{q+1}^N$$
$$\times \mathcal{M} \left(M, \mathsf{N}_{\text{ind},t}, \tau_q^{-1} \Gamma_{q+1}^{i-c_{\tilde{n}}+4}, \tilde{\tau}_q^{-1} \Gamma_{q+1}^{-1} \right). \tag{8.19}$$

Finally, we have that

$$\text{supp}_t(\mathring{R}_q) \subset [T_1, T_2]$$
$$\Rightarrow \quad \text{supp}_t(w_{q+1,\tilde{n},\tilde{p}}) \subset \left[T_1 - (\lambda_q \delta_q^{1/2})^{-1}, T_2 + (\lambda_q \delta_q^{1/2})^{-1} \right]. \tag{8.20}$$

Remark 8.3. By choosing $r = 2$, $r_2 = 1$, and $r_1 = \infty$ in (8.19) and recalling that (9.56) and (9.60b) give

$$\delta_{q+1,\tilde{n},\tilde{p}}^{1/2} \leq \Gamma_{q+1}^{-2} \delta_{q+1}^{1/2}, \qquad \mathsf{N}_{\text{fin},\tilde{n}} - \mathsf{N}_{\text{cut},t} - \mathsf{N}_{\text{cut},x} - 2\mathsf{N}_{\text{dec}} - 9 \geq 14\mathsf{N}_{\text{ind},v},$$

we may sum over $\tilde{n}$ and $\tilde{p}$ in (8.19) and use the extra negative factor of Γ_{q+1} to absorb any implicit constants. Finally, from (9.42), we have that the cost of a sharp material derivative in (8.19) is sufficient to meet the bounds in (7.3). Then we have verified (7.11), (7.18), and (7.25) at levels $\tilde{n} = 0$, $1 \leq \tilde{n} < n_{\max}$, and $\tilde{n} = n_{\max}$, respectively, and (7.3).

Proof of Corollary 8.2. Recalling the definition of $w_{(\xi)}$ from (8.4) and (8.13), we aim to prove the first estimate by applying Remark A.9, with $f = a_{(\xi)} \nabla \Phi_{(i,k)}^{-1}$, $\mathcal{C}_f = \left| \text{supp} \left(\eta_{i,j,k,q,\tilde{n},\tilde{p},l} \right) \right|^{\frac{1}{r}} \delta_{q+1,\tilde{n},\tilde{p}}^{1/2} \Gamma_{q+1}^{j+2}$, $\Phi = \Phi_{(i,k)}$, $v = v_{\ell_q}$, $\lambda = \Gamma_{q+1} \lambda_{q,\tilde{n},\tilde{p}}$, $\zeta = \tilde{\zeta} = \lambda_{q+1}$, $\mathcal{C}_\varphi = r_{q+1,\tilde{n}}^{2/r-1}$, $\mu = \lambda_{q,\tilde{n}} = \lambda_{q+1} r_{q+1,\tilde{n}}$, $\nu = \tau_q^{-1} \Gamma_{q+1}^{i-c_{\tilde{n}}+3}$, $\tilde{\nu} = \tilde{\tau}_q^{-1} \Gamma_{q+1}^{-1}$, $g = \mathbb{W}_{\xi,q+1,\tilde{n}}$, $N_t = \mathsf{N}_{\text{ind},t}$, and $N_\circ = \mathsf{N}_{\text{fin},\tilde{n}} - \mathsf{N}_{\text{cut},t} - \mathsf{N}_{\text{cut},x} - 4$. From (8.14) and Corollary 6.27, we have that for $N + M \leq \mathsf{N}_{\text{fin},\tilde{n}} - \mathsf{N}_{\text{cut},t} - \mathsf{N}_{\text{cut},x} - 4$,

$$\left\| D^N D_{t,q}^M a_{(\xi)} \right\|_{L^r} \lesssim \left| \text{supp} \left(\eta_{(i,j,k)} \right) \right|^{\frac{1}{r}} \delta_{q+1,\tilde{n},\tilde{p}}^{1/2} \Gamma_{q+1}^{j+2} \left(\Gamma_{q+1} \lambda_{q,\tilde{n},\tilde{p}} \right)^N$$
$$\times \mathcal{M} \left(M, \mathsf{N}_{\text{ind},t}, \tau_q^{-1} \Gamma_{q+1}^{i-c_{\tilde{n}}+3}, \tilde{\tau}_q^{-1} \Gamma_{q+1}^{-1} \right), \tag{8.21}$$

$$\left\| D^N D_{t,q}^M (D\Phi_{(i,k)})^{-1} \right\|_{L^\infty(\text{supp}(\psi_{i,q} \tilde{\chi}_{i,k,q}))} \leq \tilde{\lambda}_q^N \mathcal{M} \left(M, \mathsf{N}_{\text{ind},t}, \Gamma_{q+1}^{i-c_0} \tau_q^{-1}, \tilde{\tau}_q^{-1} \Gamma_{q+1}^{-1} \right), \tag{8.22}$$

$$\left\| D^N \Phi_{(i,k)} \right\|_{L^\infty(\text{supp}(\psi_{i,q} \tilde{\chi}_{i,k,q}))} \lesssim \Gamma_{q+1}^{-1} \tilde{\lambda}_q^{N-1}, \tag{8.23}$$

$$\left\| D^N \Phi_{(i,k)}^{-1} \right\|_{L^\infty(\text{supp}(\psi_{i,q}\widetilde{\chi}_{i,k,q}))} \lesssim \Gamma_{q+1}^{-1}\widetilde{\lambda}_q^{N-1}, \tag{8.24}$$

showing that (A.30), (A.31), and (A.32) are satisfied. Recall that $\mathbb{W}_{\xi,q+1,\widetilde{n}}$ is periodic to scale:

$$\lambda_{q,\widetilde{n}}^{-1} = (\lambda_{q+1}r_{q+1,\widetilde{n}})^{-1} = \left(\lambda_q^{\left(\frac{4}{5}\right)^{\widetilde{n}+1}}\lambda_{q+1}^{1-\left(\frac{4}{5}\right)^{\widetilde{n}+1}}\right)^{-1}.$$

By (9.48) and (9.60a), we have that for all q, $\widetilde{n}$, and $\widetilde{p}$,

$$\lambda_{q+1}^4 \leq \left(\frac{\lambda_{q,\widetilde{n}}}{2\pi\sqrt{3}\Gamma_{q+1}\lambda_{q,\widetilde{n},\widetilde{p}}}\right)^{\mathsf{N}_{\text{dec}}}, \quad 2\mathsf{N}_{\text{dec}} + 4 \leq \mathsf{N}_{\text{fin},\widetilde{n}} - \mathsf{N}_{\text{cut},t} - \mathsf{N}_{\text{cut},x} - 5, \tag{8.25}$$

and so the assumptions (A.34) and (A.35) from Lemma A.5 are satisfied. From the estimates in Proposition 4.4, we have that (A.33) is satisfied with $\zeta = \widetilde{\zeta} = \lambda_{q+1}$. We may thus apply Lemma A.7, Remark A.9 to obtain that for both choices of (r, r_1, r_2) and $N + M \leq \mathsf{N}_{\text{fin},\widetilde{n}} - \mathsf{N}_{\text{cut},t} - \mathsf{N}_{\text{cut},x} - 2\mathsf{N}_{\text{dec}} - 8$,

$$\left\| D^N \left(D_{t,q}^M \left(a_{(\xi)} \nabla \Phi_{(i,k)}^{-1} \right) \mathbb{W}_{\xi,q+1,\widetilde{n}} \circ \Phi_{(i,k)} \right) \right\|_{L^r}$$

$$\lesssim \sum_{m=0}^{N} \left| \text{supp}\left(\eta_{(i,j,k)}\right) \right|^{\frac{1}{r}} \delta_{q+1,\widetilde{n},\widetilde{p}}^{1/2} \Gamma_{q+1}^{j+2} \left(\Gamma_{q+1}\lambda_{q,\widetilde{n},\widetilde{p}}\right)^{N-m}$$

$$\times \mathcal{M}\left(M, \mathsf{N}_{\text{ind},t}, \tau_q^{-1}\Gamma_{q+1}^{i-c_{\widetilde{n}}+3}, \widetilde{\tau}_q^{-1}\Gamma_{q+1}^{-1}\right) \|D^m \mathbb{W}_{\xi,q+1,\widetilde{n}}\|_{L^r}$$

$$\lesssim \sum_{m=0}^{N} \left| \text{supp}\left(\eta_{(i,j,k)}\right) \right|^{\frac{1}{r}} \delta_{q+1,\widetilde{n},\widetilde{p}}^{1/2} \Gamma_{q+1}^{j+2} \left(\Gamma_{q+1}\lambda_{q,\widetilde{n},\widetilde{p}}\right)^{N-m}$$

$$\times \mathcal{M}\left(M, \mathsf{N}_{\text{ind},t}, \tau_q^{-1}\Gamma_{q+1}^{i-c_{\widetilde{n}}+3}, \widetilde{\tau}_q^{-1}\Gamma_{q+1}^{-1}\right) \lambda_{q+1}^m (r_{q+1,n})^{2/r-1}$$

$$\lesssim \left| \text{supp}\left(\eta_{(i,j,k)}\right) \right|^{\frac{1}{r}} \delta_{q+1,\widetilde{n},\widetilde{p}}^{1/2} \Gamma_{q+1}^{j+2}$$

$$\times \mathcal{M}\left(M, \mathsf{N}_{\text{ind},t}, \tau_q^{-1}\Gamma_{q+1}^{i-c_{\widetilde{n}}+3}, \widetilde{\tau}_q^{-1}\Gamma_{q+1}^{-1}\right) \lambda_{q+1}^N (r_{q+1,n})^{2/r-1}.$$

Here we have used that $\lambda_{q+1} \geq \Gamma_{q+1}\lambda_{q,\widetilde{n},\widetilde{p}}$ for all $0 \leq n \leq n_{\max}$ and $1 \leq \widetilde{p} \leq p_{\max}$, and thus we have proven (8.17).

The argument for the corrector is similar, save for the fact that $D_{t,q}$ will land on $\nabla a_{(\xi)}$, and so we require an extra commutator estimate from Lemma A.14, specifically Remark A.15. Note that $D_{t,q}\Phi_{(i,k)} = 0$ gives

$$D_{t,q}^M w_{(\xi)}^{(c)} = D_{t,q}^M \left(\nabla a_{(\xi)} \times \left(\nabla \Phi_{(i,k)}^T \mathbb{U}_{\xi,q+1,\widetilde{n}} \circ \Phi_{(i,k)} \right) \right)$$

$$= \sum_{M'+M''=M} c(M', M)\left(D_{t,q}^{M'} \nabla a_{(\xi)}\right) \times \left(\left(D_{t,q}^{M''} \nabla \Phi_{(i,k)}^T\right) \mathbb{U}_{\xi,q+1,\widetilde{n}} \circ \Phi_{(i,k)}\right).$$

Using (6.60) and (8.21) shows that (A.50) and (A.51) are satisfied with $f = \nabla a_{(\xi)}$,

$$\mathcal{C}_f = \left|\operatorname{supp}\left(\eta_{i,j,k,q,\tilde{n},\tilde{p},\tilde{l}}\right)\right|^{\frac{1}{r}}\delta_{q+1,\tilde{n},\tilde{p}}^{1/2}\Gamma_{q+1}^{j+2}\Gamma_{q+1}\lambda_{q,\tilde{n},\tilde{p}},$$

$\mathcal{C}_v = \delta_q^{\frac{1}{2}}\Gamma_{q+1}^{i+1}$, $\lambda_v = \tilde{\lambda}_v = \tilde{\lambda}_q$, $\mu_v = \Gamma_{q+1}^{i-c_0}\tau_q^{-1}$, $N_t = \mathsf{N}_{\mathrm{ind},t}$, $\tilde{\mu}_v = \tilde{\tau}_q^{-1}\Gamma_{q+1}^{-1}$, $\lambda_f = \tilde{\lambda}_f = \Gamma_{q+1}\lambda_{q,\tilde{n},\tilde{p}}$, $\mu_f = \tau_q^{-1}\Gamma_{q+1}^{i-c_{\tilde{n}}+3}$, and $\tilde{\mu}_f = \tilde{\tau}_q^{-1}\Gamma_{q+1}^{-1}$. Applying Lemma A.14 (estimate (A.54)) as before, we obtain that for $N+M \leq \mathsf{N}_{\mathrm{fin},\tilde{n}}-\mathsf{N}_{\mathrm{cut},t}-\mathsf{N}_{\mathrm{cut},x}-5$,

$$\left\|D^N D_{t,q}^M \nabla a_{(\xi)}\right\|_{L^r} \lesssim \left|\operatorname{supp}\left(\eta_{(i,j,k)}\right)\right|^{\frac{1}{r}}\delta_{q+1,\tilde{n},\tilde{p}}^{1/2}\Gamma_{q+1}^{j+2}$$

$$\times (\Gamma_{q+1}\lambda_{q,\tilde{n},\tilde{p}})^{N+1}\mathcal{M}\left(M,\mathsf{N}_{\mathrm{ind},t},\tau_q^{-1}\Gamma_{q+1}^{i-c_{\tilde{n}}+3},\tilde{\tau}_q^{-1}\Gamma_{q+1}^{-1}\right). \quad (8.26)$$

In view of (8.22) and (8.25), we may apply Lemma A.7, Remark A.9, and the estimates from Proposition 4.4 to obtain that for $N + M \leq \mathsf{N}_{\mathrm{fin},\tilde{n}} - \mathsf{N}_{\mathrm{cut},t} - \mathsf{N}_{\mathrm{cut},x} - 2\mathsf{N}_{\mathrm{dec}} - 9$

$$\left\|D^N D_{t,q}^M \left(\nabla a_{(\xi)} \times \left(\nabla \Phi_{(i,k)}^T \mathbb{U}_{\xi,q+1,\tilde{n}} \circ \Phi_{(i,k)}\right)\right)\right\|_{L^r}$$

$$\lesssim \sum_{m=0}^N \left|\operatorname{supp}\left(\eta_{(i,j,k)}\right)\right|^{\frac{1}{r}}\delta_{q+1,\tilde{n},\tilde{p}}^{1/2}\Gamma_{q+1}^{j+2}\Gamma_{q+1}\lambda_{q,\tilde{n},\tilde{p}}\lambda_{q,\tilde{n},\tilde{p}}^{N-m}$$

$$\times \mathcal{M}\left(M,\mathsf{N}_{\mathrm{ind},t},\tau_q^{-1}\Gamma_{q+1}^{i-c_{\tilde{n}}+3},\tilde{\tau}_q^{-1}\Gamma_{q+1}^{-1}\right) \left\|D^m \mathbb{U}_{\xi,q+1,\tilde{n}}\right\|_{L^r}$$

$$\lesssim \lambda_{q+1}^{m-1} \sum_{m=0}^N \left|\operatorname{supp}\left(\eta_{(i,j,k)}\right)\right|^{\frac{1}{r}}\delta_{q+1,\tilde{n},\tilde{p}}^{1/2}\Gamma_{q+1}^{j+2}\Gamma_{q+1}\lambda_{q,\tilde{n},\tilde{p}}\lambda_{q,\tilde{n},\tilde{p}}^{N-m}$$

$$\times \mathcal{M}\left(M,\mathsf{N}_{\mathrm{ind},t},\tau_q^{-1}\Gamma_{q+1}^{i-c_{\tilde{n}}+3},\tilde{\tau}_q^{-1}\Gamma_{q+1}^{-1}\right)(r_{q+1,n})^{\frac{2}{r}-1}$$

$$\lesssim \frac{\Gamma_{q+1}\lambda_{q,\tilde{n},\tilde{p}}}{\lambda_{q+1}}\lambda_{q+1}^N\left|\operatorname{supp}\left(\eta_{(i,j,k)}\right)\right|^{\frac{1}{r}}\delta_{q+1,\tilde{n},\tilde{p}}^{1/2}\Gamma_{q+1}^{j+2}$$

$$\times \mathcal{M}\left(M,\mathsf{N}_{\mathrm{ind},t},\tau_q^{-1}\Gamma_{q+1}^{i-c_{\tilde{n}}+3},\tilde{\tau}_q^{-1}\Gamma_{q+1}^{-1}\right)(r_{q+1,n})^{2/r-1}, \quad (8.27)$$

proving (8.18).

The final estimate (8.19) follows from the first two after recalling that $\psi_{i,q}$ may overlap with $\psi_{i+1,q}$, so that on the support of $\psi_{i,q}$, we will have to appeal to (8.14) at level $i + 1$. Then, we sum over $\tilde{l}$ and appeal to the bound (6.147). Next, we may sum on j, the index which we recall from Lemma 6.35 is bounded independently of q, and $\tilde{p}$, k. The powers of Γ_{q+1}^j cancel out since $rr_2 = 1$. Next, we sum over $\tilde{p}$, which is bounded independently of q, and recall that the parameter k, although not bounded independently of q, corresponds to a partition of unity, so that the number of cutoff functions which may overlap at any fixed point *is* finite and bounded independently of q. $\square$

8.3   IDENTIFICATION OF ERROR TERMS

In this section, we identify the error terms arising from the addition of $w_{q+1,\tilde{n}} = \sum_{\tilde{p}=1}^{p_{max}} w_{q+1,\tilde{n},\tilde{p}}$. After doing so, we can write down the Euler-Reynolds system satisfied by $v_{q,\tilde{n}}$, in turn verifying at level $\tilde{n}$ the conclusions (7.10), (7.17), and (7.24) of Propositions 7.3, 7.4, and 7.5, respectively.

8.3.1   The case $\tilde{n} = 0$

By the inductive assumption of Proposition 7.3, we have that $\operatorname{div} v_{\ell_q} = 0$, and

$$\partial_t v_{\ell_q} + \operatorname{div}(v_{\ell_q} \otimes v_{\ell_q}) + \nabla p_{\ell_q} = \operatorname{div} \mathring{R}_{\ell_q} + \operatorname{div} \mathring{R}_q^{\mathrm{comm}}.$$

Adding $w_{q+1,0}$ as defined in (8.6), we obtain that $v_{q,0} := v_{\ell_q} + w_{q+1,0}$ solves

$$\begin{aligned}
\partial_t v_{q,0} &+ \operatorname{div}(v_{q,0} \otimes v_{q,0}) + \nabla p_{\ell_q} \\
&= (\partial_t + v_{\ell_q} \cdot \nabla) w_{q+1,0} + w_{q+1,0} \cdot \nabla v_{\ell_q} \\
&\quad + \operatorname{div}(w_{q+1,0} \otimes w_{q+1,0}) + \operatorname{div} \mathring{R}_{\ell_q} + \operatorname{div} \mathring{R}_q^{\mathrm{comm}} \\
&:= \mathcal{T}_0 + \mathcal{N}_0 + \mathcal{O}_0 + \operatorname{div} \mathring{R}_{\ell_q} + \operatorname{div} \mathring{R}_q^{\mathrm{comm}}. \quad\quad (8.28)
\end{aligned}$$

For a fixed $\tilde{n}$, throughout this section we will consider sums over indices

$$(\xi, i, j, k, \tilde{p}, \vec{l}),$$

where the direction vector ξ takes on one of the finitely many values in Proposition 4.4, $0 \le i \le i_{\max}(q)$ indexes the velocity cutoffs (there are finitely many such values; cf. (6.50)), $0 \le j \le j_{\max}(q, \tilde{n}, \tilde{p})$ indexes the stress cutoffs (there are finitely many such values; cf. (6.129)), the parameter k indexes the time cutoffs defined in (6.96) (the number of values of k is q-dependent, but this is irrelevant because they form a partition of unity; cf. (6.94)), the parameter $1 \le \tilde{p} \le p_{\max}$ indexes which component of $\mathring{R}_{q+1,\tilde{n},\tilde{p}}$ we are working with (there are finitely many such values; cf. (9.3)), and, lastly, $\vec{l}$ indexes the checkerboard cutoffs from Definition 6.39 (again, the number of such indexes is q-dependent, but this is acceptable because they form a partition of unity; cf. (6.139)). For brevity of notation, we denote sums over such indexes as

$$\sum_{\xi, i, j, k, \tilde{p}, \vec{l}}.$$

Moreover, we shall denote as

$$\sum_{\neq \{\xi, i, j, k, \tilde{p}, \vec{l}\}} \quad\quad (8.29)$$

the *double-summation* over indexes $(\xi, i, j, k, \widetilde{p}, \vec{l})$ and $(\xi^*, i^*, j^*, k^*, p^*, \vec{l}^*)$ which belong to the set

$$
\Big\{ (\xi, i, j, k, \widetilde{p}, \vec{l}, \xi^*), (i^*, j^*, k^*, p^*, \vec{l}^*)
$$
$$
: \xi \neq \xi^* \vee i \neq i^* \vee j \neq j^* \vee k \neq k^* \vee \widetilde{p} \neq p^* \vee \vec{l} \neq \vec{l}^* \Big\}, \tag{8.30}
$$

although we remind the reader that at the current stage, $\widetilde{n} = 0$, the sum over $\widetilde{p}$ is superfluous since $w_{q+1,0} = w_{q+1,0,1}$. For the sake of consistency between $w_{q+1,0}$ and $w_{q+1,\widetilde{n}}$ for $1 \leq \widetilde{n} \leq n_{\max}$, we shall include the index $\widetilde{p}$ throughout this section. Expanding out the oscillation error $\mathcal{O}_0$, we have that

$$
\mathcal{O}_0 = \sum_{\xi, i, j, k, \widetilde{p}, \vec{l}} \operatorname{div} \left(\operatorname{curl} \left(a_{(\xi)} \nabla \Phi_{(i,k)}^T \mathbb{U}_{\xi, q+1, 0} \circ \Phi_{(i,k)} \right) \right.
$$
$$
\left. \otimes \operatorname{curl} \left(a_{(\xi)} \nabla \Phi_{(i,k)}^T \mathbb{U}_{\xi, q+1, 0} \circ \Phi_{(i,k)} \right) \right)
$$
$$
+ \sum_{\neq \{\xi, i, j, k, \widetilde{p}, \vec{l}\}} \operatorname{div} \left(\operatorname{curl} \left(a_{(\xi)} \nabla \Phi_{(i,k)}^T \mathbb{U}_{\xi, q+1, 0} \circ \Phi_{(i,k)} \right) \right.
$$
$$
\left. \otimes \operatorname{curl} \left(a_{(\xi^*)} \nabla \Phi_{(i^*, k^*)}^T \mathbb{U}_{\xi^*, q+1, 0} \circ \Phi_{(i^*, k^*)} \right) \right)
$$
$$
:= \operatorname{div} \mathcal{O}_{0,1} + \operatorname{div} \mathcal{O}_{0,2}. \tag{8.31}
$$

In Section 8.7, we will show that $\mathcal{O}_{0,2}$ is a Type 2 oscillation error so that

$$
\mathcal{O}_{0,2} = 0.
$$

Recalling identity (4.14) and the notation (9.65), we further split $\mathcal{O}_{0,1}$ as

$$
\operatorname{div} \mathcal{O}_{0,1} = \sum_{\xi, i, j, k, \widetilde{p}, \vec{l}} \operatorname{div} \left(\left(a_{(\xi)} \nabla \Phi_{(i,k)}^{-1} \mathbb{W}_{\xi, q+1, 0} \circ \Phi_{(i,k)} \right) \right.
$$
$$
\left. \otimes \left(a_{(\xi)} \nabla \Phi_{(i,k)}^{-1} \mathbb{W}_{\xi, q+1, 0} \circ \Phi_{(i,k)} \right) \right)
$$
$$
+ 2 \sum_{\xi, i, j, k, \widetilde{p}, \vec{l}} \operatorname{div} \left(\left(a_{(\xi)} \nabla \Phi_{(i,k)}^{-1} \mathbb{W}_{\xi, q+1, 0} \circ \Phi_{(i,k)} \right) \right.
$$
$$
\left. \otimes_s \left(\nabla a_{(\xi)} \times \left(\nabla \Phi_{(i,k)}^T \mathbb{U}_{\xi, q+1, 0} \circ \Phi_{(i,k)} \right) \right) \right)
$$
$$
+ \sum_{\xi, i, j, k, \widetilde{p}, \vec{l}} \operatorname{div} \left(\left(\nabla a_{(\xi)} \times \left(\nabla \Phi_{(i,k)}^T \mathbb{U}_{\xi, q+1, 0} \circ \Phi_{(i,k)} \right) \right) \right.
$$
$$
\left. \otimes \left(\nabla a_{(\xi)} \times \left(\nabla \Phi_{(i,k)}^T \mathbb{U}_{\xi, q+1, 0} \circ \Phi_{(i,k)} \right) \right) \right)
$$
$$
:= \operatorname{div} \left(\mathcal{O}_{0,1,1} + \mathcal{O}_{0,1,2} + \mathcal{O}_{0,1,3} \right). \tag{8.32}
$$

Aside from $\mathcal{O}_{0,1,1}$, each of these terms is a divergence corrector error and will therefore be estimated in Section 8.8.

Recall by Propositions 4.3, 4.4 and by (8.3) that $\mathbb{W}_{\xi,q+1,0}$ is periodized to scale $(\lambda_{q+1}r_{q+1,0})^{-1} = \lambda_{q,0}^{-1}$. Using the definition of $\mathbb{P}_{[q,n,p]}$ and (7.8), we have that[4]

$$
\mathbb{W}_{\xi,q+1,0} \otimes \mathbb{W}_{\xi,q+1,0}
$$
$$
= \fint_{\mathbb{T}^3} \mathbb{W}_{\xi,q+1,0} \otimes \mathbb{W}_{\xi,q+1,0}
$$
$$
+ \mathbb{P}_{\geq \lambda_{q,0}} \left(\sum_{n=1}^{n_{\max}} \sum_{p=1}^{p_{\max}} \mathbb{P}_{[q,n,p]} + \mathbb{P}_{[q,n_{\max},p_{\max}+1]} \right) (\mathbb{W}_{\xi,q+1,0} \otimes \mathbb{W}_{\xi,q+1,0}) \, .
$$

Combining this observation with identity (4.15) from Proposition 4.4, and with the definition of the $a_{(\xi)}$ in (8.2), we further split $\mathcal{O}_{0,1,1}$ as

$$
\mathrm{div}\,(\mathcal{O}_{0,1,1})
$$
$$
= \sum_{\xi,i,j,k,\widetilde{p},\vec{l}} \mathrm{div}\left(a_{(\xi)}^2 \nabla \Phi_{(i,k)}^{-1} \left(\fint_{\mathbb{T}^3} \mathbb{W}_{\xi,q+1,0} \otimes \mathbb{W}_{\xi,q+1,0}(\Phi_{(i,k)}) \right) \nabla \Phi_{(i,k)}^{-T} \right)
$$
$$
+ \sum_{\xi,i,j,k,\widetilde{p},\vec{l}} \mathrm{div}\left(a_{(\xi)}^2 \nabla \Phi_{(i,k)}^{-1} \mathbb{P}_{\geq \lambda_{q,0}} \left(\sum_{n=1}^{n_{\max}} \sum_{p=1}^{p_{\max}} \mathbb{P}_{[q,n,p]} + \mathbb{P}_{[q,n_{\max},p_{\max}+1]} \right) \right.
$$
$$
\left. \times (\mathbb{W} \otimes \mathbb{W})_{\xi,q+1,0}(\Phi_{(i,k)}) \nabla \Phi_{(i,k)}^{-T} \right)
$$
$$
= \mathrm{div}\left(\sum_{i,j,k,\widetilde{p},\vec{l}} \sum_{\xi} \delta_{q+1,0,\widetilde{p}} \Gamma_{q+1}^{2j+4} \eta_{(i,j,k)}^2 \gamma_{\xi}^2 \left(\frac{R_{q,0,\widetilde{p},j,i,k}}{\delta_{q+1,0,\widetilde{p}} \Gamma_{q+1}^{2j+4}} \right) \nabla \Phi_{(i,k)}^{-1} \, (\xi \otimes \xi) \, \nabla \Phi_{(i,k)}^{-T} \right)
$$
$$
+ \sum_{\xi,i,j,k,\widetilde{p},\vec{l}} \nabla a_{(\xi)}^2 \nabla \Phi_{(i,k)}^{-1} \mathbb{P}_{\geq \lambda_{q,0}} \left(\sum_{n=1}^{n_{\max}} \sum_{p=1}^{p_{\max}} \mathbb{P}_{[q,n,p]} + \mathbb{P}_{[q,n_{\max},p_{\max}+1]} \right)
$$
$$
\times (\mathbb{W} \otimes \mathbb{W})_{\xi,q+1,0}(\Phi_{(i,k)}) \nabla \Phi_{(i,k)}^{-T}
$$
$$
+ \sum_{\xi,i,j,k,\widetilde{p},\vec{l}} a_{(\xi)}^2 (\nabla \Phi_{(i,k)}^{-1})_{\alpha\theta} \mathbb{P}_{\geq \lambda_{q,0}} \left(\sum_{n=1}^{n_{\max}} \sum_{p=1}^{p_{\max}} \mathbb{P}_{[q,n,p]} + \mathbb{P}_{[q,n_{\max},p_{\max}+1]} \right)
$$
$$
\times (\mathbb{W}^\theta \mathbb{W}^\gamma)_{\xi,q+1,0}(\Phi_{(i,k)}) \partial_\alpha (\nabla \Phi_{(i,k)}^{-1})_{\zeta\gamma} \, . \tag{8.33}
$$

By Proposition 4.1, equation (4.1), and the definition (8.1), we may rewrite the

---

[4]The case $\widetilde{n} = 0$ is exceptional in the sense that the minimum frequency of $\mathbb{P}_{\geq \lambda_{q,0}}$ and the minimum frequency of $\mathbb{P}_{[q,1,0]}$ are in fact both equal to $\lambda_{q,0} = \lambda_{q,1,0} = \lambda_q^{\frac{4}{5}} \lambda_{q+1}^{\frac{1}{5}}$ from (9.27) and (9.22). For the sake of consistency with the $\widetilde{n} \geq 1$ cases, we will include the superfluous $\mathbb{P}_{\geq \lambda_{q,0}}$ in the calculations in this section.

first term on the right side of the above display as

$$\text{div} \sum_{i,j,k,\widetilde{p},\vec{l}} \sum_{\xi} \delta_{q+1,0,\widetilde{p}} \Gamma_{q+1}^{2j+4} \eta_{(i,j,k)}^2 \gamma_{\xi}^2 \left(\frac{R_{q,0,\widetilde{p},j,i,k}}{\delta_{q+1,0,\widetilde{p}} \Gamma_{q+1}^{2j+4}} \right) \nabla\Phi_{(i,k)}^{-1} (\xi \otimes \xi) \nabla\Phi_{(i,k)}^{-T}$$

$$= \text{div} \sum_{i,j,k,\vec{l}} \eta_{(i,j,k)}^2 \left(\delta_{q+1,0,1} \Gamma_{q+1}^{2j+4} \text{Id} - \mathring{R}_{\ell_q} \right)$$

$$= -\text{div} \sum_{i,j,k,\vec{l}} \eta_{(i,j,k)}^2 \mathring{R}_{\ell_q} + \nabla \left(\sum_{i,j,k,\vec{l}} \eta_{(i,j,k)}^2 \delta_{q+1,0,1} \Gamma_{q+1}^{2j+4} \right)$$

$$:= -\text{div} \left(\mathring{R}_{\ell_q} \right) + \nabla\pi. \tag{8.34}$$

In the last equality of the above display we have used the fact that by (6.142) we have

$$\mathring{R}_{\ell_q} = \sum_{i,j,k,\vec{l}} \eta_{(i,j,k)}^2 \mathring{R}_{\ell_q}. \tag{8.35}$$

We apply Proposition A.18 to the remaining two terms from (8.33) to define for $1 \leq n \leq n_{\max}$ and $1 \leq p \leq p_{\max}$ [5]

$$\mathring{H}_{q,n,p}^0 := \mathcal{H} \Bigg(\sum_{\xi,i,j,k,\widetilde{p},\vec{l}} \nabla a_{(\xi)}^2 \nabla\Phi_{(i,k)}^{-1} \mathbb{P}_{\geq \lambda_{q,0}} \mathbb{P}_{[q,n,p]}$$

$$\times (\mathbb{W}_{\xi,q+1,0} \otimes \mathbb{W}_{\xi,q+1,0})(\Phi_{(i,k)}) \nabla\Phi_{(i,k)}^{-T}$$

$$+ \sum_{\xi,i,j,k,\widetilde{p},\vec{l}} a_{(\xi)}^2 (\nabla\Phi_{(i,k)}^{-1})_{\alpha\theta} \mathbb{P}_{\geq \lambda_{q,0}} \mathbb{P}_{[q,n,p]}$$

$$\times (\mathbb{W}_{\xi,q+1,0}^\theta \mathbb{W}_{\xi,q+1,0}^\gamma)(\Phi_{(i,k)}) \partial_\alpha (\nabla\Phi_{(i,k)}^{-1})_{\zeta\gamma} \Bigg). \tag{8.36}$$

The last terms from (8.33) with $\mathbb{P}_{[q,n_{\max},p_{\max}+1]}$ will be absorbed into $\mathring{R}_{q+1}$, whereas the terms in (8.36) correspond to the error terms in (7.15).

Before amalgamating the preceding calculations, we pause to calculate the means of various terms to which the inverse divergence operator from Proposition A.18 will be applied. Examining the equality

$$\partial_t v_{q,0} + \text{div}\,(v_{q,0} \otimes v_{q,0}) + \nabla p_{\ell_q} = \mathcal{T}_0 + \mathcal{N}_0 + \mathcal{O}_0 + \text{div}\,\mathring{R}_{\ell_q} + \text{div}\,\mathring{R}_q^{\text{comm}} \tag{8.37}$$

and recalling the definitions of $\mathcal{T}_0$, $\mathcal{N}_0$, and $\mathcal{O}_0$, we see immediately that every

---

[5] Recall that $\mathcal{H}$ is the local portion of the inverse divergence operator. The pressure and the nonlocal portion will be accounted for shortly. We will check in Section 8.6 that these errors are of the form required by the inverse divergence operator as well as check the associated estimates.

term can be written as the divergence of a tensor except for $\partial_t v_{q,0}$ and $\mathcal{T}_0$. Note, however, that $v_{q,0} = v_{\ell_q} + w_{q+1,0}$, that $\int_{\mathbb{T}^3} \partial_t v_{\ell_q} = 0$ (by integrating in space (5.2)), and that $w_{q+1,0}$ is the curl of a vector field; cf. (8.13). This shows that $\int_{\mathbb{T}^3} \partial_t v_{q,0} = 0$, and thus $\int_{\mathbb{T}^3} \mathcal{T}_0 = 0$ as well. Therefore, we may use (A.72) and (A.78) to write

$$\mathcal{T}_0 = \operatorname{div}\left((\mathcal{H} + \mathcal{R}^*)\,\mathcal{T}_0\right) + \nabla P.$$

We can now combine the calculations of (8.28), (8.31), (8.32), (8.33), (8.34) (8.35), and (8.36) and let the notation $\nabla\pi$ change from line to line to incorporate all the pressure terms to write that

$$\partial_t v_{q,0} + \operatorname{div}\,(v_{q,0} \otimes v_{q,0}) + \nabla p_{\ell_q}$$

$$= \mathcal{T}_0 + \mathcal{N}_0 + \mathcal{O}_0 + \operatorname{div}\mathring{R}_{\ell_q} + \operatorname{div}\mathring{R}_q^{\mathrm{comm}}$$

$$= \mathcal{T}_0 + \mathcal{N}_0 + \operatorname{div}\,(\mathcal{O}_{0,1}) + \operatorname{div}\,(\mathcal{O}_{0,2}) + \operatorname{div}\mathring{R}_{\ell_q} + \operatorname{div}\mathring{R}_q^{\mathrm{comm}}$$

$$= \mathcal{T}_0 + \mathcal{N}_0 + \operatorname{div}\left(\mathring{R}_{\ell_q} + \mathcal{O}_{0,1,1}\right) + \operatorname{div}\,(\mathcal{O}_{0,1,2} + \mathcal{O}_{0,1,3} + \mathcal{O}_{0,2}) + \operatorname{div}\mathring{R}_q^{\mathrm{comm}}$$

$$= \mathcal{T}_0 + \mathcal{N}_0 - \nabla\pi$$

$$+ \operatorname{div}\,(\mathcal{H} + \mathcal{R}^*)\Bigg[\sum_{\xi,i,j,k,\widetilde{p},\vec{l}} \nabla a_{(\xi)}^2 \nabla\Phi_{(i,k)}^{-1}$$

$$\times \mathbb{P}_{\geq \lambda_{q,0}}\left(\sum_{n=1}^{n_{\max}}\sum_{p=1}^{p_{\max}} \mathbb{P}_{[q,n,p]} + \mathbb{P}_{[q,n_{\max},p_{\max}+1]}\right)$$

$$\times (\mathbb{W}_{\xi,q+1,0} \otimes \mathbb{W}_{\xi,q+1,0})(\Phi_{(i,k)})\nabla\Phi_{(i,k)}^{-T}$$

$$+ \sum_{\xi,i,j,k,\widetilde{p},\vec{l}} a_{(\xi)}^2 (\nabla\Phi_{(i,k)}^{-1})_{\alpha\theta}$$

$$\times \mathbb{P}_{\geq \lambda_{q,0}}\left(\sum_{n=1}^{n_{\max}}\sum_{p=1}^{p_{\max}} \mathbb{P}_{[q,n,p]} + \mathbb{P}_{[q,n_{\max},p_{\max}+1]}\right)$$

$$\times (\mathbb{W}^\theta \mathbb{W}^\gamma)_{\xi,q+1,0}(\Phi_{(i,k)})\partial_\alpha(\nabla\Phi_{(i,k)}^{-1})_{\zeta\gamma}\Bigg] \qquad (8.38)$$

$$+ \operatorname{div}\,(\mathcal{O}_{0,1,2} + \mathcal{O}_{0,1,3} + \mathcal{O}_{0,2}) + \operatorname{div}\mathring{R}_q^{\mathrm{comm}}\,.$$

After separating out the local $\mathcal{H}$ from the nonlocal $\mathcal{R}^*$ parts of the inverse divergence operator in the last two terms of the above, we may rewrite

$$\partial_t v_{q,0} + \operatorname{div}\,(v_{q,0} \otimes v_{q,0}) + \nabla p_{\ell_q} + \nabla\pi$$

$$= \operatorname{div}\Bigg[\underbrace{(\mathcal{H} + \mathcal{R}^*)\,(\mathcal{T}_0)}_{\text{transport}} + \underbrace{(\mathcal{H} + \mathcal{R}^*)\,(\mathcal{N}_0)}_{\text{Nash}} + \mathring{R}_q^{\mathrm{comm}} \qquad (8.39)$$

$$+ (\mathcal{H} + \mathcal{R}^*)\left(\sum_{\xi,i,j,k,\widetilde{p},\vec{l}} \nabla a_{(\xi)}^2 \nabla\Phi_{(i,k)}^{-1}\mathbb{P}_{[q,n_{\max},p_{\max}+1]}\right.$$

$$\times (\mathbb{W}_{\xi,q+1,0} \otimes \mathbb{W}_{\xi,q+1,0})(\Phi_{(i,k)}) \nabla \Phi_{(i,k)}^{-T} \quad (8.40)$$

$$+ \sum_{\xi,i,j,k,\widetilde{p},\vec{l}} a_{(\xi)}^2 (\nabla \Phi_{(i,k)}^{-1})_{\alpha\theta} \mathbb{P}_{[q,n_{\max},p_{\max}+1]}$$

$$\times (\mathbb{W}_{\xi,q+1,0}^\theta \mathbb{W}_{\xi,q+1,0}^\gamma)(\Phi_{(i,k)}) \partial_\alpha (\nabla \Phi_{(i,k)}^{-1})_{\varsigma\gamma} \Big) \quad (8.41)$$

$$+ \mathcal{R}^* \Bigg(\sum_{\xi,i,j,k,\widetilde{p},\vec{l}} \nabla a_{(\xi)}^2 \nabla \Phi_{(i,k)}^{-1} \mathbb{P}_{\geq \lambda_{q,0}} \left(\sum_{n=1}^{n_{\max}} \sum_{p=1}^{p_{\max}} \mathbb{P}_{[q,n,p]} \right)$$

$$\times (\mathbb{W}_{\xi,q+1,0} \otimes \mathbb{W}_{\xi,q+1,0})(\Phi_{(i,k)}) \nabla \Phi_{(i,k)}^{-T} \quad (8.42)$$

$$+ \sum_{\xi,i,j,k,\widetilde{p},\vec{l}} a_{(\xi)}^2 (\nabla \Phi_{(i,k)}^{-1})_{\alpha\theta} \mathbb{P}_{\geq \lambda_{q,0}} \left(\sum_{n=1}^{n_{\max}} \sum_{p=1}^{p_{\max}} \mathbb{P}_{[q,n,p]} \right)$$

$$\times (\mathbb{W}_{\xi,q+1,0}^\theta \mathbb{W}_{\xi,q+1,0}^\gamma)(\Phi_{(i,k)}) \partial_\alpha (\nabla \Phi_{(i,k)}^{-1})_{\varsigma\gamma} \Big) \quad (8.43)$$

$$+ \underbrace{\mathcal{O}_{0,1,2} + \mathcal{O}_{0,1,3}}_{\text{divergence corrector}} + \underbrace{\mathcal{O}_{0,2}}_{\text{Type 2}} \Bigg] \quad (8.44)$$

$$+ \text{div}\, \mathcal{H} \Bigg(\sum_{\xi,i,j,k,\widetilde{p},\vec{l}} \nabla a_{(\xi)}^2 \nabla \Phi_{(i,k)}^{-1} \mathbb{P}_{\geq \lambda_{q,0}} \left(\sum_{n=1}^{n_{\max}} \sum_{p=1}^{p_{\max}} \mathbb{P}_{[q,n,p]} \right)$$

$$\times (\mathbb{W}_{\xi,q+1,0} \otimes \mathbb{W}_{\xi,q+1,0})(\Phi_{(i,k)}) \nabla \Phi_{(i,k)}^{-T} \quad (8.45)$$

$$+ \sum_{\xi,i,j,k,\widetilde{p},\vec{l}} a_{(\xi)}^2 (\nabla \Phi_{(i,k)}^{-1})_{\alpha\theta} \mathbb{P}_{\geq \lambda_{q,0}} \left(\sum_{n=1}^{n_{\max}} \sum_{p=1}^{p_{\max}} \mathbb{P}_{[q,n,p]} \right)$$

$$\times (\mathbb{W}_{\xi,q+1,0}^\theta \mathbb{W}_{\xi,q+1,0}^\gamma)(\Phi_{(i,k)}) \partial_\alpha (\nabla \Phi_{(i,k)}^{-1})_{\varsigma\gamma} \Big) \quad (8.46)$$

$$:= \text{div}\, (\mathring{R}_{q+1}^0) + \text{div} \left(\sum_{n=1}^{n_{\max}} \sum_{p=1}^{p_{\max}} \mathring{H}_{q,n,p}^0 \right) + \text{div}\, \mathring{R}_q^{\text{comm}},$$

thus verifying (7.10) from Proposition 7.3, after condensing the terms from (8.40), (8.41), (8.42), and (8.43) into $\mathring{R}_{q+1}^0$, and using (8.36) to place the terms from (8.45) and (8.46) into $\mathring{H}_{q,n,p}^0$.

8.3.2   The case $1 \leq \widetilde{n} \leq n_{\max} - 1$

From (7.17), we assume that $v_{q,\widetilde{n}-1}$ is divergence-free and is a solution to

$$\partial_t v_{q,\widetilde{n}-1} + \text{div}\, (v_{q,\widetilde{n}-1} \otimes v_{q,\widetilde{n}-1}) + \nabla p_{q,\widetilde{n}-1}$$

$$= \text{div}\left(\mathring{R}_{q+1}^{\tilde{n}-1}\right) + \text{div}\left(\sum_{n=\tilde{n}}^{n_{\max}}\sum_{p=1}^{p_{\max}}\sum_{n'=0}^{\tilde{n}-1}\mathring{H}_{q,n,p}^{n'}\right) + \text{div}\,\mathring{R}_q^{\text{comm}}.$$

Now using the definition of $\mathring{R}_{q,\tilde{n},\tilde{p}}$ from (8.7) and adding $w_{q+1,\tilde{n}}$ as defined in (8.13), we have that $v_{q,\tilde{n}} := v_{q,\tilde{n}-1} + w_{q+1,\tilde{n}} = v_{\ell_q} + \sum_{0 \le n' \le \tilde{n}-1} w_{q+1,n'} + w_{q+1,\tilde{n}}$ solves

$$\partial_t v_{q,\tilde{n}} + \text{div}\,(v_{q,\tilde{n}} \otimes v_{q,\tilde{n}}) + \nabla p_{q,\tilde{n}-1}$$

$$= \text{div}\left(\mathring{R}_{q+1}^{\tilde{n}-1}\right) + \text{div}\left(\sum_{n=\tilde{n}+1}^{n_{\max}}\sum_{p=1}^{p_{\max}}\sum_{n'=0}^{\tilde{n}-1}\mathring{H}_{q,n,p}^{n'}\right) + \text{div}\,\mathring{R}_q^{\text{comm}}$$

$$+ (\partial_t + v_{\ell_q} \cdot \nabla)w_{q+1,\tilde{n}} + w_{q+1,\tilde{n}} \cdot \nabla v_{\ell_q}$$

$$+ \sum_{n' \le \tilde{n}-1}\text{div}\,(w_{q+1,\tilde{n}} \otimes w_{q+1,n'} + w_{q+1,n'} \otimes w_{q+1,\tilde{n}})$$

$$+ \text{div}\left(w_{q+1,\tilde{n}} \otimes w_{q+1,\tilde{n}} + \sum_{\tilde{p}=1}^{p_{\max}}\mathring{R}_{q,\tilde{n},\tilde{p}}\right). \tag{8.47}$$

The first term on the right-hand side is $\mathring{R}_{q+1}^{\tilde{n}-1}$, which satisfies the same estimates as $\mathring{R}_{q+1}^{\tilde{n}}$ by (7.20) and will thus be absorbed into $\mathring{R}_{q+1}^{\tilde{n}}$ (these estimates do not change in $\tilde{n}$ save for implicit constants). The second term, save for the fact that the sum is over $n' \le \tilde{n}-1$ rather than $n' \le \tilde{n}$ and is therefore missing the terms $\mathring{H}_{q,n,p}^{\tilde{n}}$, matches (7.17) at level $\tilde{n}$ (i.e., replacing every instance of $\tilde{n}-1$ with $\tilde{n}$). As before, we apply the inverse divergence operators from Proposition A.18 to the transport and Nash errors to obtain

$$(\partial_t + v_{\ell_q} \cdot \nabla)w_{q+1,\tilde{n}} + w_{q+1,\tilde{n}} \cdot \nabla v_{\ell_q} + \nabla\pi$$
$$= \text{div}\left((\mathcal{H} + \mathcal{R}^*)\left((\partial_t + v_{\ell_q} \cdot \nabla)w_{q+1,\tilde{n}} + w_{q+1,\tilde{n}} \cdot \nabla v_{\ell_q}\right)\right),$$

and these errors are absorbed into $\mathring{R}_{q+1}^{\tilde{n}}$ or the new pressure. We will show in Section 8.7 that the interaction of $w_{q+1,\tilde{n}}$ with previous terms $w_{q+1,n'}$ is a Type 2 oscillation error so that

$$\sum_{n' \le \tilde{n}-1}(w_{q+1,\tilde{n}} \otimes w_{q+1,n'} + w_{q+1,n'} \otimes w_{q+1,\tilde{n}}) = 0. \tag{8.48}$$

So to verify (7.17) at level $\tilde{n}$, only the analysis of

$$\text{div}\left(w_{q+1,\tilde{n}} \otimes w_{q+1,\tilde{n}} + \sum_{\tilde{p}=1}^{p_{\max}}\mathring{R}_{q,\tilde{n},\tilde{p}}\right)$$

remains. Reusing the notations from $(8.29)^6$ and writing out the self-interaction of $w_{q+1,\widetilde{n}}$ yields

$$\text{div} \left(w_{q+1,\widetilde{n}} \otimes w_{q+1,\widetilde{n}} \right)$$

$$= \sum_{\xi,i,j,k,\widetilde{p},\vec{l}} \text{div} \left(\text{curl} \left(a_{(\xi)} \nabla \Phi_{(i,k)}^T \mathbb{U}_{\xi,q+1,\widetilde{n}} \right) \otimes \text{curl} \left(a_{(\xi)} \nabla \Phi_{i,k}^T \mathbb{U}_{\xi,q+1,\widetilde{n}} \right) \right)$$

$$+ \sum_{\neq \{\xi,i,j,k,\widetilde{p},\vec{l}\}} \text{div} \left(\text{curl} \left(a_{(\xi)} \nabla \Phi_{(i,k)}^T \mathbb{U}_{\xi,q+1,\widetilde{n}} \right) \otimes \text{curl} \left(a_{(\xi')} \nabla \Phi_{(i',k')}^T \mathbb{U}_{\xi',q+1,\widetilde{n}} \right) \right)$$

$$:= \text{div}\, \mathcal{O}_{\widetilde{n},1} + \text{div}\, \mathcal{O}_{\widetilde{n},2}. \tag{8.49}$$

As before, we will show that $\mathcal{O}_{\widetilde{n},2}$ is a Type 2 oscillation error so that

$$\mathcal{O}_{\widetilde{n},2} = 0.$$

Splitting $\mathcal{O}_{\widetilde{n},1}$ gives

$$\text{div}\, \mathcal{O}_{\widetilde{n},1} = \sum_{\xi,i,j,k,\widetilde{p},\vec{l}} \text{div} \left(\left(a_{(\xi)} \nabla \Phi_{(i,k)}^{-1} \mathbb{W}_{\xi,q+1,\widetilde{n}} \circ \Phi_{(i,k)} \right) \right.$$

$$\left. \otimes \left(a_{(\xi)} \nabla \Phi_{(i,k)}^{-1} \mathbb{W}_{\xi,q+1,\widetilde{n}} \circ \Phi_{(i,k)} \right) \right)$$

$$+ 2 \sum_{\xi,i,j,k,\widetilde{p},\vec{l}} \text{div} \left(\left(a_{(\xi)} \nabla \Phi_{(i,k)}^{-1} \mathbb{W}_{\xi,q+1,\widetilde{n}} \circ \Phi_{(i,k)} \right) \right.$$

$$\left. \otimes_s \left(\nabla a_{(\xi)} \times \left(\nabla \Phi_{(i,k)}^T \mathbb{U}_{\xi,q+1,\widetilde{n}} \circ \Phi_{(i,k)} \right) \right) \right)$$

$$+ \sum_{\xi,i,j,k,\widetilde{p},\vec{l}} \text{div} \left(\left(\nabla a_{(\xi)} \times \left(\nabla \Phi_{(i,k)}^T \mathbb{U}_{\xi,q+1,\widetilde{n}} \circ \Phi_{(i,k)} \right) \right) \right.$$

$$\left. \otimes \left(\nabla a_{(\xi)} \times \left(\nabla \Phi_{(i,k)}^T \mathbb{U}_{\xi,q+1,\widetilde{n}} \circ \Phi_{(i,k)} \right) \right) \right)$$

$$:= \text{div} \left(\mathcal{O}_{\widetilde{n},1,1} + \mathcal{O}_{\widetilde{n},1,2} + \mathcal{O}_{\widetilde{n},1,3} \right). \tag{8.50}$$

The last two of these terms are again divergence corrector errors and will therefore be absorbed into $\mathring{R}_{q+1}^{\widetilde{n}}$ and estimated in Section 8.8. So the only terms remaining from (8.47) are $\mathcal{O}_{\widetilde{n},1,1}$ and $\sum_{\widetilde{p}=1}^{p_{\max}} \mathring{R}_{q,\widetilde{n},\widetilde{p}}$, which are analyzed in a fashion similar to the $\widetilde{n} = 0$ case, save for the fact that summation over $\widetilde{p}$ is now crucial.

Recall—cf. (8.10)—that $\mathbb{W}_{\xi,q+1,\widetilde{n}}$ is periodized to scale $(\lambda_{q+1} r_{q+1,\widetilde{n}})^{-1} =$

---

^{6}In a slight abuse of notation, notice that the admissible values of $\vec{l}$ have changed, since these parameters describe the checkerboard cutoff functions at scale $\lambda_{q,\widetilde{n},1}^{-1}$ and thus depend on $\widetilde{n}$.

$\lambda_{q,\tilde{n}}^{-1}$. Using (7.8), we have that

$$
\mathbb{W}_{\xi,q+1,\tilde{n}} \otimes \mathbb{W}_{\xi,q+1,\tilde{n}}
$$

$$
= \fint_{\mathbb{T}^3} \mathbb{W}_{\xi,q+1,\tilde{n}} \otimes \mathbb{W}_{\xi,q+1,\tilde{n}}
$$

$$
+ \mathbb{P}_{\geq \lambda_{q,\tilde{n}}} \left(\sum_{n=\tilde{n}+1}^{n_{\max}} \sum_{p=1}^{p_{\max}} \mathbb{P}_{[q,n,p]} + \mathbb{P}_{[q,n_{\max},p_{\max}+1]} \right) \left(\mathbb{W}_{\xi,q+1,\tilde{n}} \otimes \mathbb{W}_{\xi,q+1,\tilde{n}} \right).
$$

Combining this division with identity (4.15) from Proposition 4.4, we further split $\mathcal{O}_{\tilde{n},1,1}$ as

$$
\mathrm{div}\,(\mathcal{O}_{\tilde{n},1,1})
$$

$$
= \sum_{\xi,i,j,k,\tilde{p},\vec{l}} \mathrm{div}\left[a_{(\xi)}^2 \nabla\Phi_{(i,k)}^{-1} \left(\fint_{\mathbb{T}^3} \mathbb{W}_{\xi,q+1,\tilde{n}} \otimes \mathbb{W}_{\xi,q+1,\tilde{n}}(\Phi_{(i,k)}) \right) \nabla\Phi_{(i,k)}^{-T} \right]
$$

$$
+ \sum_{\xi,i,j,k,\tilde{p},\vec{l}} \mathrm{div}\left[a_{(\xi)}^2 \nabla\Phi_{(i,k)}^{-1} \mathbb{P}_{\geq \lambda_{q,\tilde{n}}} \left(\sum_{n=\tilde{n}+1}^{n_{\max}} \sum_{p=1}^{p_{\max}} \mathbb{P}_{[q,n,p]} + \mathbb{P}_{[q,n_{\max},p_{\max}+1]} \right) \right.
$$

$$
\left. \times (\mathbb{W} \otimes \mathbb{W})_{\xi,q+1,\tilde{n}}(\Phi_{(i,k)}) \nabla\Phi_{(i,k)}^{-T} \right]
$$

$$
= \mathrm{div}\left[\sum_{\xi,i,j,k,\tilde{p},\vec{l}} \delta_{q+1,\tilde{n},\tilde{p}} \Gamma_{q+1}^{2j+4} \eta_{(i,j,k)}^2 \gamma_{\xi}^2 \left(\frac{R_{q,\tilde{n},\tilde{p},j,i,k}}{\delta_{q+1,\tilde{n},\tilde{p}}\Gamma_{q+1}^{2j+4}} \right) \nabla\Phi_{(i,k)}^{-1} \left(\xi \otimes \xi \right) \nabla\Phi_{(i,k)}^{-T} \right]
$$

$$
+ \sum_{\xi,i,j,k,\tilde{p},\vec{l}} \nabla a_{(\xi)}^2 \nabla\Phi_{(i,k)}^{-1} \mathbb{P}_{\geq \lambda_{q,\tilde{n}}} \left(\sum_{n=\tilde{n}+1}^{n_{\max}} \sum_{p=1}^{p_{\max}} \mathbb{P}_{[q,n,p]} + \mathbb{P}_{[q,n_{\max},p_{\max}+1]} \right)
$$

$$
\times (\mathbb{W} \otimes \mathbb{W})_{\xi,q+1,\tilde{n}}(\Phi_{(i,k)}) \nabla\Phi_{(i,k)}^{-T}
$$

$$
+ \sum_{\xi,i,j,k,\tilde{p},\vec{l}} a_{(\xi)}^2 (\nabla\Phi_{(i,k)}^{-1})_{\alpha\theta} \mathbb{P}_{\geq \lambda_{q,\tilde{n}}} \left(\sum_{n=\tilde{n}+1}^{n_{\max}} \sum_{p=1}^{p_{\max}} \mathbb{P}_{[q,n,p]} + \mathbb{P}_{[q,n_{\max},p_{\max}+1]} \right)
$$

$$
\times (\mathbb{W}^\theta \mathbb{W}^\gamma)_{\xi,q+1,\tilde{n}}(\Phi_{(i,k)}) \partial_\alpha (\nabla\Phi_{(i,k)}^{-1})_{\zeta\gamma}. \tag{8.51}
$$

By Proposition 4.1, equation (4.1), and identity (8.8), we obtain that

$$
\mathrm{div} \sum_{i,j,k,\tilde{p},\vec{l}} \sum_{\xi} \delta_{q+1,\tilde{n},\tilde{p}} \Gamma_{q+1}^{2j+4} \eta_{(i,j,k)}^2 \gamma_\xi^2 \left(\frac{R_{q,\tilde{n},\tilde{p},j,i,k}}{\delta_{q+1,\tilde{n},\tilde{p}}\Gamma_{q+1}^{2j+4}} \right) \nabla\Phi_{(i,k)}^{-1} \left(\xi \otimes \xi \right) \nabla\Phi_{(i,k)}^{-T}
$$

$$
= \mathrm{div} \sum_{i,j,k,\tilde{p},\vec{l}} \eta_{(i,j,k)}^2 \left(\delta_{q+1,\tilde{n},\tilde{p}} \Gamma_{q+1}^{2j+4} \mathrm{Id} - \sum_{\tilde{p}=1}^{p_{\max}} \mathring{R}_{q,\tilde{n},\tilde{p}} \right)
$$

$$= -\text{div} \sum_{i,j,k,\vec{l}} \sum_{\widetilde{p}=1}^{p_{\max}} \eta^2_{(i,j,k)} \mathring{R}_{q,\widetilde{n},\widetilde{p}} + \nabla \left(\sum_{i,j,k,\vec{l}} \eta^2_{(i,j,k)} \delta_{q+1,\widetilde{n},\widetilde{p}} \Gamma^{2j+4}_{q+1} \right)$$

$$:= -\text{div} \sum_{\widetilde{p}=1}^{p_{\max}} \mathring{R}_{q,\widetilde{n},\widetilde{p}} + \nabla \pi \, , \tag{8.52}$$

where in the last equality we have appealed to (6.142). We can finally apply Proposition A.18 to the remaining terms in (8.51) for $\widetilde{n} + 1 \leq n \leq n_{\max}$ and $1 \leq p \leq p_{\max}$, to define

$$\mathring{H}^{\widetilde{n}}_{q,n,p} := \mathcal{H} \Bigg[\sum_{\xi,i,j,k,\widetilde{p}} \nabla a^2_{(\xi)} \nabla \Phi^{-1}_{(i,k)} \mathbb{P}_{\geq \lambda_{q,\widetilde{n}}} \mathbb{P}_{[q,n,p]}$$

$$\times (\mathbb{W}_{\xi,q+1,\widetilde{n}} \otimes \mathbb{W}_{\xi,q+1,\widetilde{n}})(\Phi_{(i,k)}) \nabla \Phi^{-T}_{(i,k)}$$

$$+ \sum_{\xi,i,j,k,\widetilde{p}} a^2_{(\xi)} (\nabla \Phi^{-1}_{(i,k)})_{\alpha\theta} \mathbb{P}_{\geq \lambda_{q,\widetilde{n}}} \mathbb{P}_{[q,n,p]}$$

$$\times (\mathbb{W}^\theta_{\xi,q+1,\widetilde{n}} \mathbb{W}^\gamma_{\xi,q+1,\widetilde{n}})(\Phi_{(i,k)}) \partial_\alpha (\nabla \Phi^{-1}_{(i,k)})_{\zeta\gamma} \Bigg]. \tag{8.53}$$

As before, the terms from (8.51) with $\mathbb{P}_{[q,n_{\max},p_{\max}+1]}$ will be absorbed into $\mathring{R}^{\widetilde{n}}_{q+1}$. We will show shortly that the terms $\mathring{H}^{\widetilde{n}}_{q,n,p}$ in (8.53) are precisely the terms needed to make (8.47) match (7.17) at level $\widetilde{n}$. As before, any nonlocal inverse divergence terms will be absorbed into $\mathring{R}^{\widetilde{n}}_{q+1}$.

Recall from (7.9) that $\mathring{R}^{\widetilde{n}}_{q+1}$ will include $\mathring{R}^{\widetilde{n}-1}_{q+1}$ in addition to error terms arising from the addition of $w_{q+1,\widetilde{n}}$ which are small enough to be absorbed in $\mathring{R}_{q+1}$. Then to check (7.17), we return to (8.47) and use (8.49), (8.50), (8.51), (8.52), and (8.53) to write

$$\partial_t v_{q,\widetilde{n}} + \text{div} \, (v_{q,\widetilde{n}} \otimes v_{q,\widetilde{n}}) + \nabla p_{q,\widetilde{n}-1}$$

$$= \text{div} \left(\sum_{n=\widetilde{n}+1}^{n_{\max}} \sum_{p=1}^{p_{\max}} \sum_{n'=0}^{\widetilde{n}-1} \mathring{H}^{n'}_{q,n,p} \right) + \text{div} \left(\mathring{R}^{\widetilde{n}-1}_{q+1} \right) + \text{div} \, \mathring{R}^{\text{comm}}_q$$

$$+ (\partial_t + v_{\ell_q} \cdot \nabla) w_{q+1,\widetilde{n}} + w_{q+1,\widetilde{n}} \cdot \nabla v_{\ell_q}$$

$$+ \sum_{n' \leq \widetilde{n}-1} \text{div} \, (w_{q+1,\widetilde{n}} \otimes w_{q+1,n'} + w_{q+1,n'} \otimes w_{q+1,\widetilde{n}})$$

$$+ \text{div} \left(w_{q+1,\widetilde{n}} \otimes w_{q+1,\widetilde{n}} + \sum_{\widetilde{p}=1}^{p_{\max}} \mathring{R}_{q,\widetilde{n},\widetilde{p}} \right)$$

$$= \text{div} \, \mathring{R}^{\text{comm}}_q + \text{div} \left(\sum_{n=\widetilde{n}+1}^{n_{\max}} \sum_{p=1}^{p_{\max}} \sum_{n'=0}^{\widetilde{n}-1} \mathring{H}^{n'}_{q,n,p} \right)$$

$$+ \operatorname{div} \left(\mathring{R}_{q+1}^{\tilde{n}-1} + (\mathcal{H} + \mathcal{R}^*) \left(\partial_t w_{q+1,\tilde{n}} + v_{\ell_q} \cdot \nabla w_{q+1,\tilde{n}} \right) \right.$$

$$+ (\mathcal{H} + \mathcal{R}^*) \left(w_{q+1,\tilde{n}} \cdot \nabla v_{\ell_q} \right)$$

$$\left. + \sum_{n' \le \tilde{n}-1} \left(w_{q+1,\tilde{n}} \otimes w_{q+1,n'} + w_{q+1,n'} \otimes w_{q+1,\tilde{n}} \right) \right)$$

$$+ \operatorname{div} \left(\mathcal{O}_{\tilde{n},1,2} + \mathcal{O}_{\tilde{n},1,3} + \mathcal{O}_{\tilde{n},2} \right) + \nabla \pi + \operatorname{div} \left(\mathcal{O}_{\tilde{n},1,1} + \sum_{\tilde{p}=1}^{p_{\max}} \mathring{R}_{q,\tilde{n},\tilde{p}} \right)$$

$$= \operatorname{div} \mathring{R}_q^{\mathrm{comm}} + \operatorname{div} \left(\sum_{n=\tilde{n}+1}^{n_{\max}} \sum_{p=1}^{p_{\max}} \sum_{n'=0}^{\tilde{n}-1} \mathring{H}_{q,n,p}^{n'} \right)$$

$$+ \operatorname{div} \left(\mathring{R}_{q+1}^{\tilde{n}-1} + (\mathcal{H} + \mathcal{R}^*) \left(\partial_t w_{q+1,\tilde{n}} + v_{\ell_q} \cdot \nabla w_{q+1,\tilde{n}} \right) \right.$$

$$+ (\mathcal{H} + \mathcal{R}^*) \left(w_{q+1,\tilde{n}} \cdot \nabla v_{\ell_q} \right)$$

$$\left. + \sum_{n' \le \tilde{n}-1} \left(w_{q+1,\tilde{n}} \otimes w_{q+1,n'} + w_{q+1,n'} \otimes w_{q+1,\tilde{n}} \right) \right)$$

$$+ \operatorname{div} \left(\mathcal{O}_{\tilde{n},1,2} + \mathcal{O}_{\tilde{n},1,3} + \mathcal{O}_{\tilde{n},2} \right) + \nabla \pi$$

$$+ \operatorname{div} \left(\mathcal{H} + \mathcal{R}^* \right) \left(\sum_{\xi,i,j,k,\tilde{p},\vec{l}} \nabla a_{(\xi)}^2 \nabla \Phi_{(i,k)}^{-1} \right.$$

$$\times \mathbb{P}_{\ge \lambda_{q,\tilde{n}}} \left(\sum_{n=\tilde{n}+1}^{n_{\max}} \sum_{p=1}^{p_{\max}} \mathbb{P}_{[q,n,p]} + \mathbb{P}_{[q,n_{\max},p_{\max}+1]} \right)$$

$$\times (\mathbb{W} \otimes \mathbb{W})_{\xi,q+1,\tilde{n}} (\Phi_{(i,k)}) \nabla \Phi_{(i,k)}^{-T}$$

$$+ \sum_{\xi,i,j,k,\tilde{p},\vec{l}} a_{(\xi)}^2 (\nabla \Phi_{(i,k)}^{-1})_{\alpha\theta}$$

$$\times \mathbb{P}_{\ge \lambda_{q,\tilde{n}}} \left(\sum_{n=\tilde{n}+1}^{n_{\max}} \sum_{p=1}^{p_{\max}} \mathbb{P}_{[q,n,p]} + \mathbb{P}_{[q,n_{\max},p_{\max}+1]} \right)$$

$$\left. \times (\mathbb{W}^\theta \mathbb{W}^\gamma)_{\xi,q+1,\tilde{n}} (\Phi_{(i,k)}) \partial_\alpha (\nabla \Phi_{(i,k)}^{-1})_{\zeta\gamma} \right). \tag{8.54}$$

In order to check which contributions go into $\mathring{R}_{q+1}^{\tilde{n}}$ and which go into $\mathring{H}_{q,n,p}^{\tilde{n}}$, we further decompose the above as

$$\partial_t v_{q,\tilde{n}} + \operatorname{div} \left(v_{q,\tilde{n}} \otimes v_{q,\tilde{n}} \right) + \nabla \pi$$

$$= \operatorname{div} \mathring{R}_q^{\mathrm{comm}} + \operatorname{div} \left(\sum_{n=\tilde{n}+1}^{n_{\max}} \sum_{p=1}^{p_{\max}} \sum_{n'=0}^{\tilde{n}-1} \mathring{H}_{q,n,p}^{n'} \right)$$

$$+ \operatorname{div}\left(\mathring{R}_{q+1}^{\widetilde{n}-1} + \underbrace{(\mathcal{H} + \mathcal{R}^*)\left(\partial_t w_{q+1,\widetilde{n}} + v_{\ell_q} \cdot \nabla w_{q+1,\widetilde{n}}\right)}_{\text{transport}}\right. \tag{8.55}$$

$$+ \underbrace{(\mathcal{H} + \mathcal{R}^*)\left(w_{q+1,\widetilde{n}} \cdot \nabla v_{\ell_q}\right)}_{\text{Nash}} \tag{8.56}$$

$$\left.+ \underbrace{\sum_{n' \leq \widetilde{n}-1}\left(w_{q+1,\widetilde{n}} \otimes w_{q+1,n'} + w_{q+1,n'} \otimes w_{q+1,\widetilde{n}}\right)}_{\text{Type 2}}\right) \tag{8.57}$$

$$+ \operatorname{div}\left(\underbrace{\mathcal{O}_{\widetilde{n},1,2} + \mathcal{O}_{\widetilde{n},1,3}}_{\text{divergence corrector}} + \underbrace{\mathcal{O}_{\widetilde{n},2}}_{\text{Type 2}}\right) \tag{8.58}$$

$$+ \operatorname{div}\left[(\mathcal{H} + \mathcal{R}^*)\left(\sum_{\xi,i,j,k,\widetilde{p},\vec{l}} \nabla a_{(\xi)}^2 \nabla \Phi_{(i,k)}^{-1} \mathbb{P}_{[q,n_{\max},p_{\max}+1]}\right.\right.$$

$$\times (\mathbb{W}_{\xi,q+1,\widetilde{n}} \otimes \mathbb{W}_{\xi,q+1,\widetilde{n}})(\Phi_{(i,k)})\nabla \Phi_{(i,k)}^{-T} \tag{8.59}$$

$$+ \sum_{\xi,i,j,k,\widetilde{p},\vec{l}} a_{(\xi)}^2 (\nabla \Phi_{(i,k)}^{-1})_{\alpha\theta} \mathbb{P}_{[q,n_{\max},p_{\max}+1]}$$

$$\left.\times (\mathbb{W}_{\xi,q+1,\widetilde{n}}^\theta \mathbb{W}_{\xi,q+1,\widetilde{n}}^\gamma)(\Phi_{(i,k)})\partial_\alpha (\nabla \Phi_{(i,k)}^{-1})_{\zeta\gamma}\right) \tag{8.60}$$

$$+ \mathcal{R}^*\left(\sum_{\xi,i,j,k,\widetilde{p},\vec{l}} \nabla a_{(\xi)}^2 \nabla \Phi_{(i,k)}^{-1} \mathbb{P}_{\geq \lambda_{q,\widetilde{n}}}\left(\sum_{n=\widetilde{n}+1}^{n_{\max}}\sum_{p=1}^{p_{\max}} \mathbb{P}_{[q,n,p]} + \mathbb{P}_{[q,n_{\max},p_{\max}+1]}\right)\right.$$

$$\times (\mathbb{W} \otimes \mathbb{W})_{\xi,q+1,\widetilde{n}}(\Phi_{(i,k)})\nabla \Phi_{(i,k)}^{-T} \tag{8.61}$$

$$+ \sum_{\xi,i,j,k,\widetilde{p},\vec{l}} a_{(\xi)}^2 (\nabla \Phi_{(i,k)}^{-1})_{\alpha\theta} \mathbb{P}_{\geq \lambda_{q,\widetilde{n}}}\left(\sum_{n=\widetilde{n}+1}^{n_{\max}}\sum_{p=1}^{p_{\max}} \mathbb{P}_{[q,n,p]} + \mathbb{P}_{[q,n_{\max},p_{\max}+1]}\right)$$

$$\left.\left.\times (\mathbb{W}^\theta \mathbb{W}^\gamma)_{\xi,q+1,\widetilde{n}}(\Phi_{(i,k)})\partial_\alpha (\nabla \Phi_{(i,k)}^{-1})_{\zeta\gamma}\right)\right] \tag{8.62}$$

$$+ \operatorname{div}\left[\mathcal{H}\left(\sum_{\xi,i,j,k,\widetilde{p},\vec{l}} \nabla a_{(\xi)}^2 \nabla \Phi_{(i,k)}^{-1} \mathbb{P}_{\geq \lambda_{q,\widetilde{n}}}\left(\sum_{n=\widetilde{n}+1}^{n_{\max}}\sum_{p=1}^{p_{\max}} \mathbb{P}_{[q,n,p]}\right)\right.\right.$$

$$\times (\mathbb{W} \otimes \mathbb{W})_{\xi,q+1,\widetilde{n}}(\Phi_{(i,k)})\nabla \Phi_{(i,k)}^{-T} \tag{8.63}$$

$$+ \sum_{\xi,i,j,k,\widetilde{p},\vec{l}} a_{(\xi)}^2 (\nabla \Phi_{(i,k)}^{-1})_{\alpha\theta} \mathbb{P}_{\geq \lambda_{q,\widetilde{n}}}\left(\sum_{n=\widetilde{n}+1}^{n_{\max}}\sum_{p=1}^{p_{\max}} \mathbb{P}_{[q,n,p]}\right)$$

$$\left.\left.\times (\mathbb{W}^\theta \mathbb{W}^\gamma)_{\xi,q+1,\widetilde{n}}(\Phi_{(i,k)})\partial_\alpha (\nabla \Phi_{(i,k)}^{-1})_{\zeta\gamma}\right)\right] \tag{8.64}$$

$$= \operatorname{div} \mathring{R}_q^{\text{comm}} + \operatorname{div} \mathring{R}_{q+1}^{\widetilde{n}} + \operatorname{div} \sum_{n=\widetilde{n}+1}^{n_{\max}}\sum_{p=1}^{p_{\max}}\sum_{n'=0}^{\widetilde{n}} \mathring{H}_{q,n,p}^{n'},$$

so that the terms in (8.55), (8.56), (8.57), (8.58), (8.59), (8.60), (8.61), and (8.62) are placed into $\mathring{R}^{\tilde{n}}_{q+1}$, while the terms in (8.63) and (8.64), and the triple sum of $\mathring{H}^{n'}_{q,n,p}$ terms in the first line, are incorporated into the new triple sum of $\mathring{H}^{n'}_{q,n,p}$ terms. Note that we have implicitly used in the above equalities that $\left(\partial_t + v_{\ell_q} \cdot \nabla\right) w_{q+1,\tilde{n}}$ has zero mean, which can be deduced in the same fashion as for the case $\tilde{n} = 0$.

8.3.3 The case $\tilde{n} = n_{\max}$

From (7.24), we assume that $v_{q,n_{\max}-1}$ is divergence-free and is a solution to

$$\partial_t v_{q,n_{\max}-1} + \text{div}\left(v_{q,n_{\max}-1} \otimes v_{q,n_{\max}-1}\right) + \nabla p_{q,n_{\max}-1}$$
$$= \text{div}\left(\mathring{R}^{n_{\max}-1}_{q+1}\right) + \text{div}\left(\sum_{n'=0}^{n_{\max}-1} \sum_{p=1}^{p_{\max}} \mathring{H}^{n'}_{q,n_{\max},p}\right) + \text{div}\,\mathring{R}^{\text{comm}}_q.$$

Now using the definition of $\mathring{R}_{q,n_{\max},p}$ from (8.7) and adding $w_{q+1,n_{\max}}$ as defined in (8.13), we have that $v_{q+1} := v_{q,n_{\max}-1} + w_{q+1,n_{\max}}$ solves

$$\partial_t v_{q+1} + \text{div}\left(v_{q+1} \otimes v_{q+1}\right) + \nabla p_{q,n_{\max}-1}$$
$$= \text{div}\,\mathring{R}^{\text{comm}}_q + \text{div}\left(\mathring{R}^{n_{\max}-1}_{q+1}\right) + (\partial_t + v_{\ell_q} \cdot \nabla)w_{q+1,n_{\max}} + w_{q+1,n_{\max}} \cdot \nabla v_{\ell_q}$$
$$+ \sum_{n' \leq n_{\max}-1} \text{div}\left(w_{q+1,n_{\max}} \otimes w_{q+1,n'} + w_{q+1,n'} \otimes w_{q+1,n_{\max}}\right)$$
$$+ \text{div}\left(w_{q+1,n_{\max}} \otimes w_{q+1,n_{\max}} + \sum_{p=1}^{p_{\max}} \mathring{R}_{q,n_{\max},p}\right). \tag{8.65}$$

We absorb the term $\text{div}\left(\mathring{R}^{n_{\max}-1}_{q+1}\right)$ into $\mathring{R}_{q+1}$ immediately. We will then show that, up to a pressure term,

$$(\mathcal{H} + \mathcal{R}^*)\left((\partial_t + v_{\ell_q} \cdot \nabla)\, w_{q+1,n_{\max}}\right), \qquad (\mathcal{H} + \mathcal{R}^*)\left(w_{q+1,n_{\max}} \cdot \nabla v_{\ell_q}\right)$$

can be absorbed into $\mathring{R}_{q+1}$ in Sections 8.4 and 8.5, respectively. We will show in 8.7 that the interaction of $w_{q+1,n_{\max}}$ with previous perturbations $w_{q+1,n'}$ will satisfy

$$\sum_{n' \leq n_{\max}-1}\left(w_{q+1,n_{\max}} \otimes w_{q+1,n'} + w_{q+1,n'} \otimes w_{q+1,n_{\max}}\right) = 0. \tag{8.66}$$

Thus it remains to analyze

$$\text{div}\left(w_{q+1,n_{\max}} \otimes w_{q+1,n_{\max}} + \sum_{p=1}^{p_{\max}} \mathring{R}_{q,n_{\max}}\right)$$

from (8.65). Reusing the notations from (8.29)–(8.30), we can write out the self-interaction of $w_{q+1,n_{\max}}$ as

$$
\begin{aligned}
\operatorname{div}&\left(w_{q+1,n_{\max}} \otimes w_{q+1,n_{\max}}\right) \\
= &\sum_{\xi,i,j,k,p,\vec{l}} \operatorname{div}\left(\operatorname{curl}\left(a_{(\xi)}\nabla\Phi_{(i,k)}^T \mathbb{U}_{\xi,q+1,n_{\max}}\right) \otimes \operatorname{curl}\left(a_{(\xi)}\nabla\Phi_{i,k}^T \mathbb{U}_{\xi,q+1,n_{\max}}\right)\right) \\
&+ \sum_{\neq\{\xi,i,j,k,p,\vec{l}\}} \operatorname{div}\left(\operatorname{curl}\left(a_{(\xi)}\nabla\Phi_{(i,k)}^T \mathbb{U}_{\xi,q+1,n_{\max}}\right) \otimes \operatorname{curl}\left(a_{(\xi')}\nabla\Phi_{(i',k')}^T \mathbb{U}_{\xi',q+1,n_{\max}}\right)\right) \\
:= &\operatorname{div}\mathcal{O}_{n_{\max},1} + \operatorname{div}\mathcal{O}_{n_{\max},2}.
\end{aligned}
\tag{8.67}
$$

As before, we will show in Section 8.7 that $\mathcal{O}_{n_{\max},2}$ is a Type 2 oscillation error and so

$$
\mathcal{O}_{n_{\max},2} = 0.
$$

Splitting $\mathcal{O}_{n_{\max},1}$ gives

$$
\begin{aligned}
\operatorname{div}\mathcal{O}_{n_{\max},1} = &\sum_{\xi,i,j,k,p,\vec{l}} \operatorname{div}\left(\left(a_{(\xi)}\nabla\Phi_{(i,k)}^{-1} \mathbb{W}_{\xi,q+1,n_{\max}} \circ \Phi_{(i,k)}\right) \right. \\
&\qquad\qquad\qquad \left. \otimes \left(a_{(\xi)}\nabla\Phi_{(i,k)}^{-1} \mathbb{W}_{\xi,q+1,n_{\max}} \circ \Phi_{(i,k)}\right)\right) \\
&+ 2 \sum_{\xi,i,j,k,p,\vec{l}} \operatorname{div}\left(\left(a_{(\xi)}\nabla\Phi_{(i,k)}^{-1} \mathbb{W}_{\xi,q+1,n_{\max}} \circ \Phi_{(i,k)}\right)\right. \\
&\qquad\qquad\qquad \left. \otimes_s \left(\nabla a_{(\xi)} \times \left(\nabla\Phi_{(i,k)}^T \mathbb{U}_{\xi,q+1,n_{\max}} \circ \Phi_{(i,k)}\right)\right)\right) \\
&+ \sum_{\xi,i,j,k,p,\vec{l}} \operatorname{div}\left(\left(\nabla a_{(\xi)} \times \left(\nabla\Phi_{(i,k)}^T \mathbb{U}_{\xi,q+1,n_{\max}} \circ \Phi_{(i,k)}\right)\right)\right. \\
&\qquad\qquad\qquad \left. \otimes \left(\nabla a_{(\xi)} \times \left(\nabla\Phi_{(i,k)}^T \mathbb{U}_{\xi,q+1,n_{\max}} \circ \Phi_{(i,k)}\right)\right)\right) \\
:= &\operatorname{div}\left(\mathcal{O}_{n_{\max},1,1} + \mathcal{O}_{n_{\max},1,2} + \mathcal{O}_{n_{\max},1,3}\right).
\end{aligned}
\tag{8.68}
$$

The last two of these three terms are again divergence corrector errors and will therefore be absorbed into $\mathring{R}_{q+1}$ and estimated in Section 8.8.

Recall—cf. (8.3)—that $\mathbb{W}_{\xi,q+1,n_{\max}}$ is periodized to scale $(\lambda_{q+1}r_{q+1,n_{\max}})^{-1} = \lambda_{q,n_{\max}}^{-1}$. Combining this observation with (4.15) from Proposition 4.4 and (7.8), we further split $\mathcal{O}_{n_{\max},1,1}$ as[7]

$$
\operatorname{div}\left(\mathcal{O}_{n_{\max},1,1}\right)
$$

---

[7] In this case, the projection $\mathbb{P}_{\geq\lambda_{q,n_{\max}}}$ has a greater minimum frequency than the projection $\mathbb{P}_{[q,n_{\max},p_{\max}+1]}$; cf. (9.28), (9.22), and (7.6). For the sake of consistency, we write $\mathbb{P}_{\geq\lambda_{q,n_{\max}}}\mathbb{P}_{[q,n_{\max},p_{\max}+1]}$ throughout this section.

$$= \sum_{\xi,i,j,k,p,\vec{l}} \mathrm{div}\left[a_{(\xi)}^2 \nabla\Phi_{(i,k)}^{-1} \left(\fint_{\mathbb{T}^3} \mathbb{W}_{\xi,q+1,n_{\max}} \otimes \mathbb{W}_{\xi,q+1,n_{\max}}(\Phi_{(i,k)}) \right) \nabla\Phi_{(i,k)}^{-T} \right]$$

$$+ \sum_{\xi,i,j,k,p,\vec{l}} \mathrm{div}\left[a_{(\xi)}^2 \nabla\Phi_{(i,k)}^{-1} \mathbb{P}_{\geq\lambda_{q,n_{\max}}} \mathbb{P}_{[q,n_{\max},p_{\max}+1]} \right.$$

$$\left. \times (\mathbb{W}_{\xi,q+1,n_{\max}} \otimes \mathbb{W}_{\xi,q+1,n_{\max}})(\Phi_{(i,k)}) \nabla\Phi_{(i,k)}^{-T} \right]$$

$$= \mathrm{div} \sum_{\xi,i,j,k,p,\vec{l}} \delta_{q+1,n_{\max},p} \Gamma_{q+1}^{2j+4} \eta_{(i,j,k)}^2 \gamma_\xi^2 \left(\frac{R_{q,n_{\max},p,j,i,k}}{\delta_{q+1,n_{\max},p} \Gamma_{q+1}^{2j+4}} \right)$$

$$\times \nabla\Phi_{(i,k)}^{-1}(\xi\otimes\xi)\nabla\Phi_{(i,k)}^{-T}$$

$$+ \sum_{\xi,i,j,k,p,\vec{l}} \nabla a_{(\xi)}^2 \nabla\Phi_{(i,k)}^{-1} \mathbb{P}_{\geq\lambda_{q,n_{\max}}} \mathbb{P}_{[q,n_{\max},p_{\max}+1]}$$

$$\times (\mathbb{W}_{\xi,q+1,n_{\max}} \otimes \mathbb{W}_{\xi,q+1,n_{\max}})(\Phi_{(i,k)}) \nabla\Phi_{(i,k)}^{-T}$$

$$+ \sum_{\xi,i,j,k,p,\vec{l}} a_{(\xi)}^2 (\nabla\Phi_{(i,k)}^{-1})_{\alpha\theta} \mathbb{P}_{\geq\lambda_{q,n_{\max}}} \mathbb{P}_{[q,n_{\max},p_{\max}+1]}$$

$$\times (\mathbb{W}_{\xi,q+1,n_{\max}}^\theta \mathbb{W}_{\xi,q+1,n_{\max}}^\gamma)(\Phi_{(i,k)}) \partial_\alpha(\nabla\Phi_{(i,k)}^{-1})_{\zeta\gamma}. \quad (8.69)$$

By (4.1) from Proposition 4.1 and (8.8), we obtain that

$$\mathrm{div} \sum_{\xi,i,j,k,p,\vec{l}} \delta_{q+1,n_{\max},p} \Gamma_{q+1}^{2j+4} \eta_{(i,j,k)}^2 \gamma_\xi^2 \left(\frac{R_{q,n_{\max},p,j,i,k}}{\delta_{q+1,n_{\max},p} \Gamma_{q+1}^{2j+4}} \right)$$

$$\times \nabla\Phi_{(i,k)}^{-1}(\xi\otimes\xi)\nabla\Phi_{(i,k)}^{-T}$$

$$= \mathrm{div} \sum_{i,j,k,p,\vec{l}} \eta_{(i,j,k)}^2 \left(\delta_{q+1,n_{\max},p} \Gamma_{q+1}^{2j+4} \mathrm{Id} - \mathring{R}_{q,n_{\max},p} \right)$$

$$= -\mathrm{div} \sum_{i,j,k,\vec{l}} \sum_{p=1}^{p_{\max}} \eta_{(i,j,k)}^2 \mathring{R}_{q,n_{\max},p} + \nabla\left(\sum_{i,j,k,p,\vec{l}} \eta_{(i,j,k)}^2 \delta_{q+1,n_{\max},p} \Gamma_{q+1}^{2j+4} \right)$$

$$:= -\mathrm{div} \sum_{p=1}^{p_{\max}} \mathring{R}_{q,n_{\max},p} + \nabla\pi, \quad (8.70)$$

where in the last line we have used (6.142). We can apply Proposition A.18 to the remaining two terms in (8.69) to produce the terms

$$(\mathcal{H} + \mathcal{R}^*)\left[\sum_{\xi,i,j,k,p,\vec{l}} \nabla a_{(\xi)}^2 \nabla\Phi_{(i,k)}^{-1} \mathbb{P}_{\geq\lambda_{q,n_{\max}}} \mathbb{P}_{[q,n_{\max},p_{\max}+1]} \right.$$

$$\times (\mathbb{W}\otimes\mathbb{W})_{\xi,q+1,n_{\max}}(\Phi_{(i,k)}) \nabla\Phi_{(i,k)}^{-T}$$

$$+ \sum_{\xi,i,j,k,p,\vec{l}} a^2_{(\xi)} (\nabla\Phi^{-1}_{(i,k)})_{\alpha\theta} \mathbb{P}_{\geq\lambda_{q,n_{\max}}} \mathbb{P}_{[q,n_{\max},p_{\max}+1]}$$

$$\times (\mathbb{W}^\theta \mathbb{W}^\gamma)_{\xi,q+1,n_{\max}} (\Phi_{(i,k)}) \partial_\alpha (\nabla\Phi^{-1}_{(i,k)})_{\zeta\gamma} \Big], \quad (8.71)$$

which will be absorbed into $\mathring{R}_{q+1}$ and estimated in Section 8.6.

Before combining the previous steps, we remind the reader that at this point, $\mathring{R}_{q+1}$ will be fully defined, and will include $\mathring{R}^{n_{\max}-1}_{q+1}$, all the error terms arising from the addition of $w_{q+1,n_{\max}}$, and $\mathring{R}^{\mathrm{comm}}_q$. Then from (8.65), (8.66), (8.67), (8.68), (8.69), (8.70), and (8.71), we can finally write that

$$\partial_t v_{q+1} + \mathrm{div}\,(v_{q+1} \otimes v_{q+1}) + \nabla p_{q,n_{\max}-1}$$

$$= \mathrm{div}\,\mathring{R}^{\mathrm{comm}}_q + \mathrm{div}\left(\mathring{R}^{n_{\max}-1}_{q+1}\right) + (\partial_t + v_{\ell_q}\cdot\nabla)w_{q+1,n_{\max}} + w_{q+1,n_{\max}}\cdot\nabla v_{\ell_q}$$

$$+ \sum_{n'\leq n_{\max}-1} \mathrm{div}\,(w_{q+1,n_{\max}} \otimes w_{q+1,n'} + w_{q+1,n'} \otimes w_{q+1,n_{\max}})$$

$$+ \mathrm{div}\left(w_{q+1,n_{\max}} \otimes w_{q+1,n_{\max}} + \sum_{p=1}^{p_{\max}} \mathring{R}_{q,n_{\max},p}\right)$$

$$= \mathrm{div}\,\mathring{R}^{\mathrm{comm}}_q + \mathrm{div}\left[\mathring{R}^{n_{\max}-1}_{q+1} + (\mathcal{H}+\mathcal{R}^*)\left(\partial_t w_{q+1,n_{\max}} + v_{\ell_q}\cdot\nabla w_{q+1,n_{\max}}\right)\right.$$

$$+ (\mathcal{H}+\mathcal{R}^*)\left(w_{q+1,n_{\max}}\cdot\nabla v_{\ell_q}\right)$$

$$\left.+ \sum_{n'\leq n_{\max}-1}\left(w_{q+1,n_{\max}} \otimes w_{q+1,n'} + w_{q+1,n'} \otimes w_{q+1,n_{\max}}\right)\right]$$

$$+ \mathrm{div}\left[\mathcal{O}_{n_{\max},1,1} + \mathcal{O}_{n_{\max},1,2} + \mathcal{O}_{n_{\max},1,3} + \mathcal{O}_{n_{\max},2} + \sum_{p=1}^{p_{\max}} \mathring{R}_{q,n_{\max},p}\right] + \nabla\pi$$

$$= \mathrm{div}\,\mathring{R}^{\mathrm{comm}}_q + \mathrm{div}\left[\mathring{R}^{n_{\max}-1}_{q+1} + \underbrace{(\mathcal{H}+\mathcal{R}^*)\left(\partial_t w_{q+1,n_{\max}} + v_{\ell_q}\cdot\nabla w_{q+1,n_{\max}}\right)}_{\text{transport}}\right.$$

$$(8.72)$$

$$+ \underbrace{(\mathcal{H}+\mathcal{R}^*)\left(w_{q+1,n_{\max}}\cdot\nabla v_{\ell_q}\right)}_{\text{Nash}}$$

$$\left.+ \underbrace{\sum_{n'\leq n_{\max}-1}\left(w_{q+1,n_{\max}} \otimes w_{q+1,n'} + w_{q+1,n'} \otimes w_{q+1,n_{\max}}\right)}_{\text{Type 2}}\right]$$

$$(8.73)$$

$$+ \mathrm{div}\left[\underbrace{\mathcal{O}_{n_{\max},1,2} + \mathcal{O}_{n_{\max},1,3}s}_{\text{divergence corrector}} + \underbrace{\mathcal{O}_{n_{\max},2}}_{\text{Type 2}}\right] + \nabla\pi \qquad (8.74)$$

$$+ \operatorname{div}\left[(\mathcal{H} + \mathcal{R}^*)\left(\sum_{\xi,i,j,k,p,\vec{l}} \nabla a_{(\xi)}^2 \nabla \Phi_{(i,k)}^{-1} \mathbb{P}_{\geq \lambda_{q,n_{\max}}} \mathbb{P}_{[q,n_{\max},p_{\max}+1]}\right.\right.$$

$$\left. \times (\mathbb{W} \otimes \mathbb{W})_{\xi,q+1,n_{\max}}(\Phi_{(i,k)}) \nabla \Phi_{(i,k)}^{-T} \right. \tag{8.75}$$

$$+ \sum_{\xi,i,j,k,p} a_{(\xi)}^2 (\nabla \Phi_{(i,k)}^{-1})_{\alpha\theta} \mathbb{P}_{\geq \lambda_{q,n_{\max}}} \mathbb{P}_{[q,n_{\max},p_{\max}+1]}$$

$$\left.\left. \times (\mathbb{W}^\theta \mathbb{W}^\gamma)_{\xi,q+1,n_{\max}}(\Phi_{(i,k)}) \partial_\alpha (\nabla \Phi_{(i,k)}^{-1})_{\zeta\gamma}\right)\right] \tag{8.76}$$

$$= \operatorname{div}(\mathring{R}_{q+1}) + \nabla\pi,$$

where the terms in (8.75) and (8.76) are Type 1 errors. This concludes the proof after again noting that $\left(\partial_t + v_{\ell_q} \cdot \nabla\right) w_{q+1,\tilde{n}}$ has zero mean.

8.4   TRANSPORT ERRORS

Lemma 8.4. *For all $0 \leq \tilde{n} \leq n_{\max}$, the transport errors satisfy*

$$D_{t,q} w_{q+1,\tilde{n}} = \partial_t w_{q+1,\tilde{n}} + v_{\ell_q} \cdot \nabla w_{q+1,\tilde{n}}$$
$$= \operatorname{div} \circ (\mathcal{H} + \mathcal{R}^*)\left(\partial_t w_{q+1,\tilde{n}} + v_{\ell_q} \cdot \nabla w_{q+1,\tilde{n}}\right) + \nabla p_{\tilde{n}}$$

with the estimates

$$\left\|\psi_{i,q} D^N D_{t,q}^M \left((\mathcal{H} + \mathcal{R}^*)\left(\partial_t w_{q+1,\tilde{n}} + v_{\ell_q} \cdot \nabla w_{q+1,\tilde{n}}\right)\right)\right\|_{L^1}$$
$$\lesssim \delta_{q+2} \Gamma_{q+1}^{-C_R-1} \lambda_{q+1}^N \mathcal{M}\left(M, \mathsf{N}_{\mathrm{ind},t}, \tau_q^{-1}\Gamma_{q+1}^{i+1}, \Gamma_{q+1}^{-1}\tilde{\tau}_q^{-1}\right)$$

for all $N, M \leq 3\mathsf{N}_{\mathrm{ind},v}$.

Proof of Lemma 8.4. The transport errors are given in (8.39), (8.55), and (8.72). Writing out the transport error, we have that

$$\left(\partial_t + v_{\ell_q} \cdot \nabla\right) w_{q+1,\tilde{n}}$$

$$= \left(\partial_t + v_{\ell_q} \cdot \nabla\right)\left(\sum_{i,j,k,\tilde{p},\vec{l},\xi} \operatorname{curl}\left(a_{\xi,i,j,k,q,\tilde{n},\tilde{p},\vec{l}} \nabla \Phi_{(i,k)}^T \mathbb{U}_{\xi,q+1,\tilde{n}} \circ \Phi_{(i,k)}\right)\right)$$

$$= \sum_{i,j,k,\tilde{p},\vec{l},\xi} \left(\partial_t + v_{\ell_q} \cdot \nabla\right)\left(a_{(\xi)} \nabla \Phi_{(i,k)}^{-1}\right) \mathbb{W}_{\xi,q+1,\tilde{n}} \circ \Phi_{(i,k)}$$

$$+ \sum_{i,j,k,\tilde{p},\vec{l},\xi} \left(\left(\partial_t + v_{\ell_q} \cdot \nabla\right) \nabla a_{(\xi)}\right) \times \left(\nabla \Phi_{(i,k)} \mathbb{U}_{\xi,q+1,\tilde{n}} \circ \Phi_{(i,k)}\right)$$

$$+ \sum_{i,j,k,\widetilde{p},\vec{l},\xi} \nabla a_{(\xi)} \times \left(\left((\partial_t + v_{\ell_q} \cdot \nabla) \nabla \Phi_{(i,k)} \right) \mathbb{U}_{\xi,q+1,\widetilde{n}} \circ \Phi_{(i,k)} \right). \qquad (8.77)$$

Due to the fact that the second two terms arise from the addition of the corrector defined in (8.5) and (8.12), and the fact that the bounds for the corrector in (8.18) are stronger than that of the principal part of the perturbation, we shall completely estimate only the first term and simply indicate the setup for the second and third. Before applying Proposition A.18, recall that the inverse divergence of (8.77) needs to be estimated on the support of a cutoff $\psi_{i,q}$ in order to verify (7.13), and (7.20), and (7.27). Recall from the identification of the error terms (cf. (8.37) and the subsequent argument) that for all $\widetilde{n}$, $(\partial_t + v_{\ell_q} \cdot \nabla) w_{q+1,\widetilde{n}}$ has zero mean. Thus, although each individual term in the final equality in (8.77) may not have zero mean, we can safely apply $\mathcal{H}$ and $\mathcal{R}^*$ to each term and estimate the outputs while ignoring the last term in (A.78).

We will apply Proposition A.18, specifically Remark A.19, to each summand in the first term on the right side of (8.77), with the following choices. We set $v = v_{\ell_q}$, and $D_t = D_{t,q} = \partial_t + v_{\ell_q} \cdot \nabla$ as usual. We set $N_* = M_* = \lfloor 1/2 \left(N_{\mathrm{fin},\widetilde{n}} - N_{\mathrm{cut},t} - N_{\mathrm{cut},x} - 5 \right) \rfloor$, with N_{dec} and d satisfying (9.60a). We define

$$G = (\partial_t + v_{\ell_q} \cdot \nabla)(a_{(\xi)} \nabla \Phi_{(i,k)}^{-1})\xi,$$

with $\lambda = \Gamma_{q+1} \lambda_{q,\widetilde{n},\widetilde{p}}$, $\nu = \tau_q^{-1} \Gamma_{q+1}^{i-\mathsf{c}_{\widetilde{n}}+3}$, $M_t = N_{\mathrm{ind},t}$, $\widetilde{\nu} = \widetilde{\tau}_q^{-1} \Gamma_{q+1}^{-1}$, and

$$\mathcal{C}_G = \left| \mathrm{supp}\, (\eta_{i,j,k,q,\widetilde{n},\widetilde{p},\vec{l}}) \right| \delta_{q+1,\widetilde{n},1}^{1/2} \Gamma_{q+1}^{i-\mathsf{c}_{\widetilde{n}}+j+5} \tau_q^{-1},$$

which is the correct amplitude in view of (8.14) with $r = 1$, $r_1 = r_2 = 2$, and (6.114). Thus, we have that

$$\left\| D^N D_{t,q}^M G \right\|_{L^1} \lesssim \mathcal{C}_G \left(\lambda_{q,\widetilde{n},\widetilde{p}} \Gamma_{q+1} \right)^N \mathcal{M} \left(M, N_{\mathrm{ind},t} - 1, \tau_q^{-1} \Gamma_{q+1}^{i-\mathsf{c}_{\widetilde{n}}+3}, \widetilde{\tau}_q^{-1} \Gamma_{q+1}^{-1} \right),$$
$$(8.78)$$

for all $N, M \leq \lfloor 1/2 \left(N_{\mathrm{fin},\widetilde{n}} - N_{\mathrm{cut},t} - N_{\mathrm{cut},x} - 5 \right) \rfloor$ after using (9.42) and (9.52), and so (A.66) is satisfied. We set $\Phi = \Phi_{i,k}$ and $\lambda' = \widetilde{\lambda}_q$. Appealing as usual to Corollary 6.27 and (6.60), we have that (A.67) and (A.68) are satisfied.

Referring to (1) from Proposition 4.4, we set $\varrho = \varrho_{\xi,\lambda_{q+1},r_{q+1,\widetilde{n}}}$ and $\vartheta = \vartheta_{\xi,\lambda_{q+1},r_{q+1,\widetilde{n}}}$. Setting $\zeta = \lambda_{q+1}$, we have that (1) is satisfied. Setting $\mu = \lambda_{q+1} r_{q+1,\widetilde{n}} = \lambda_{q,\widetilde{n}}$ and referring to (2) from Proposition 4.4, we have that (2) is satisfied. Setting $\Lambda = \zeta = \lambda_{q+1}$ and $C_* = r_{q+1,\widetilde{n}}$ and referring to (4.11) and (4.12) from Proposition 4.4, we have that (A.69) is satisfied. (A.70) is immediate from the definitions. Referring to (9.48), we have that (A.71) is satisfied. Thus, we conclude from (A.73) with α_R as in (9.53), that for $N, M \leq \lfloor 1/2 \left(N_{\mathrm{fin},\widetilde{n}} - N_{\mathrm{cut},t} - N_{\mathrm{cut},x} - 5 \right) \rfloor - \mathsf{d}$,

$$\left\| D^N D_{t,q}^M \left(\mathcal{H} \left((\partial_t + v_{\ell_q} \cdot \nabla)(a_{(\xi)} \nabla \Phi_{(i,k)}^{-1})\xi \right) \right) \right\|_{L^1}$$

$$= \left\| D^N D_{t,q}^M \left(\mathcal{H} \left(G_\varrho \circ \Phi \right) \right) \right\|_{L^1}$$

$$\lesssim \left| \mathrm{supp}\, (\eta_{i,j,k,q,\tilde{n},\tilde{p},\vec{l}}) \right| \delta_{q+1,\tilde{n},1}^{1/2} \Gamma_{q+1}^{i-c_{\tilde{n}}+j+6} \tau_q^{-1} r_{q+1,\tilde{n}}$$

$$\times \lambda_{q+1}^{-1} \lambda_{q+1}^N \mathcal{M} \left(M, \mathsf{N}_{\mathrm{ind},t}, \tau_q^{-1} \Gamma_{q+1}^i, \tilde{\tau}_q^{-1} \Gamma_{q+1}^{-1} \right),$$

after appealing to (9.42). From (9.60c), these bounds are valid for all $N, M \leq 3\mathsf{N}_{\mathrm{ind},v}$. The bound obtained above is next summed over $(i,j,k,\tilde{p},\tilde{n},\vec{l})$. First, we treat the sum over $\vec{l}$. By noting that (6.147) with $r_1 = 2$ and $r_2 = 2$ and (9.42) imply

$$\sum_{\vec{l}} \left| \mathrm{supp}\, (\eta_{i,j,k,q,\tilde{n},\tilde{p},\vec{l}}) \right| \Gamma_{q+1}^{i-c_{\tilde{n}}+j+6} \leq \Gamma_{q+1}^{-2\left(\frac{i}{2}+\frac{j}{2}\right)+\frac{c_b}{2}+2} \Gamma_{q+1}^{i-c_{\tilde{n}}+j+6} = \Gamma_{q+1}^{\frac{c_b}{2}+3},$$

we conclude that

$$\left\| D^N D_{t,q}^M \left(\mathcal{H} \left(\partial_t w_{q+1,\tilde{n}} + v_{\ell_q} \cdot \nabla w_{q+1,\tilde{n}} \right) \right) \right\|_{L^1(\mathrm{supp}\, \psi_{i,q})}$$

$$\lesssim \sum_{i'=i-1}^{i+1} \sum_{j,k,\tilde{p},\xi} \Gamma_{q+1}^{\frac{c_b}{2}+3} \delta_{q+1,\tilde{n},1}^{1/2} \tau_q^{-1} r_{q+1,\tilde{n}}$$

$$\times \lambda_{q+1}^{-1} \lambda_{q+1}^N \mathcal{M} \left(M, \mathsf{N}_{\mathrm{ind},t}, \tau_q^{-1} \Gamma_{q+1}^{i'}, \tilde{\tau}_q^{-1} \Gamma_{q+1}^{-1} \right)$$

$$\lesssim \Gamma_{q+1}^{4+\frac{c_b}{2}} \delta_{q+1,\tilde{n},1}^{\frac{1}{2}} \tau_q^{-1} r_{q+1,\tilde{n}} \lambda_{q+1}^{-1} \lambda_{q+1}^N \mathcal{M} \left(M, \mathsf{N}_{\mathrm{ind},t}, \tau_q^{-1} \Gamma_{q+1}^{i+1}, \tilde{\tau}_q^{-1} \Gamma_{q+1}^{-1} \right)$$

$$\lesssim \Gamma_{q+1}^{-C_R-1} \delta_{q+2} \lambda_{q+1}^N \mathcal{M} \left(M, \mathsf{N}_{\mathrm{ind},t}, \tau_q^{-1} \Gamma_{q+1}^{i+1}, \tilde{\tau}_q^{-1} \Gamma_{q+1}^{-1} \right) \qquad (8.79)$$

after also using (9.57).

To finish the proof for the first term in (8.77), we must provide a matching estimate for the $\mathcal{R}^*$ portion. Following again the parameter choices in Remark A.19, we set $N_\circ = M_\circ = 3\mathsf{N}_{\mathrm{ind},v}$. As in the argument from Lemma 8.6, we have that (A.75), (A.76), and (A.77) are satisfied, this time with $\zeta = \lambda_{q+1}$. Thus we achieve the estimate in (A.79). Summing over $\vec{l}$ loses a factor less than λ_{q+1}^3, while summing over the other indices costs a constant independent of q. This completes the estimate for the first term from (8.77).

For the second and third terms, we explain how to identify G and ϱ in order to give an idea of how to obtain similar estimates. Using 1 from Proposition 4.4 and the vector calculus identity $\mathrm{curl} \circ \mathrm{curl} = \nabla \circ \mathrm{div} - \Delta$, we obtain that

$$\mathbb{U}_{\xi,q+1,\tilde{n}} = \mathrm{curl} \left(\xi \lambda_{q+1}^{-2d} \Delta^{d-1} \left(\vartheta_{\xi,\lambda_{q+1},r_{q+1,\tilde{n}}} \right) \right)$$

$$= \lambda_{q+1}^{-2d} \xi \times \nabla \left(\Delta^{d-1} \left(\vartheta_{\xi,\lambda_{q+1},r_{q+1,\tilde{n}}} \right) \right). \qquad (8.80)$$

With a little massaging, one can now rewrite the second and third terms in (8.77) in the form $G_\varrho \circ \Phi_{(i,k)}$. Since both terms have traded a spatial derivative on $\mathbb{U}_{\xi,q+1,\tilde{n}}$ for a spatial derivative on $a_{(\xi)}$, inducing a gain, one can easily show that the estimates for these terms will be even stronger than those for the first term.

Notice that we have set $N_* = M_* = \lfloor 1/2 \left(N_{\mathrm{fin},\tilde{n}} - N_{\mathrm{cut},t} - N_{\mathrm{cut},x} - 7\right)\rfloor$ since we have lost a spatial derivative on $a_{(\xi)}$. We omit the rest of the details. $\qquad\square$

8.5 NASH ERRORS

Lemma 8.5. *For all $0 \le \tilde{n} \le n_{\max}$, the Nash errors satisfy*

$$w_{q+1,\tilde{n}} \cdot \nabla v_{\ell_q} = \mathrm{div}\left((\mathcal{H} + \mathcal{R}^*)\, w_{q+1,\tilde{n}} \cdot \nabla v_{\ell_q}\right) + \nabla p_{\tilde{n}}$$

with

$$\left\| \psi_{i,q} D^k D_{t,q}^m \left((\mathcal{H} + \mathcal{R}^*)\, w_{q+1,\tilde{n}} \cdot \nabla v_{\ell_q}\right) \right\|_{L^1}$$
$$\lesssim \delta_{q+2} \Gamma_{q+1}^{-\mathsf{C_R}-1} \lambda_{q+1}^N \mathcal{M}\left(M, \mathsf{N}_{\mathrm{ind},t}, \tau_q^{-1} \Gamma_{q+1}^{i+1}, \Gamma_{q+1}^{-1} \tilde{\tau}_q^{-1}\right)$$

for all $N, M \le 3\mathsf{N}_{\mathrm{ind},v}$.

Proof of Lemma 8.5. The estimates are similar to those in Lemma 8.4. Writing out the Nash error, we have that

$$w_{q+1,\tilde{n}} \cdot \nabla v_{\ell_q} = \sum_{i-1 \le i' \le i+1} \sum_{j,k,\widetilde{p},\vec{l},\xi} \mathrm{curl}\left(a_{\xi,i,j,k,q,\tilde{n}} \nabla \Phi_{(i,k)}^T \mathbb{U}_{\xi,q+1,\tilde{n}} \circ \Phi_{(i,k)}\right)$$

$$= \left(\sum_{i,j,k,\widetilde{p},\vec{l},\xi} \nabla a_{(\xi)} \times \left(\Phi_{(i,k)}^T \mathbb{U}_{\xi,q+1,\tilde{n}} \circ \Phi_{(i,k)}\right)\right) \cdot \nabla v_{\ell_q}$$

$$+ \left(\sum_{i,j,k,\widetilde{p},\vec{l},\xi} a_{(\xi)} \nabla \Phi_{(i,k)}^{-1} \mathbb{W}_{\xi,q+1,\tilde{n}} \circ \Phi_{(i,k)}\right) \cdot \nabla v_{\ell_q}. \qquad (8.81)$$

Due to the fact that the first term arises from the addition of the corrector defined in (8.5) and (8.12), and the fact that the bounds for the corrector in (8.18) are stronger than that of the principal part of the perturbation, we shall completely estimate only the second term and simply indicate the setup for the first. Before applying Proposition A.18, recall that the inverse divergence of (8.77) needs to be estimated on the support of a cutoff $\psi_{i,q}$ in order to verify (7.5), (7.13), and (7.20). Note that the Nash error can be written as $\mathrm{div}\left(w_{q+1,\tilde{n}} \cdot v_{\ell_q}\right)$ and so has zero mean. Thus, although each individual term in the final equality in (8.81) may not have zero mean, we can safely apply $\mathcal{H}$ and $\mathcal{R}^*$ to each term and estimate the outputs while ignoring the last term in (A.78).

We will apply Proposition A.18 to the second term with the following choices. We set $v = v_{\ell_q}$, and $D_t = D_{t,q} = \partial_t + v_{\ell_q} \cdot \nabla$ as usual. We set $N_* = M_* =$

$\lfloor 1/2\, (N_{\mathrm{fin},\tilde{n}} - N_{\mathrm{cut},x} - N_{\mathrm{cut},t} - 4)\rfloor$, with N_{dec} and d satisfying (9.60a). We define

$$G = a_{(\xi)} \nabla \Phi^{-1}_{(i,k)} \xi \cdot \nabla v_{\ell_q}$$

and set

$$\mathcal{C}_G = \left| \mathrm{supp}\,(\eta_{i,j,k,q,\tilde{n},\tilde{p},\vec{l}}) \right| \delta^{1/2}_{q+1,\tilde{n},1} \Gamma^{i-c_{\tilde{n}}+j+5}_{q+1} \tau^{-1}_q,$$

$\lambda = \Gamma_{q+1}\lambda_{q,\tilde{n},\tilde{p}}$, $\nu = \tau^{-1}_q \Gamma^{i-c_{\tilde{n}}+3}_{q+1}$, $M_t = N_{\mathrm{ind},t}$, and $\tilde{\nu} = \tilde{\tau}^{-1}_q \Gamma^{-1}_{q+1}$. From (8.14) with $r = 1$ and $r_1 = r_2 = 2$, (6.114), and (6.60), we have that for $N, M \leq \lfloor 1/2\, (N_{\mathrm{fin},\tilde{n}} - N_{\mathrm{cut},x} - N_{\mathrm{cut},t} - 4)\rfloor$

$$\left\| D^N D^M_{t,q} G \right\|_{L^1} \lesssim \mathcal{C}_G \left(\Gamma_{q+1}\lambda_{q,\tilde{n},\tilde{p}}\right)^N \mathcal{M}\left(M, N_{\mathrm{ind},t}, \tau^{-1}_q \Gamma^{i-c_{\tilde{n}}+3}_{q+1}, \tilde{\tau}^{-1}_q \Gamma^{-1}_{q+1}\right), \tag{8.82}$$

and so (A.66) is satisfied. Note that we have used (9.39) when converting the $\delta^{1/2}_q \tilde{\lambda}_q$ to a τ^{-1}_q. Setting $\Phi = \Phi_{(i,k)}$ and $\lambda' = \tilde{\lambda}_q$, we have that (A.67) and (A.68) are satisfied as usual. The choices of ϱ, ϑ, ζ, μ, Λ, and $\mathcal{C}_*$ are identical to those of the transport error (both terms contain $\mathbb{W}_{\xi,q+1,\tilde{n}} \circ \Phi_{(i,k)}$), and so we have that (1)–(2), (A.69), (A.70), and (A.71) are satisfied as well. Since the bound (8.82) is identical to that of (8.78), we obtain an estimate identical to (8.79). The argument for the $\mathcal{R}^*$ portion follows analogously to that for the first term from the transport error. Finally, after using (8.80) again, one may obtain similar estimates for the first term in (8.81), concluding the proof. □

8.6 TYPE 1 OSCILLATION ERRORS

The Type 1 oscillation errors are defined in the three parameter regimes $\tilde{n} = 0$, $1 \leq \tilde{n} \leq n_{\max} - 1$, and $\tilde{n} = n_{\max}$. In the case $\tilde{n} = 0$, Type 1 oscillation errors stem from the term identified in (8.38), which we recall is

$$(\mathcal{H} + \mathcal{R}^*)\Bigg[\sum_{\xi,i,j,k,\tilde{p},\vec{l}} \nabla a^2_{(\xi)} \nabla \Phi^{-1}_{(i,k)} \mathbb{P}_{\geq \lambda_{q,0}} \left(\sum_{n=1}^{n_{\max}} \sum_{p=1}^{p_{\max}} \mathbb{P}_{[q,n,p]} + \mathbb{P}_{[q,n_{\max},p_{\max}+1]}\right)$$

$$\times (\mathbb{W}_{\xi,q+1,0} \otimes \mathbb{W}_{\xi,q+1,0})(\Phi_{(i,k)}) \nabla \Phi^{-T}_{(i,k)}$$

$$+ \sum_{\xi,i,j,k,\tilde{p},\vec{l}} a^2_{(\xi)} (\nabla \Phi^{-1}_{(i,k)})_{\alpha\theta} \mathbb{P}_{\geq \lambda_{q,0}} \left(\sum_{n=1}^{n_{\max}} \sum_{p=1}^{p_{\max}} \mathbb{P}_{[q,n,p]} + \mathbb{P}_{[q,n_{\max},p_{\max}+1]}\right)$$

$$\times (\mathbb{W}^{\theta}_{\xi,q+1,0} \mathbb{W}^{\gamma}_{\xi,q+1,0})(\Phi_{(i,k)}) \partial_\alpha (\nabla \Phi^{-1}_{(i,k)}) \zeta\gamma \Bigg]. \tag{8.83}$$

This sum is divided into the terms identified in (8.40), (8.41), (8.42), (8.43), (8.45), and (8.46). The errors defined in (8.45) and (8.46) are $\mathring{H}^0_{q,n,p}$ errors and will be corrected by later perturbations $w_{q+1,n,p}$, while the others will be immediately absorbed into $\mathring{R}_{q+1}^0$.

In the case $1 \leq \tilde{n} \leq n_{\max} - 1$, Type 1 oscillation errors stem from the term identified in (8.54):

$$
(\mathcal{H} + \mathcal{R}^*) \Bigg[\sum_{\xi,i,j,k,\widetilde{p},\vec{l}} \nabla a_{(\xi)}^2 \nabla \Phi_{(i,k)}^{-1} \mathbb{P}_{\geq \lambda_{q,\tilde{n}}} \Bigg(\sum_{n=\tilde{n}+1}^{n_{\max}} \sum_{p=1}^{p_{\max}} \mathbb{P}_{[q,n,p]} + \mathbb{P}_{[q,n_{\max},p_{\max}+1]} \Bigg)
$$
$$
\times (\mathbb{W}_{\xi,q+1,\tilde{n}} \otimes \mathbb{W}_{\xi,q+1,\tilde{n}})(\Phi_{(i,k)}) \nabla \Phi_{(i,k)}^{-T}
$$
$$
+ \sum_{\xi,i,j,k,\widetilde{p},\vec{l}} a_{(\xi)}^2 (\nabla \Phi_{(i,k)}^{-1})_{\alpha\theta} \mathbb{P}_{\geq \lambda_{q,\tilde{n}}} \Bigg(\sum_{n=\tilde{n}+1}^{n_{\max}} \sum_{p=1}^{p_{\max}} \mathbb{P}_{[q,n,p]} + \mathbb{P}_{[q,n_{\max},p_{\max}+1]} \Bigg)
$$
$$
\times (\mathbb{W}_{\xi,q+1,\tilde{n}}^\theta \mathbb{W}_{\xi,q+1,\tilde{n}}^\gamma)(\Phi_{(i,k)}) \partial_\alpha (\nabla \Phi_{(i,k)}^{-1})_{\varsigma\gamma} \Bigg]. \tag{8.84}
$$

This sum is divided into the terms identified in (8.59), (8.60), (8.61), (8.62), (8.63), and (8.64). As before, the last two terms are $H_{q,n,p}^{\tilde{n}}$ errors and will be corrected by later perturbations, while the others are absorbed into $\mathring{R}_{q+1}^{\tilde{n}}$.

In the case $\tilde{n} = n_{\max}$, Type 1 oscillation errors are identified in (8.75) and (8.76) as

$$
(\mathcal{H} + \mathcal{R}^*) \Bigg[\sum_{\xi,i,j,k,p,\vec{l}} \nabla a_{(\xi)}^2 \nabla \Phi_{(i,k)}^{-1} \mathbb{P}_{\geq \lambda_{q,n_{\max}}} \mathbb{P}_{[q,n_{\max},p_{\max}+1]}
$$
$$
\times (\mathbb{W}_{\xi,q+1,n_{\max}} \otimes \mathbb{W}_{\xi,q+1,n_{\max}})(\Phi_{(i,k)}) \nabla \Phi_{(i,k)}^{-T}
$$
$$
+ \sum_{\xi,i,j,k,p,\vec{l}} a_{(\xi)}^2 (\nabla \Phi_{(i,k)}^{-1})_{\alpha\theta} \mathbb{P}_{\geq \lambda_{q,n_{\max}}} \mathbb{P}_{[q,n_{\max},p_{\max}+1]}
$$
$$
\times (\mathbb{W}_{\xi,q+1,n_{\max}}^\theta \mathbb{W}_{\xi,q+1,n_{\max}}^\gamma)(\Phi_{(i,k)}) \partial_\alpha (\nabla \Phi_{(i,k)}^{-1})_{\varsigma\gamma} \Bigg]. \tag{8.85}
$$

These errors are completely absorbed into $\mathring{R}_{q+1}$.

To prove the desired estimates on these error terms, we will first analyze a single term of the form

$$
(\mathcal{H} + \mathcal{R}^*) \Bigg[\sum_{\xi,i,j,k,\widetilde{p},\vec{l}} \nabla a_{(\xi)}^2 \nabla \Phi_{(i,k)}^{-1} \mathbb{P}_{\geq \lambda_{q,\tilde{n}}} \mathbb{P}_{[q,n,p]}
$$
$$
\times (\mathbb{W}_{\xi,q+1,\tilde{n}} \otimes \mathbb{W}_{\xi,q+1,\tilde{n}})(\Phi_{(i,k)}) \nabla \Phi_{(i,k)}^{-T}
$$
$$
+ \sum_{\xi,i,j,k,\widetilde{p},\vec{l}} a_{(\xi)}^2 (\nabla \Phi_{(i,k)}^{-1})_{\alpha\theta} \mathbb{P}_{\geq \lambda_{q,\tilde{n}}} \mathbb{P}_{[q,n,p]}
$$
$$
\times (\mathbb{W}_{\xi,q+1,\tilde{n}}^\theta \mathbb{W}_{\xi,q+1,\tilde{n}}^\gamma)(\Phi_{(i,k)}) \partial_\alpha (\nabla \Phi_{(i,k)}^{-1})_{\varsigma\gamma} \Bigg]
$$

$$=: (\mathcal{H} + \mathcal{R}^*)\, \mathcal{O}_{\tilde{n},\tilde{p},n,p}. \tag{8.86}$$

The estimates in Lemma 8.6 for this term on the support of a cutoff function $\psi_{i,q}$ will depend on $\tilde{n}$ and $\tilde{p}$, which range from $0 \le \tilde{n} \le n_{\max}$ and $1 \le \tilde{p} \le p_{\max}$, respectively, and n and p, which range from $\tilde{n}+1 \le n \le n_{\max}$ and $1 \le p \le p_{\max}$, with the additional endpoint case $n = n_{\max}$, $p = p_{\max} + 1$. We then use this general estimate to specify in Remark 8.7 how the terms corresponding to various values of n, $\tilde{n}$, p, and $\tilde{p}$ are absorbed into either higher order stresses $\mathring{H}^{\tilde{n}}_{q,n,p}$ or $\mathring{R}^{\tilde{n}}_{q+1}$, and eventually $\mathring{R}_{q+1}$.

Lemma 8.6. *The terms $\mathcal{O}_{\tilde{n},\tilde{p},n,p}$ defined in (8.86) satisfy the following.*

1. *For the special case $n = n_{\max}$, $p = p_{\max} + 1$ and for all $0 \le \tilde{n} \le n_{\max}$, $1 \le \tilde{p} \le p_{\max}$, as well as for all cases $0 \le \tilde{n} < n \le n_{\max}$, $1 \le p, \tilde{p} \le p_{\max}$, the nonlocal portion of the inverse divergence satisfies*

$$\left\| D^N D_{t,q}^M \left(\mathcal{R}^* \mathcal{O}_{\tilde{n},\tilde{p},n,p} \right) \right\|_{L^1(\mathbb{T}^3)} \le \frac{\delta_{q+2}}{\lambda_{q+1}} \lambda_{q+1}^N \tau_q^{-M} \tag{8.87}$$

   for all $N, M \le 3\mathsf{N}_{\mathrm{ind},v}$.

2. *For $n = n_{\max}$, $p = p_{\max} + 1$, all $0 \le \tilde{n} \le n_{\max}$, and $1 \le \tilde{p} \le p_{\max}$, the high frequency, local portion of the inverse divergence satisfies*

$$\left\| D^N D_{t,q}^M \left(\mathcal{H} \mathcal{O}_{\tilde{n},\tilde{p},n_{\max},p_{\max}+1} \right) \right\|_{L^1(\mathrm{supp}\,\psi_{i,q})}$$
$$\lesssim \Gamma_{q+1}^{-\mathsf{C}_{\mathcal{R}}} \Gamma_{q+1}^{-1} \delta_{q+2} \lambda_{q+1}^N \mathcal{M}\left(M, \mathsf{N}_{\mathrm{ind},t}, \tau_q^{-1}\Gamma_{q+1}^{i-\mathsf{c}_{\tilde{n}}+4}, \Gamma_{q+1}^{-1}\widetilde{\tau}_q^{-1}\right) \tag{8.88}$$

   for all $N, M \le 3\mathsf{N}_{\mathrm{ind},v}$.

3. *For $0 \le \tilde{n} < n \le n_{\max}$ and $1 \le p, \tilde{p} \le p_{\max}$, the medium frequency, local portion of the inverse divergence satisfies*

$$\left\| D^N D_{t,q}^M \left(\mathcal{H} \mathcal{O}_{\tilde{n},\tilde{p},n,p} \right) \right\|_{L^1 \mathrm{supp}\,(\psi_{i,q})}$$
$$\lesssim \delta_{q+1,n,p} \lambda_{q,n,p}^N \mathcal{M}\left(M, \mathsf{N}_{\mathrm{ind},t}, \tau_q^{-1}\Gamma_{q+1}^{i-\mathsf{c}_{\tilde{n}}+4}, \Gamma_{q+1}^{-1}\widetilde{\tau}_q^{-1}\right) \tag{8.89}$$

   for all $N + M \le \mathsf{N}_{\mathrm{fin},n}$.

Remark 8.7. Note that after appealing to $\tilde{n} \le n - 1$, (9.35), and (9.42), (8.89) matches (7.15), (7.22), and (7.29), or equivalently (6.118). In addition, after appealing again to $\tilde{n} \le n - 1$, (9.35), and (9.42), (8.87) and (8.88) are sufficient to meet (7.13), (7.20), and (7.27).

Proof of Lemma 8.6. The first step is to use item (1) and (4.15) from Proposition 4.4 to rewrite (8.86) as

$$(\mathcal{H} + \mathcal{R}^*)\left[\sum_{\xi,i,j,k,\tilde{p},\vec{l}} \nabla a_{(\xi)}^2 \nabla \Phi_{(i,k)}^{-1} \mathbb{P}_{\ge \lambda_{q,\tilde{n}}} \mathbb{P}_{[q,n,p]} \right.$$

$$\times \, (\mathbb{W}_{\xi,q+1,\widetilde{n}} \otimes \mathbb{W}_{\xi,q+1,\widetilde{n}})(\Phi_{(i,k)}) \nabla \Phi_{(i,k)}^{-T}$$

$$+ \sum_{\xi,i,j,k,\widetilde{p},\vec{l}} a_{(\xi)}^2 (\nabla \Phi_{(i,k)}^{-1})_{\theta\alpha} \mathbb{P}_{\geq \lambda_{q,\widetilde{n}}} \mathbb{P}_{[q,n,p]}$$

$$\times \, (\mathbb{W}_{\xi,q+1,\widetilde{n}}^{\theta} \mathbb{W}_{\xi,q+1,\widetilde{n}}^{\gamma})(\Phi_{(i,k)}) \partial_{\alpha} (\nabla \Phi_{(i,k)}^{-1})_{\gamma\kappa} \Bigg]$$

$$= (\mathcal{H} + \mathcal{R}^*) \Bigg[\sum_{\xi,i,j,k,\widetilde{p},\vec{l}} \mathbb{P}_{\geq \lambda_{q,\widetilde{n}}} \mathbb{P}_{[q,n,p]} \left(\left(\varrho_{\xi,\lambda_{q+1},r_{q+1,\widetilde{n}}} \right)^2 \right) (\Phi_{(i,k)})$$

$$\times \left(\partial_{\alpha} a_{(\xi)}^2 (\nabla \Phi_{(i,k)}^{-1})_{\gamma\kappa} \xi^{\theta} \xi^{\gamma} (\nabla \Phi_{(i,k)}^{-T})_{\theta\alpha} \right.$$

$$\left. + a_{(\xi)}^2 (\nabla \Phi_{(i,k)})^{-1})_{\theta\alpha} \xi^{\theta} \xi^{\gamma} \partial_{\alpha} (\nabla \Phi_{(i,k)}^{-1})_{\gamma\kappa} \right) \Bigg]. \qquad (8.90)$$

Next, we must identify the functions and the values of the parameters which will be used in the application of Proposition A.18, specifically Remark A.19. We first address the bounds required in (A.66), (A.67), and (A.68), which we can treat simultaneously for items (1), (2), and (3). Afterwards, we split the proof into two parts. First, we set $n = n_{\max}$, $p = p_{\max} + 1$ and prove (8.87) for *only* these specific values of n and p, as we simultaneously prove (8.88). Next, we consider $n < n_{\max}$ and prove (8.87) in the remaining cases, as we simultaneously prove (8.89).

Returning to (A.66), we will verify that this inequality holds with $v = v_{\ell_q}$, $D_t = D_{t,q} = \partial_t + v_{\ell_q} \cdot \nabla$, and $N_* = M_* = \lfloor N^\sharp/2 \rfloor$, where $N^\sharp = N_{\text{fin},\widetilde{n}} - N_{\text{cut},t} - N_{\text{cut},x} - 5$. In order to verify the assumption $N_* - d \geq 2N_{\text{dec}} + 4$, we use that N_{dec} and d satisfy (9.60a), which gives that

$$2N_{\text{dec}} + 4 \leq \lfloor 1/2 \left(N_{\text{fin},\widetilde{n}} - N_{\text{cut},t} - N_{\text{cut},x} - 5 \right) - d \rfloor. \qquad (8.91)$$

Denoting the κth component of the below vector field G by G_κ, we fix a value of $(\xi, i, j, k, \widetilde{p}, \vec{l})$ and set

$$G_\kappa = \partial_\alpha a_{(\xi)}^2 \left(\nabla \Phi_{(i,k)}^{-1} \right)_{\gamma\kappa} \xi^\theta \xi^\gamma \left(\nabla \Phi_{(i,k)}^{-T} \right)_{\alpha\theta}$$

$$+ a_{(\xi)}^2 \left(\nabla \Phi_{(i,k)} \right)^{-1})_{\alpha\theta} \xi^\theta \xi^\gamma \partial_\alpha \left(\nabla \Phi_{(i,k)}^{-1} \right)_{\kappa\gamma}. \qquad (8.92)$$

We now establish (A.66)–(A.68) with the parameter choices

$$\mathcal{C}_G = |\text{supp}\, (\eta_{i,j,k,q,\widetilde{n},\widetilde{p},\vec{l}})| \Gamma_{q+1}^{2j-3-C_b} \Gamma_q^{-C_R} \delta_{q+1} \widetilde{\lambda}_q \prod_{n' \leq \widetilde{n}} \left(f_{q,n'} \Gamma_{q+1}^{8+C_b} \right), \qquad (8.93)$$

$\lambda = \lambda_{q,\widetilde{n},\widetilde{p}} \Gamma_{q+1}$, $M_t = N_{\text{ind},t}$, $\nu = \tau_q^{-1} \Gamma_{q+1}^{i-c_{\widetilde{n}}+4}$, $\widetilde{\nu} = \widetilde{\tau}_q^{-1} \Gamma_{q+1}^{-1}$, and $\lambda' = \widetilde{\lambda}_q$. Applying Lemma 8.1 and estimate (8.26) with $r = 2$, $r_2 = 1$, $r_1 = \infty$, and the bounds (6.113) and (6.114), we see that

$$
\left\| D^N D_{t,q}^M \left(\partial_\alpha a_{(\xi)}^2 \left(\nabla \Phi_{(i,k)}^{-1} \right)_{\gamma\kappa} \xi^\theta \xi^\gamma \left(\nabla \Phi_{(i,k)}^{-T} \right)_{\alpha\theta} \right) \right\|_{L^1}
$$

$$
\lesssim |\mathrm{supp}\,(\eta_{i,j,k,q,\widetilde{n},\widetilde{p},\widetilde{l}})| \Gamma_{q+1}^{2j+5} \lambda_{q,\widetilde{n},\widetilde{p}} \delta_{q+1,\widetilde{n},\widetilde{p}}
$$
$$
\times (\Gamma_{q+1} \lambda_{q,\widetilde{n},\widetilde{p}})^N \mathcal{M} \left(M, \mathsf{N}_{\mathrm{ind},t}, \tau_q^{-1} \Gamma_{q+1}^{i-\mathsf{c}_{\widetilde{n}}+3}, \widetilde{\tau}_q^{-1} \Gamma_{q+1}^{-1} \right)
$$

$$
\lesssim |\mathrm{supp}\,(\eta_{i,j,k,q,\widetilde{n},\widetilde{p},\widetilde{l}})| \Gamma_{q+1}^{2j-2-\mathsf{C}_b}
$$
$$
\times \Gamma_{q+1}^{-1} \Gamma_q^{-\mathsf{C}_R} \delta_{q+1} \widetilde{\lambda}_q \prod_{n' \le \widetilde{n}} \left(f_{q,n'} \Gamma_{q+1}^{8+\mathsf{C}_b} \right)
$$
$$
\times (\Gamma_{q+1} \lambda_{q,\widetilde{n},\widetilde{p}})^N \mathcal{M} \left(M, \mathsf{N}_{\mathrm{ind},t}, \tau_q^{-1} \Gamma_{q+1}^{i-\mathsf{c}_{\widetilde{n}}+3}, \widetilde{\tau}_q^{-1} \Gamma_{q+1}^{-1} \right) \qquad (8.94)
$$

holds for all $N, M \le \lfloor 1/2\,(\mathsf{N}_{\mathrm{fin},\widetilde{n}} - \mathsf{N}_{\mathrm{cut},t} - \mathsf{N}_{\mathrm{cut},x} - 5) \rfloor$. To achieve the last inequality, we have used the definition of $\delta_{q+1,\widetilde{n},\widetilde{p}}$ in (9.34) and the definition of $f_{q,\widetilde{n}}$ in (9.31) to rewrite

$$
\delta_{q+1,\widetilde{n},\widetilde{p}} \lambda_{q,\widetilde{n},\widetilde{p}} \Gamma_{q+1}^{7+\mathsf{C}_b} = \Gamma_{q+1}^{-1} \Gamma_q^{-\mathsf{C}_R} \delta_{q+1} \widetilde{\lambda}_q \prod_{n' \le \widetilde{n}} \left(f_{q,n'} \Gamma_{q+1}^{8+\mathsf{C}_b} \right).
$$

For the second half of G_κ, we can appeal to (6.113) and (6.114), and use that $\widetilde{\lambda}_q \le \lambda_{q,\widetilde{n},\widetilde{p}}$ for all $\widetilde{n}$ and $\widetilde{p}$ to deduce that

$$
\left\| D^N D_{t,q}^M \partial_\alpha \left(\nabla \Phi_{i,k}^{-1} \right)_{\gamma\kappa} \right\|_{L^\infty(\mathrm{supp}\,\psi_{i,q}\widetilde{\chi}_{i,k,q})}
$$
$$
\le (\Gamma_{q+1} \lambda_{q,\widetilde{n},\widetilde{p}})^{N+1} \mathcal{M} \left(M, \mathsf{N}_{\mathrm{ind},t}, \tau_q^{-1} \Gamma_{q+1}^{i-\mathsf{c}_0}, \widetilde{\tau}_q^{-1} \Gamma_{q+1}^{-1} \right)
$$

for $N, M \le \lfloor 1/2\,(\mathsf{N}_{\mathrm{fin},\widetilde{n}} - \mathsf{N}_{\mathrm{cut},t} - \mathsf{N}_{\mathrm{cut},x} - 5) \rfloor$. Combining these estimates shows that

$$
\left\| D^N D_{t,q}^M G_\kappa \right\|_{L^1} \lesssim \mathcal{C}_G \, (\Gamma_{q+1} \lambda_{q,\widetilde{n},\widetilde{p}})^N \mathcal{M} \left(M, \mathsf{N}_{\mathrm{ind},t}, \tau_q^{-1} \Gamma_{q+1}^{i-\mathsf{c}_{\widetilde{n}}+3}, \widetilde{\tau}_q^{-1} \Gamma_{q+1}^{-1} \right)
$$
$$
\tag{8.95}
$$

for $N, M \le \lfloor 1/2\,(\mathsf{N}_{\mathrm{fin},\widetilde{n}} - \mathsf{N}_{\mathrm{cut},t} - \mathsf{N}_{\mathrm{cut},x} - 5) \rfloor$, showing that (A.66) has been satisfied.

We set the flow in Proposition A.18 as $\Phi = \Phi_{i',k}$, which by definition satisfies $D_{t,q}\Phi_{i',k} = 0$. Appealing to (6.109) and (6.112), we have that (A.67) is satisfied. From (6.60), the choice of ν from earlier, and (9.39), we have that $Dv = Dv_{\ell_q}$ satisfies the bound (A.68).

Proof of item (2) and of item (1) when $n = n_{\max}$, $p = p_{\max} + 1$. We first assume that $\widetilde{n} < n_{\max}$. In this case, we have that the minimum frequency $\lambda_{q,n_{\max}+1,0}$ of $\mathbb{P}_{[q,n_{\max},p_{\max}+1]}$ is larger than the minimum frequency $\lambda_{q,\widetilde{n}}$ of $\mathbb{P}_{\ge \lambda_{q,\widetilde{n}}}$ from (9.28) and (9.22). We therefore can discard $\mathbb{P}_{\ge \lambda_{q,\widetilde{n}}}$ from (8.90) and with the goal of satisfying verifying (1)–(3) of Proposition A.18, we set

$$\zeta = \lambda_{q,n_{\max}+1,0}, \qquad \mu = \lambda_{q,\tilde{n}}, \qquad \Lambda = \lambda_{q+1}, \tag{8.96}$$

and

$$\varrho = \mathbb{P}_{[q,n_{\max},p_{\max}+1]}\left(\left(\varrho_{\xi,\lambda_{q+1},r_{q+1,\tilde{n}}}\right)^2\right), \tag{8.97a}$$

$$\vartheta = \lambda_{q,n_{\max}+1,0}^{2d}\Delta^{-d}\mathbb{P}_{[q,n_{\max},p_{\max}+1]}\left(\varrho_{\xi,\lambda_{q+1},r_{q+1,\tilde{n}}}^2\right), \tag{8.97b}$$

where we recall that $\varrho_{\xi,\lambda,r}$ is defined via Propositions 4.3 and 4.4. We then have immediately that

$$\varrho = \mathbb{P}_{[q,n_{\max},p_{\max}+1]}\left(\left(\varrho_{\xi,\lambda_{q+1},r_{q+1,\tilde{n}}}\right)^2\right)$$
$$= \lambda_{q,n_{\max}+1,0}^{-2d}\Delta^d\lambda_{q,n_{\max}+1,0}^{2d}\Delta^{-d}\left(\mathbb{P}_{[q,n_{\max},p_{\max}+1]}\left(\varrho_{\xi,\lambda_{q+1},r_{q+1,\tilde{n}}}^2\right)\right)$$
$$= \lambda_{q,n_{\max}+1,0}^{-2d}\Delta^d\vartheta, \tag{8.98}$$

and so (1) from Proposition A.18 is satisfied. By property (1) of Proposition 4.3, we have that the functions ϱ and ϑ defined in (8.97) are both periodic to scale $(\lambda_{q+1}r_{q+1,\tilde{n}})^{-1} = \lambda_{q,\tilde{n}}^{-1}$, and so (2) is satisfied. The estimates in (A.69) follow with $\mathcal{C}_* = 1$ from standard Littlewood-Paley arguments (see also the discussion in part (b) of Remark A.21) and item (5) from Proposition 4.4. Note that in the case $N = 2d$ in (A.69), the inequality is weakened by a factor of $\lambda_{q+1}^{\alpha_R}$, for an arbitrary $\alpha_R > 0$; thus, (2) is satisfied. At this stage let us fix a value for this parameter α_R: we choose it to be sufficiently small (with respect to b and ε_Γ) to ensure that the loss $\lambda_{q+1}^{\alpha_R}$ may be absorbed by the spare negative factor of Γ_{q+1} in the definition of $\mathcal{C}_G$, as is postulated in (9.53). From (9.19), (9.22), (9.26), and (9.29), we have that

$$\widetilde{\lambda}_q \leq \lambda_{q,\tilde{n},\tilde{p}} \ll \lambda_{q,\tilde{n}} \leq \lambda_{q,n_{\max}+1,0} \leq \lambda_{q+1},$$

and so (A.70) is satisfied. From (9.48) we have that

$$\lambda_{q+1}^4 \leq \left(\frac{\lambda_{q,\tilde{n}}}{2\pi\sqrt{3}\Gamma_{q+1}\lambda_{q,\tilde{n},\tilde{p}}}\right)^{N_{\text{dec}}}$$

if N_{dec} is chosen large enough, and so (A.71) is satisfied. Applying the estimate (A.73) with α as in (9.53), recalling the value for $\mathcal{C}_G$ in (8.93), using (6.19) and (6.147) with $r_1 = \infty$ and $r_2 = 1$, we obtain that

$$\left\|D^N D_{t,q}^M\left(\mathcal{HO}_{\tilde{n},\tilde{p},n_{\max},p_{\max}+1}\right)\right\|_{L^1(\text{supp}\,\psi_{i,q})}$$

$$\lesssim \sum_{i'=i-1}^{i+1} \sum_{\xi,j,k,\vec{l}} \Lambda^{\alpha_R} |\text{supp}\,(\eta_{i,j,k,q,\widetilde{n},\widetilde{p},\vec{l}})| \Gamma_{q+1}^{2j-3-C_b} \Gamma_q^{-C_R}$$

$$\times \delta_{q+1}\widetilde{\lambda}_q \prod_{n'\leq\widetilde{n}} \left(f_{q,n'}\Gamma_{q+1}^{8+C_b}\right) C_* \zeta^{-1} \mathcal{M}\,(N,1,\zeta,\Lambda)\,\mathcal{M}\,(M,M_t,\nu,\widetilde{\nu})$$

$$\lesssim \Gamma_{q+1}\left(\Gamma_{q+1}^{-1}\Gamma_q^{-C_R}\delta_{q+1}\widetilde{\lambda}_q \prod_{n'\leq\widetilde{n}} \left(f_{q,n'}\Gamma_{q+1}^{8+C_b}\right)\right)$$

$$\times \lambda_{q,n_{\max}+1,0}^{-1}\lambda_{q+1}^N \mathcal{M}\left(M,N_{\text{ind},t},\tau_q^{-1}\Gamma_{q+1}^{i-c_{\widetilde{n}}+4},\widetilde{\tau}_q^{-1}\Gamma_{q+1}^{-1}\right)$$

$$\lesssim \Gamma_{q+1}^{-C_R}\Gamma_{q+1}^{-1}\delta_{q+2}\lambda_{q+1}^N \mathcal{M}\left(M,N_{\text{ind},t},\tau_q^{-1}\Gamma_{q+1}^{i-c_{\widetilde{n}}+4},\widetilde{\tau}_q^{-1}\Gamma_{q+1}^{-1}\right), \tag{8.99}$$

for $N, M \leq \lfloor 1/2\,(N_{\text{fin},\widetilde{n}} - N_{\text{cut},t} - N_{\text{cut},x} - 5)\rfloor - \mathsf{d}$. In the last inequality, we have used the parameter estimate (9.54), which directly implies

$$\Gamma_q^{-C_R}\delta_{q+1}\widetilde{\lambda}_q \prod_{n'\leq\widetilde{n}} \left(f_{q,n'}\Gamma_{q+1}^{8+C_b}\right) \lambda_{q,n_{\max}+1,0}^{-1} \leq \Gamma_{q+1}^{-C_R}\Gamma_{q+1}^{-1}\delta_{q+2}. \tag{8.100}$$

Then, after using (9.60c), which gives that for all $\widetilde{n}$ we have

$$\lfloor 1/2\,(N_{\text{fin},\widetilde{n}} - N_{\text{cut},t} - N_{\text{cut},x} - 5)\rfloor - \mathsf{d} \geq 3N_{\text{ind},v}, \tag{8.101}$$

the range of derivatives allowed in (8.99) is exactly as needed in (8.88), thereby proving this bound.

Continuing to follow the parameter choices in Remark A.19, we set $N_\circ = M_\circ = 3N_{\text{ind},v}$, and as before $N^\sharp = N_{\text{fin},\widetilde{n}} - N_{\text{cut},t} - N_{\text{cut},x} - 5$. From (9.60d), we have that the condition $N_\circ \leq N^\sharp/4$ is satisfied. The inequalities (A.75) and (A.76) follow from the discussion in Remark A.19. The inequality in (A.77) follows from (9.43), (9.55), the fact that $\lambda = \Gamma_{q+1}\lambda_{q,\widetilde{n},\widetilde{p}} \leq \Gamma_{q+1}\lambda_{q,\widetilde{n},p_{\max}}$, and $\zeta = \lambda_{q,n_{\max}+1,0} > \lambda_{q,n_{\max}-1} \geq \lambda_{q,\widetilde{n}}$, as in the discussion in Remark A.19. Having satisfied these assumptions, we may now appeal to the estimate in (A.79), which gives (8.87) for the case $\widetilde{n} < n = n_{\max}$, $p = p_{\max} + 1$, and any value of $\widetilde{p}$.

Recall we began this case by assuming that $\widetilde{n} < n_{\max}$. In the case $\widetilde{n} = n_{\max}$ and $1 \leq \widetilde{p} \leq p_{\max}$, we have from (9.22) and (9.29) that $\lambda_{q,n_{\max}} > \lambda_{q,n_{\max}+1,0}$, and so

$$\mathbb{P}_{[q,n_{\max},p_{\max}+1]}\mathbb{P}_{\geq\lambda_{q,\widetilde{n}}} = \mathbb{P}_{\geq\lambda_{q,n_{\max}}}.$$

Then we can set $\zeta = \mu = \lambda_{q,n_{\max}}$. The only change is that (8.100) becomes stronger, since $\lambda_{q,n_{\max}} > \lambda_{q,n_{\max}+1,0}$, and so the desired estimates follow by arguing as before. We omit further details.

Proof of item (3) and of item (1) when $p \neq p_{\max} + 1$ and $n \leq n_{\max}$. Note that in both of these cases we have $\widetilde{n} < n$. We first point that that we may assume that n and p are such that $\lambda_{q,\widetilde{n}} < \lambda_{q,n,p}$. If not, then $\mathbb{P}_{\geq\lambda_{q,\widetilde{n}}}\mathbb{P}_{[q,n,p]} = 0$, and so the estimate is trivially satisfied. We then set

$$\zeta = \max\left\{\lambda_{q,\tilde{n}}, \lambda_{q,n,p-1}\right\}, \qquad \mu = \lambda_{q,\tilde{n}}, \qquad \Lambda = \lambda_{q,n,p}, \tag{8.102}$$

and

$$\varrho = \mathbb{P}_{\geq \lambda_{q,\tilde{n}}} \mathbb{P}_{[q,n,p]}\left(\left(\varrho_{\xi,\lambda_{q+1},r_{q+1,\tilde{n}}}\right)^2\right), \tag{8.103a}$$

$$\vartheta = \zeta^{2d}\Delta^{-d}\mathbb{P}_{\geq \lambda_{q,\tilde{n}}} \mathbb{P}_{[q,n,p]}\left(\varrho_{\xi,\lambda_{q+1},r_{q+1,\tilde{n}}}^2\right). \tag{8.103b}$$

We then have from the discussion part (b) of Remark A.21 that

$$\varrho = \mathbb{P}_{\geq \lambda_{q,\tilde{n}}} \mathbb{P}_{[q,n,p]}\left(\left(\varrho_{\xi,\lambda_{q+1},r_{q+1,\tilde{n}}}\right)^2\right)$$
$$= \zeta^{-2d}\Delta^{d}\zeta^{2d}\Delta^{-d}\left(\mathbb{P}_{\geq \lambda_{q,\tilde{n}}} \mathbb{P}_{[q,n,p]}\left(\varrho_{\xi,\lambda_{q+1},r_{q+1,\tilde{n}}}^2\right)\right)$$
$$= \zeta^{-2d}\Delta^{d}\vartheta, \tag{8.104}$$

and so (1) from Proposition A.18 is satisfied. By property (1) of Proposition 4.3, ϱ and ϑ are both periodic to scale $(\lambda_{q+1}r_{q+1,\tilde{n}})^{-1} = \lambda_{q,\tilde{n}}^{-1}$, and so (2) is satisfied. The estimates in (A.69) follow with $\mathcal{C}_* = 1$ from the discussion in part (b) of Remark A.21. Note that in the case $N = 2d$ in (A.69), the inequality is weakened by a factor of $\lambda_{q+1}^{\alpha_R}$, and so (2) is satisfied. Here we again use α_R as in (9.53), so this loss will be absorbed using a factor of Γ_{q+1}. From (9.19), (9.26), (9.29), and (9.22), and the assumption that $\lambda_{q,\tilde{n}} < \lambda_{q,n,p}$, we have that

$$\tilde{\lambda}_q \leq \lambda_{q,\tilde{n},\tilde{p}} \ll \lambda_{q,\tilde{n}} \leq \max\left\{\lambda_{q,\tilde{n}}, \lambda_{q,n,p-1}\right\} \leq \lambda_{q,n,p},$$

and so, since $\Lambda \leq \lambda_{q+1}$, (A.70) is satisfied. From (9.48) we have that

$$\lambda_{q+1}^4 \leq \left(\frac{\lambda_{q,\tilde{n}}}{2\pi\sqrt{3}\Gamma_{q+1}\lambda_{q,\tilde{n},\tilde{p}}}\right)^{N_{\text{dec}}},$$

and so (A.71) is satisfied. Applying the estimate (A.73) for the parameter range in Remark A.19, recalling that (8.92) includes the indicator function of $\mathrm{supp}\,(\psi_{i,q})$, recalling the definition of $\mathcal{C}_G$ in (8.93), using (6.19) and (6.147) with $r_1 = \infty$ and $r_2 = 1$, and using $\zeta^{-1} \leq \lambda_{q,n,p-1}^{-1}$, we have that

$$\left\|D^N D_{t,q}^M\left(\mathcal{H}\mathcal{O}_{\tilde{n},\tilde{p},n,p}\right)\right\|_{L^1(\mathrm{supp}\,\psi_{i,q})}$$

$$\lesssim \sum_{i'=i-1}^{i+1}\sum_{\xi,j,k,\vec{l}} \Lambda^{\alpha_R}|\mathrm{supp}\,(\eta_{i,j,k,q,\tilde{n},\tilde{p},\vec{l}})|\Gamma_{q+1}^{2j-3-C_b}\Gamma_q^{-C_R}$$

$$\times \delta_{q+1}\tilde{\lambda}_q \prod_{n'\leq\tilde{n}}\left(f_{q,n'}\Gamma_{q+1}^{8+C_b}\right)\mathcal{C}_*\zeta^{-1}\mathcal{M}\left(N,1,\zeta,\Lambda\right)\mathcal{M}\left(M,M_t,\nu,\tilde{\nu}\right)$$

$$\lesssim \Gamma_{q+1}\Gamma_{q+1}^{-1}\Gamma_q^{-C_R}\delta_{q+1}\widetilde{\lambda}_q \prod_{n'\leq\widetilde{n}}\left(f_{q,n'}\Gamma_{q+1}^{8+C_b}\right)$$

$$\times \lambda_{q,n,p-1}^{-1}\lambda_{q,n,p}^N \mathcal{M}\left(M,\mathsf{N}_{\mathrm{ind},t},\tau_q^{-1}\Gamma_{q+1}^{i-c_{\widetilde{n}}+4},\widetilde{\tau}_q^{-1}\Gamma_{q+1}^{-1}\right)$$

$$\lesssim \delta_{q+1,n,p}\lambda_{q,n,p}^N \mathcal{M}\left(M,\mathsf{N}_{\mathrm{ind},t},\tau_q^{-1}\Gamma_{q+1}^{i-c_{\widetilde{n}}+4},\widetilde{\tau}_q^{-1}\Gamma_{q+1}^{-1}\right). \tag{8.105}$$

In the last inequality, we have used the fact that since $n < \widetilde{n}$, by (9.34) we have

$$\Gamma_q^{-C_R}\delta_{q+1}\widetilde{\lambda}_q \prod_{n'\leq\widetilde{n}}\left(f_{q,n'}\Gamma_{q+1}^{8+C_b}\right)\lambda_{q,n,p-1}^{-1} \leq \delta_{q+1,n,p} \tag{8.106}$$

for all $N, M \leq \lfloor 1/2\left(\mathsf{N}_{\mathrm{fin},\widetilde{n}} - \mathsf{N}_{\mathrm{cut},t} - \mathsf{N}_{\mathrm{cut},x} - 5\right)\rfloor - \mathsf{d}$. Then after using (9.61), which gives that for all $\widetilde{n} < n$

$$\lfloor 1/2\left(\mathsf{N}_{\mathrm{fin},\widetilde{n}} - \mathsf{N}_{\mathrm{cut},t} - \mathsf{N}_{\mathrm{cut},x} - 5\right)\rfloor - \mathsf{d} \geq \mathsf{N}_{\mathrm{fin},n}, \tag{8.107}$$

we have achieved (8.89).

Continuing to follow the parameter choices in Remark A.19, we set $N_\circ = M_\circ = 3\mathsf{N}_{\mathrm{ind},v}$, and as before $N^\sharp = \mathsf{N}_{\mathrm{fin},\widetilde{n}} - \mathsf{N}_{\mathrm{cut},t} - \mathsf{N}_{\mathrm{cut},x} - 5$. From (9.60d), we have that the condition $N_\circ \leq N^\sharp/4$ is satisfied. The inequalities (A.75) and (A.76) follow from the discussion in Remark A.19. The inequality in (A.77) follows from (9.55) and the fact that $\lambda = \Gamma_{q+1}\lambda_{q,\widetilde{n},\widetilde{p}} \leq \Gamma_{q+1}\lambda_{q,\widetilde{n},p_{\max}}$ and $\zeta = \max\{\lambda_{q,\widetilde{n}}, \lambda_{q,n,p-1}\} \geq \lambda_{q,\widetilde{n}}$. We then achieve the concluded estimate in (A.79), which gives (8.87) for the case $p \neq p_{\max} + 1$, $n \leq n_{\max}$, and any values of $\widetilde{n}, \widetilde{p}$ with $\widetilde{n} < n$. $\qquad\square$

8.7   TYPE 2 OSCILLATION ERRORS

In order to show that the Type 2 errors (previously identified in (8.44), (8.57), (8.58), (8.73), (8.74)) vanish, we will apply Proposition 4.8 on the support of a specific cutoff function

$$\eta = \eta_{i,j,k,q,n,p,\vec{l}} = \psi_{i,q}\chi_{i,k,q}\overline{\chi}_{q,n,p}\omega_{i,j,q,n,p}\zeta_{i,q,k,n,\vec{l}}.$$

Before we apply the proposition, we first estimate in Lemma 8.8 the number of cutoff functions η^* which may overlap with η, with an eye towards keeping track of all the pipes that we will have to dodge in order to successfully place pipes on η. The next three Lemmas ((8.9)–(8.11)) are technical in nature and are necessary in order to apply Lemma 4.7. Specifically, we show that given η, η^* and a fixed time t^*, one may find a convex set which contains the intersection of the supports of η and η^* at t^*. The time t^* will be the time at which the pipes on η^* are *straight*, and combined with the convexity, Lemma 4.7 may be applied. The upshot of this is that the pipes belonging to η^* only undergo mild deformations on the support of η. This allows us to finally apply Proposition 4.8 to place pipes on η which dodge all pipes originating from overlapping cutoff functions η^*. We remark that since $\overline{\chi}_{q,n,p}$ depends only on n and p, which are indices already encoded in $\omega_{i,j,q,n,p}$, throughout this section we will suppress the dependence of the cumulative cutoff function η on $\overline{\chi}_{q,n,p}$ (defined in (6.104)), as it does not affect any of the estimates.

8.7.1 Preliminary estimates

Lemma 8.8 (Keeping track of overlap). *Given a cutoff function* $\eta_{i,j,k,q,n,p,\vec{l}}$, *consider the set of all tuples* $\left(i^*, j^*, k^*, n^*, p^*, \vec{l}^*\right)$ *such that the cutoff function* $\eta_{i^*,j^*,k^*,q,n^*,p^*,\vec{l}^*}$ *satisfies:*

1. *$n^* \leq n$,*
2. *there exists (x,t) such that*

$$\eta_{i,j,k,q,n,p,\vec{l}}(x,t)\eta_{i^*,j^*,k^*,q,n^*,p^*,\vec{l}^*}(x,t) \neq 0. \tag{8.108}$$

Then the cardinality of the set of all such tuples is bounded above by $C_\eta \Gamma_{q+1}$, where the constant C_η depends only on $n_{\max}$, $p_{\max}$, $j_{\max}$, and dimensional constants. In particular, due to (9.2), (9.3), and (6.129), C_η is independent of q and the values of the other parameters indexing the cutoff functions.

Proof of Lemma 8.8. Recall that the cutoff functions are defined by

$$\eta_{i,j,k,q,n,p,\vec{l}}(x,t) = \psi_{i,q}(x,t)\chi_{i,k,q}(t)\overline{\chi}_{q,n,p}(t)\omega_{i,j,q,n,p}(x,t)\zeta_{i,q,k,n,\vec{l}}(x,t). \tag{8.109}$$

As noted in the outline of this section, we will suppress the dependence on $\overline{\chi}_{q,n,p}$, since the n and p indices are already accounted for in $\omega_{i,j,q,n,p}$. The proof proceeds by first counting the number of combinations (i^*, k^*) for which it is possible that there exists (x,t) such that

$$\psi_{i,q}(x,t)\chi_{i,k,q}(t)\psi_{i^*,q}(x,t)\chi_{i^*,k^*,q}(t) \neq 0. \tag{8.110}$$

Next, for a given (i^*, k^*), we count the number of values of (j^*, n^*, p^*) such that there exists (x,t) such that

$$\omega_{i,j,q,n,p}(x,t)\omega_{i^*,j^*,q,n^*,p^*}(x,t) \neq 0. \tag{8.111}$$

Finally, for a given $(i^*, k^*, j^*, n^*, p^*)$, we count the number of triples (l^*, w^*, h^*) such that $n^* \leq n$ and there exists (x,t) such that

$$\zeta_{i,q,k,n,p,\vec{l}}(x,t)\zeta_{i^*,q,k^*,n^*,p^*,\vec{l}^*}(x,t) \neq 0. \tag{8.112}$$

Recalling the definition of $\chi_{i,k,q}$ from (6.96) and (6.98), we see that $\psi_{i,q}\chi_{i,k^*,q}$ may have nonempty overlap with $\psi_{i,q}\chi_{i,k,q}$ if and only if $k^* \in \{k-1, k, k+1\}$.

Next, from (6.19), we have that only $\psi_{i-1,q}$ and $\psi_{i+1,q}$ may overlap with $\psi_{i,q}$. Now, let $(x,t) \in \operatorname{supp} \psi_{i,q}\chi_{i,k,q}$ be given such that there exists k_{i-1} such that

$$\psi_{i,q}(x,t)\chi_{i,k,q}(t)\psi_{i-1,q}(x,t)\chi_{i-1,k_{i-1},q}(t) \neq 0.$$

From the definition of $\chi_{i-1,k_{i-1},q}$, it is immediate that the diameter of the support of $\chi_{i-1,k_{i-1},q}$ is *larger* than the diameter of the support of $\chi_{i,k,q}$. It follows that there can be at most three values of k^* (one of which is k_{i-1}) such that $\chi_{i-1,k^*,q}$ has nonempty overlap with $\chi_{i,k,q}$. Finally, let $(x,t) \in \operatorname{supp} \psi_{i,q}\chi_{i,k,q}$ be given such that there exists k_{i+1} such that

$$\psi_{i,q}(x,t)\chi_{i,k,q}(t)\psi_{i+1,q}(x,t)\chi_{i+1,k_{i+1},q}(t) \neq 0.$$

From the definition of $\chi_{i+1,k^*,q}$, there exists a constant $\mathcal{C}_\chi$ depending on χ but not i, q, or k^* such that for all $|k'| \geq \mathcal{C}_\chi\Gamma_{q+1}$

$$\chi_{i+1,k_{i+1}+k',q}(t)\chi_{i,k,q}(t) = 0$$

for all $t \in \mathbb{R}$. Therefore, the number of k^* such that $\chi_{i+1,k^*,q}$ may have nonempty overlap with $\chi_{i,k,q}$ is no more than $2\mathcal{C}_\chi\Gamma_{q+1}+1$. In summary, the number of pairs (i^*,k^*) such that (8.110) holds for some (x,t) is bounded above by

$$3 + 3 + 2\mathcal{C}_\chi\Gamma_{q+1} + 1 \leq 3\mathcal{C}_\chi\Gamma_{q+1} \tag{8.113}$$

if λ_0 is sufficiently large, where the implicit constant is independent of q or any other parameters which index the cutoff functions.

Now let (i^*,k^*) be given such that $\psi_{i^*,q}\chi_{i^*,k^*,q}$ has nonempty overlap with $\psi_{i,q}\chi_{i,k,q}$. Once values of n^*, p^*, and j^* are chosen, these three parameters along with the value of i^* uniquely determine a stress cutoff function $\omega_{i^*,j^*,q,n^*,p^*}$. Since i^* was fixed, we may let j^*, n^*, and p^* vary. Using that $j^* \leq j_{max} \leq 4b/(\varepsilon_\Gamma(b-1))$ from (6.129), $n^* \leq n_{max}$, $p^* \leq p_{max}$, where n_{max} and p_{max} are independent of q, the number of tuples (i^*,k^*,j^*,n^*,p^*) such that there exists (x,t) with

$$\psi_{i,q}(x,t)\chi_{i,k,q}(x,t)\omega_{i,j,q,n,p}(x,t)\psi_{i^*,q}(x,t)\chi_{i^*,k^*,q}(x,t)\omega_{i^*,j^*,q,n^*,p^*}(x,t) \neq 0 \tag{8.114}$$

is bounded by a dimensional constant multiplied by $\Gamma_{q+1}n_{max}p_{max}4b/(\varepsilon_\Gamma(b-1))$.

Finally, fix a tuple (i^*,k^*,j^*,n^*,p^*) such that (8.114) holds at (x,t). From (6.139), there exists $\vec{l}^* = (l^*,w^*,h^*)$ such that $\zeta_{i^*,q,k^*,n^*,\vec{l}^*}(x,t) \neq 0$. From (6.141), (6.108), and the fact that $n^* \leq n$, there exists a dimensional constant $\mathcal{C}_\zeta$ such at most $\mathcal{C}_\zeta$ of the checkerboard cutoffs neighboring $\zeta_{i^*,q,k^*,n^*,\vec{l}^*}$ can intersect the support of $\zeta_{i,q,k,n,\vec{l}}$. Since all Lagrangian trajectories originating at (x,t) follow the same velocity field v_{ℓ_q} and the checkerboard cutoffs are precomposed with Lagrangian flows, this property is preserved in time. Thus we have shown that for each tuple (i^*,k^*,j^*,n^*,p^*), the number of associated tuples (l^*,w^*,h^*) such that $\zeta_{i^*,q,k^*,n^*,\vec{l}^*}$ can have nonempty intersection with $\zeta_{i,q,k,n,\vec{l}}$ is bounded by a dimensional constant independent of q.

Combining the preceding arguments, we obtain that the number of cutoff functions $\eta_{i^*,j^*,k^*,q,n^*,p^*,\vec{l}^*}$ which may overlap nontrivially with $\eta_{i,j,k,q,n,p,\vec{l}}$ is bounded by a dimensional constant multiplied by $\Gamma_{q+1} n_{\max} p_{\max} 4b/(\varepsilon_\Gamma(b-1))$, finishing the proof.   □

Lemma 8.9. *Let* $(x,t),(y,t) \in \operatorname{supp} \psi_{i,q}$ *be such that* $\psi_{i,q}^2(x,t) \geq 1/4$ *and* $\psi_{i,q}^2(y,t) \leq 1/8$. *Then there exists a geometric constant* $C_* > 1$ *such that*

$$|x - y| \geq C_* \left(\Gamma_q \lambda_q\right)^{-1}. \tag{8.115}$$

Proof Lemma 8.9. Let $L(x,y)$ be the line segment connecting x and y. From (6.36), we have that for $z \in L(x,y)$ (in fact for all $z \in \mathbb{T}^3$),

$$|\nabla \psi_{i,q}(z)| \lesssim \psi_{i,q}^{1-\frac{1}{N_{\text{fin}}}}(z)\lambda_q \Gamma_q. \tag{8.116}$$

Thus we can write

$$\frac{1}{8} \leq \left|\psi_{i,q}^2(x,t) - \psi_{i,q}^2(y,t)\right| \leq 2\left|\psi_{i,q}(x) - \psi_{i,q}(y)\right|$$

$$\leq 2\left|\int_0^1 \nabla \psi_{i,q}(x + t(y-x)) \cdot (y-x)\, dt\right|$$

$$\leq 2|x-y|\, \|\nabla \psi_{i,q}\|_{L^\infty}$$

$$\lesssim \Gamma_q \lambda_q |x-y|,$$

and (8.115) follows.   □

Lemma 8.10. *Consider cutoff functions*

$$\eta := \eta_{i,j,k,q,n,p,\vec{l}} = \psi_{i,q}\chi_{i,k,q}\omega_{i,j,q,n,p}\zeta_{i,k,q,n,\vec{l}},$$

$$\eta^* := \eta_{i^*,j^*,k^*,q,n^*,p^*,\vec{l}^*} = \psi_{i^*,q}\chi_{i^*,k^*,q}\omega_{i^*,j^*,q,n^*,p^*}\zeta_{i^*,k^*,q,n^*,\vec{l}^*},$$

where $n^* \leq n$ *and* η *and* η^* *overlap as in Lemma 8.8. Let* $t^* \in \operatorname{supp} \chi_{i^*,k^*,q}$ *be given. Then there exists a convex set* $\Omega := \Omega(\eta, \eta^*, t^*)$ *with diameter* $\lambda_{q,n,0}^{-1}\Gamma_{q+1}$ *such that*

$$\left(\operatorname{supp} \zeta_{i,k,q,n,\vec{l}} \cap \{t = t^*\}\right) \subset \Omega \subset \operatorname{supp} \psi_{i\pm,q}. \tag{8.117}$$

Proof of Lemma 8.10. Let $(x,t_0) \in \operatorname{supp}(\eta\eta^*)$. Then there exists $i' \in \{i-1,i,i+1\}$ such that $\psi_{i',q}^2(x,t_0) \geq \frac{1}{2}$. Consider the flow $X(x,t)$ originating from (x,t_0). Then for any t such that $|t-t_0| \leq \tau_q \Gamma_{q+1}^{-i+5+c_0}$, we can apply Lemma 6.24 to deduce that $\psi_{i',q}^2(t, X(x,t)) \geq \frac{1}{4}$. By the definition of $\chi_{i^*,k^*,q}$, the fact that $i^* \in \{i-1,i,i+1\}$, the existence of $(x,t_0) \in \operatorname{supp}(\chi_{i,k,q}\chi_{i^*,k^*,q})$, and the fact that $t^* \in \operatorname{supp} \chi_{i^*,k^*,q}$, we in particular deduce that $\psi_{i',q}^2(t^*, X(x,t^*)) \geq \frac{1}{4}$.

Now, let y be such that

$$|X(x,t^*) - y| \leq \lambda_{q,n,0}^{-1}\Gamma_{q+1} \leq \widetilde{\lambda}_q^{-1} < \mathcal{C}_*\widetilde{\lambda}_q^{-1}$$

for $\mathcal{C}_*$ given in (8.115), where we have used the definitions of $\lambda_{q,n,0}$ in (9.26)–(9.28). Then, from Lemma 8.9, it cannot be the case that $\psi_{i',q}^2(t^*,y) \leq \frac{1}{8}$, and thus

$$y \in \operatorname{supp}\psi_{i',q} \cap \{t = t^*\} \subset \operatorname{supp}\psi_{i\pm,q} \cap \{t = t^*\}. \tag{8.118}$$

Since y is arbitrary, we conclude that the ball of radius $\Gamma_{q+1}\lambda_{q,n,0}^{-1}$ is contained in $\operatorname{supp}\psi_{i\pm,q} \cap \{t = t^*\}$. We let $\Omega(\eta,\eta^*,t^*)$ be precisely this ball (hence a convex set). Since $D_{t,q}\zeta_{i,k,q,n,\vec{l}} = 0$ and $(x,t_0) \in \operatorname{supp}\zeta_{i,k,q,n,\vec{l}}$, we have that $X(x,t^*) \in \operatorname{supp}\zeta_{i,k,q,n,\vec{l}} \cap \{t = t^*\}$. Then, recalling that the support of $\zeta_{i,k,q,n,\vec{l}}$ must obey the diameter bound in (6.141) on the support of $\widetilde{\chi}_{i,k,q}$, which contains the support of $\chi_{i^*,k^*,q}$ by (6.103), we conclude that

$$\operatorname{supp}\zeta_{i,k,q,n,\vec{l}} \cap \{t = t^*\} \subset \Omega. \tag{8.119}$$

Combining (8.118) and (8.119) concludes the proof of the lemma. $\qquad\square$

Lemma 8.11. *As in Lemma 8.8, consider cutoff functions*

$$\eta := \eta_{i,j,k,q,n,p,\vec{l}} = \psi_{i,q}\chi_{i,k,q}\omega_{i,j,q,n,p}\zeta_{i,k,q,n,\vec{l}},$$

$$\eta^* := \eta_{i^*,j^*,k^*,q,n^*,p^*,\vec{l}^*} = \psi_{i^*,q}\chi_{i^*,k^*,q}\omega_{i^*,j^*,q,n^*,p^*}\zeta_{i^*,k^*,q,n^*,\vec{l}^*}.$$

Let $t^* \in \operatorname{supp}\chi_{i^*,k^*,q}$ be such that $\Phi^* := \Phi_{(i^*,k^*)}$ is the identity at time t^*. Using Lemma 8.10, define $\Omega := \Omega(\eta,\eta^*,t^*)$. Define $\Omega(t) := \Omega(\eta,\eta^*,t^*,t) := X(\Omega,t)$, where $X(\cdot,t^*)$ is the identity.

1. *For $t \in \operatorname{supp}\chi_{i,k,q}$,*

$$\operatorname{supp}\eta(\cdot,t) \subset \Omega(t) \subset \operatorname{supp}\psi_{i\pm,q}. \tag{8.120}$$

2. *Let $\mathbb{W}^* \circ \Phi^* := \mathbb{W}_{\xi^*,q+1,n^*}^{i^*,j^*,k^*,n^*,\vec{l}^*} \circ \Phi_{(i^*,k^*)}$ be an intermittent pipe flow supported on η^*. Then there exists a geometric constant $\mathcal{C}_{\text{pipe}}$ such that*

$$(\operatorname{supp}\mathbb{W}^* \circ \Phi^* \cap \{t = t^*\} \cap \Omega) \subset \bigcup_{n=1}^{N} S_n,$$

where the sets S_n are cylinders concentrated around line segments A_n for $n \in \{1,...,N\}$ with

$$N \leq \mathcal{C}_{\text{pipe}}\left(\frac{\lambda_{q,n}}{\lambda_{q,n,0}\Gamma_{q+1}^{-1}}\right)^2. \tag{8.121}$$

3. $\mathbb{W}^* \circ \Phi^*(\cdot, t)$ *and the associated axes* $A_n(t)$ *and sets* $S_n(t)$ *satisfy the conclusions of Lemma 4.7 on the set* $\Omega(t)$ *for* $t \in \operatorname{supp} \chi_{i,k,q}$.

Proof of Lemma 8.11. From the previous lemma, we have that for all $y \in \Omega$, $\psi^2_{i\pm,q}(y, t^*) \geq 1/8$. Applying Lemma 6.24, we have that for all t with $|t - t^*| \leq \tau_q \Gamma_{q+1}^{-i+5+c_0}$, the Lagrangian flow originating from (y, t^*) has the property that

$$\psi^2_{i\pm,q}(t, X(y, t)) \geq 1/16. \tag{8.122}$$

Recalling from (6.102) that the diameter of the support of $\widetilde{\chi}_{i^*,k^*,q}$ is $\tau_q \Gamma_{q+1}^{-i^*+c_0}$ and that $i - 1 \leq i^* \leq i + 1$, we have that in particular the Lagrangian flow originating at (y, t^*) satisfies (8.122) for all $t \in \operatorname{supp} \widetilde{\chi}_{i^*,k^*,q}$. From (6.103), (8.122) is then satisfied in particular for all $t \in \operatorname{supp} \chi_{i,k,q}$, thus proving the second inclusion from (8.120). To prove the first inclusion, we use (8.117), the definition of $\Omega(t)$, and the equality $D_{t,q}\zeta_{i,k,q,n,\vec{l}} = 0$ to deduce that

$$\operatorname{supp} \zeta_{i,k,q,n,\vec{l}}(\cdot, t) \subset \Omega(t),$$

finishing the proof of (8.120).

To prove the second claim, recall that $\mathbb{W}^* \circ \Phi^*$ at $t = t^*$ is periodic to scale λ_{q,n^*}^{-1} for $n^* \leq n$, and the diameter of Ω is $2\lambda_{q,n,0}^{-1}\Gamma_{q+1}$ (in fact Ω is a ball). Considering the quotient of the respective diameters squared, the claim then follows after absorbing the geometric constant n_ξ^* from Proposition 4.3 into $\mathcal{C}_{\text{pipe}}$.

To see that we may apply Lemma 4.7, first note that $\Omega = \Omega(t^*)$ is convex by construction, and so the first assumption of Lemma 4.7 is met. We choose $v = v_{\ell_q}$ and X and Φ to be the associated backward and forward flows originating from $t_0 = t^*$. From (6.60), (8.120), and (9.19), we have that for $t \in \operatorname{supp} \chi_{i,k,q}$ and $x \in \Omega(t)$,

$$\left|\nabla v_{\ell_q}(x, t)\right| \lesssim \delta_q^{1/2}\widetilde{\lambda}_q \Gamma_{q+1}^{i+2} = \delta_q^{1/2}\lambda_q \Gamma_{q+1}^{i+7}, \tag{8.123}$$

and so (4.21) is satisfied with $C = i + 7$. Recall again from (6.103) that $\operatorname{supp} \widetilde{\chi}_{i^*,k^*,q}$ contains the support of $\chi_{i,k,q}$, and that from (6.102) the support of $\widetilde{\chi}_{i^*,k^*,q}$ has diameter $\tau_q \Gamma_{q+1}^{-i^*+c_0}$. We then use (9.39) and (9.19) to write that for any $t \in \operatorname{supp} \widetilde{\chi}_{i^*,k^*,q}$ we have

$$|t - t^*| \leq \tau_q \Gamma_{q+1}^{-i^*+c_0+1} \leq \tau_q \Gamma_{q+1}^{-i+c_0+2}$$

$$\leq \left(\delta_q^{1/2}\widetilde{\lambda}_q \Gamma_{q+1}^{c_0+6}\right)^{-1} \Gamma_{q+1}^{-i+c_0+2}$$

$$= \left(\delta_q^{1/2}\lambda_q \Gamma_{q+1}^{c_0+11}\right)^{-1} \Gamma_{q+1}^{-i+c_0+2}$$

$$\leq \left(\delta_q^{1/2}\lambda_q \Gamma_{q+1}^{i+9}\right)^{-1},$$

so that (4.20) is satisfied since $C + 2 = i + 9$. We can now apply Lemma 4.7, concluding the proof of the lemma. $\qquad\qquad\qquad\qquad\qquad\qquad\qquad\qquad\square$

8.7.2   Applying Proposition 4.8

Lemma 8.12. *The Type 2 oscillation errors vanish. More specifically,*

1. *when $\tilde{n} = 0$, the Type 2 errors identified in* (8.44) *vanish;*
2. *when $1 \leq \tilde{n} \leq n_{\max} - 1$, the Type 2 errors identified in* (8.57) *and* (8.58) *vanish;*
3. *when $\tilde{n} = n_{\max}$, the Type 2 errors identified in* (8.73) *and* (8.74) *vanish.*

Proof of Lemma 8.12. We first recall what the Type 2 oscillation errors are. When $\tilde{n} = 0$, the errors identified in (8.44) can be written using (8.31) as

$$\mathcal{O}_{0,2} = \sum_{\neq \{\xi,i,j,k,\widetilde{p},\vec{l}\}} \operatorname{curl}\left(a_{(\xi)} \nabla \Phi_{(i,k)}^T \mathbb{U}_{\xi,q+1,0} \circ \Phi_{(i,k)} \right)$$
$$\otimes \operatorname{curl}\left(a_{(\xi^*)} \nabla \Phi_{(i^*,k^*)}^T \mathbb{U}_{\xi^*,q+1,0} \circ \Phi_{(i^*,k^*)} \right), \tag{8.124}$$

where the notation $\neq \{\xi, i, j, k, \widetilde{p}, \vec{l}\}$ is defined in (8.30) and denotes summation over all pairs of cutoff function indices for which at least one parameter differs between the two pairs. When $1 \leq \tilde{n} \leq n_{\max}$, the Type 2 errors identified in (8.57) and (8.73) can be written as

$$2 \sum_{n' \leq \tilde{n}-1} w_{q+1,\tilde{n}} \otimes_s w_{q+1,n'}$$

$$= 2 \sum_{n^* \leq \tilde{n}-1} \sum_{\xi,i,j,k,\widetilde{p},\vec{l}} \sum_{\xi^*,i^*,j^*,k^*,p^*,\vec{l}^*} \operatorname{curl}\left(a_{(\xi)} \nabla \Phi_{(i,k)}^T \mathbb{U}_{\xi,q+1,\tilde{n}} \circ \Phi_{(i,k)} \right)$$
$$\otimes_s \operatorname{curl}\left(a_{(\xi^*)} \nabla \Phi_{(i^*,k^*)}^T \mathbb{U}_{\xi^*,q+1,n^*} \circ \Phi_{(i^*,k^*)} \right). \tag{8.125}$$

When $1 \leq \tilde{n} \leq n_{\max}$, the Type 2 errors identified in (8.58) and (8.74) can be written as

$$\sum_{\neq \{\xi,i,j,k,\widetilde{p},\vec{l}\}} \operatorname{curl}\left(a_{(\xi)} \nabla \Phi_{(i,k)}^T \mathbb{U}_{\xi,q+1,\tilde{n}} \right) \otimes \operatorname{curl}\left(a_{(\xi^*)} \nabla \Phi_{(i^*,k^*)}^T \mathbb{U}_{\xi^*,q+1,\tilde{n}} \right),$$
$$\tag{8.126}$$

where the notation $\neq \{\xi, i, j, k, \widetilde{p}, \vec{l}\}$ has been reused from (8.30). To show that the errors defined in (8.124), (8.125), and (8.126) vanish, it suffices to show the following. For pairs of cutoff functions $\eta_{i,j,k,q,\tilde{n},\widetilde{p},\vec{l}}$ and $\eta_{i^*,j^*,k^*,q,n^*,p^*,\vec{l}^*}$ satisfying the two conditions in Lemma 8.8, and vectors $\xi, \xi^* \in \Xi$,

$$\operatorname{supp}\left(\mathbb{W}_{\xi,q+1,\tilde{n}}^{i,j,k,\tilde{n},\widetilde{p},\vec{l}} \circ \Phi_{(i,k)} \right) \cap \operatorname{supp} \eta_{i,j,k,q,\tilde{n},\widetilde{p},\vec{l}}$$
$$\cap \operatorname{supp}\left(\mathbb{W}_{\xi^*,q+1,n^*}^{i^*,j^*,k^*,n^*,p^*,\vec{l}^*} \circ \Phi_{(i^*,k^*)} \right) \cap \operatorname{supp} \eta_{i^*,j^*,k^*,q,n^*,p^*,\vec{l}^*} = \emptyset. \tag{8.127}$$

The proof of this claim will proceed by fixing $\tilde{n}$, using the preliminary estimates, and applying Proposition 4.8.

Let $\tilde{n}$ be fixed and assume that $w_{q+1,n'}$ for $n' < \tilde{n}$ has been defined (when $\tilde{n} = 0$, this assumption is vacuous). In particular, placements have been chosen for all intermittent pipe flows indexed by n'. Now, consider all the cutoff functions $\eta_{i,j,k,q,\tilde{n},\tilde{p},\vec{l}}$ utilized at stage $\tilde{n}$. Since the parameters indexing the cutoff functions are countable, we may choose *any* ordering of the tuples $(i, j, k, \tilde{p}, \vec{l})$ at level $\tilde{n}$. Combined with an ordering of the direction vectors $\xi \in \Xi$, we thus have an ordering of the cutoff functions $\eta_{i,j,k,q,\tilde{n},\tilde{p},\vec{l}}$ and the associated intermittent pipe flows $\mathbb{W}_{\xi,q+1,\tilde{n}}^{i,j,k,\tilde{n},\tilde{p},\vec{l}} \circ \Phi_{(i,k)}$.

To ease notation, we will abbreviate the cutoff functions as η_z and the associated intermittent pipe flows as $(\mathbb{W} \circ \Phi)_z$, where $z \in \mathbb{N}$ corresponds to the ordering. We will apply Proposition 4.8 inductively on z such that the following two conditions hold. Our goal is to place the pipe flow $(\mathbb{W} \circ \Phi)_z$ such that

$$\operatorname{supp}(\mathbb{W} \circ \Phi)_{z'} \cap \operatorname{supp}(\mathbb{W} \circ \Phi)_z \cap \operatorname{supp} \eta_z = \emptyset, \qquad (8.128)$$

for all $z' < z$, and such that

$$\operatorname{supp} w_{q+1,n'} \cap \operatorname{supp}(\mathbb{W} \circ \Phi)_z \cap \operatorname{supp} \eta_z = \emptyset, \qquad (8.129)$$

for all $n' < \tilde{n}$. The first condition shows that all Type 2 errors such as (8.124) and (8.126) which arise from two sets of pipes both indexed by $\tilde{n}$ vanish, while the second condition shows that the Type 2 errors which arise from pipes indexed by $n' < \tilde{n}$ interacting with pipes indexed by $\tilde{n}$ vanish, such as (8.125).

Throughout the rest of the proof, z' will only ever denote an integer less than z such that η_z and $\eta_{z'}$ overlap. Although we have suppressed the indices, note that $\eta_{z'}$ and η_z both correspond to the index $\tilde{n}$. Conversely, let $\eta_{z''}$ denote a generic cutoff function indexed by n' which overlaps with η_z. By Lemma 8.8, there exists a geometric constant $\mathcal{C}_\eta$ such that the number of cutoff functions $\eta_{z'}$ or $\eta_{z''}$ which overlap with η_z is bounded above by $\mathcal{C}_\eta \Gamma_{q+1}$. Let $t_{z'} \in \operatorname{supp} \chi_{i_{z'},k_{z'},q}$ be the time for which $\Phi_{i_{z'},k_{z'},q}$ is the identity, and let $\Omega(\eta_z, \eta_{z'}, t_{z'})$ be the convex set constructed in Lemma 8.10, where we have set $t^* = t_{z'}$. Let $\Omega(\eta_z, \eta_{z'}, t_{z'}, t)$ denote the image of $\Omega(\eta_z, \eta_{z'}, t_{z'})$ under this flow, as defined in Lemma 8.11. We then have that the set

$$\operatorname{supp}(\mathbb{W} \circ \Phi)_{z'} \cap \operatorname{supp} \Omega(\eta_z, \eta_{z'}, t_{z'}) \cap \{t = t_{z'}\} \qquad (8.130)$$

is contained in the union of sets $S_n^{z'}$ concentrated around axes $A_n^{z'}$ for

$$n \leq \mathcal{C}_{\text{pipe}} \Gamma_{q+1}^2 \frac{\lambda_{q,\tilde{n}}^2}{\lambda_{q,\tilde{n},0}^2},$$

and the flowed axes $A_n^{z'}$ and pipes of $(\mathbb{W} \circ \Phi)_{z'}$ satisfy the conclusions of Lemma 4.7. Furthermore, substituting z'' for z' in the preceding discussion, all the analogous definitions and conclusions can be made for cutoff functions $\eta_{z''}$ and pipe flows $(\mathbb{W} \circ \Phi)_{z''}$.

We will apply Proposition 4.8 with the following choices. Let t_z be the time at which the flow map $\Phi_{i,k,q}$ corresponding to η_z is the identity. Set

$$\Omega = \left(\bigcup_{z' < z} \Omega\left(\eta_z, \eta_{z'}, t_{z'}, t_z\right) \right) \bigcup \left(\bigcup_{n' < \tilde{n}} \Omega\left(\eta_z, \eta_{z''}, t_{z''}, t_z\right) \right) \qquad (8.131)$$

and set

$$r_1 = \Gamma_{q+1}^{-1} \frac{\lambda_{q,\tilde{n},0}}{\lambda_{q+1}} = \begin{cases} \left(\dfrac{\lambda_q}{\lambda_{q+1}}\right)^{\left(\frac{4}{5}\right)^{\tilde{n}-1} \cdot \frac{5}{6}} \Gamma_{q+1}^{-1} & \text{if } \tilde{n} \geq 2 \\[2mm] \left(\dfrac{\lambda_q}{\lambda_{q+1}}\right)^{\frac{4}{5}} \Gamma_{q+1}^{-1} & \text{if } \tilde{n} = 1 \\[2mm] \dfrac{\lambda_q}{\lambda_{q+1}} & \text{if } \tilde{n} = 0. \end{cases} \qquad (8.132)$$

We have used here the definitions of $\lambda_{q,\tilde{n},0}$ given in (9.27), (9.26), and (9.28). Note that by (8.120), $\operatorname{supp} \eta_z(\cdot, t_z) \subset \Omega\left(\eta_z, \eta_{z'}, t_{z'}, t_z\right)$ for each $z' < z$, with the analogous inclusion holding when z' is replaced by z''. In particular, we have that $\operatorname{supp} \eta_z(\cdot, t_z) \subset \Omega$. Furthermore, we have additionally from Lemma 8.11 that Lemma 4.7 may be applied on $\Omega(t)$ for all $t \in \chi_{i,k,q}$. Thus, the diameter of $\Omega(\eta_z, \eta_{z'}, t_{z'}, t_z)$ satisfies

$$\operatorname{diam}\left(\Omega\left(\eta_z, \eta_{z'}, t_{z'}, t_z\right)\right) \leq (1 + \Gamma_{q+1}^{-1})\operatorname{diam}\left(\Omega(\eta_z, \eta_{z'}, t_{z'})\right)$$
$$= 2(1 + \Gamma_{q+1}^{-1})\lambda_{q,\tilde{n},0}^{-1}\Gamma_{q+1}. \qquad (8.133)$$

Using the fact that the diameter of the support of $\eta_z(\cdot, t_z)$ is bounded by a dimensional constant time $\lambda_{q,\tilde{n},0}^{-1}$ from (6.141) and recalling that $\operatorname{supp} \eta_z(\cdot, t_z) \subset \Omega\left(\eta_z, \eta_{z'}, t_{z'}, t_z\right)$ with the analogous conclusion holding for z'', we have that

$$\operatorname{diam}(\Omega) \leq 4(1 + \Gamma_{q+1}^{-1})\lambda_{q,\tilde{n},0}^{-1}\Gamma_{q+1} + \Gamma_{q+1}\lambda_{q,\tilde{n},0}^{-1}$$
$$\leq 6(1 + \Gamma_{q+1}^{-1})\Gamma_{q+1}\left(\lambda_{q,\tilde{n},0}\right)^{-1}$$
$$\leq 16(\lambda_{q+1}r_1)^{-1}$$

for each value of $\tilde{n}$ from (8.132), and so (4.28) is satisfied.

Now set

$$\mathcal{C}_A = \mathcal{C}_{\text{pipe}}\mathcal{C}_\eta\Gamma_{q+1}, \qquad r_2 = r_{q+1,n} \approx \left(\frac{\lambda_q}{\lambda_{q+1}}\right)^{\left(\frac{4}{5}\right)^{\tilde{n}+1}},$$

where above we have appealed to (9.23) and (9.25). By (8.121) and Lemma 8.8, the total number of pipes contained in Ω is no more than

$$\mathcal{C}_{\text{pipe}}\mathcal{C}_\eta\Gamma_{q+1}^3 \frac{\lambda_{q,\tilde{n}}^2}{\lambda_{q,\tilde{n},1}^2}.$$

Then we can write

$$\mathcal{C}_{\text{pipe}}\mathcal{C}_\eta\Gamma_{q+1}^3 \frac{\lambda_{q,\tilde{n}}^2}{\lambda_{q,\tilde{n},0}^2} = \mathcal{C}_A \frac{r_2^2}{r_1^2},$$

and so (4.29) is satisfied. Furthermore, the assumptions on the axes and the neighborhoods of the axes required by Proposition 4.8 follow from Lemma 8.11, which allows us to appeal to the conclusions of Lemma 4.7. Finally, from (9.58a), we have that for $\tilde{n} \geq 2$,

$$C_*\mathcal{C}_A r_2^4 \leq 16C_*\mathcal{C}_{\text{pipe}}\mathcal{C}_\eta\Gamma_{q+1} \left(\frac{\lambda_q}{\lambda_{q+1}}\right)^{\left(\frac{4}{5}\right)^{\tilde{n}+1}\cdot 4}$$

$$\leq \left(\frac{\lambda_q}{\lambda_{q+1}}\right)^{\left(\frac{4}{5}\right)^{\tilde{n}-1}\cdot\frac{5}{6}\cdot 3} \Gamma_{q+1}^{-3} = r_1^3, \tag{8.134}$$

showing that (4.31) is satisfied for $\tilde{n} \geq 2$. In the cases $\tilde{n} = 0$ and $\tilde{n} = 1$, the desired inequalities follow from (8.132) and (9.58b) and (9.58c), and so we have checked that (4.31) is satisfied for all $0 \leq \tilde{n} \leq n_{\text{max}}$. Then from the conclusion (4.32) of Proposition 4.8, we have that on the support of Ω, which in particular contains the support of $\eta_z(\cdot, t_z)$ from (8.120), we can choose the support of $(\mathbb{W} \circ \Phi)_z$ to be disjoint from the support of $(\mathbb{W} \circ \Phi)_{z'}$ and $(\mathbb{W} \circ \Phi)_{z''}$ for all overlapping z'' and z'. Then since $D_{t,q}(\mathbb{W} \circ \Phi)_z = D_{t,q}(\mathbb{W} \circ \Phi)_{z'} = D_{t,q}(\mathbb{W} \circ \Phi)_{z''} = 0$, (8.128) and (8.129) are satisfied, concluding the proof. □

8.8   DIVERGENCE CORRECTOR ERRORS

Lemma 8.13. *For all $0 \leq \tilde{n} \leq n_{\text{max}}$, $1 \leq \tilde{p} \leq p_{\text{max}}$, and $j \in \{2,3\}$, the divergence corrector errors $\mathcal{O}_{\tilde{n},1,j}$ satisfy*

$$\left\|\psi_{i,q}D^k D_{t,q}^m \mathcal{O}_{\tilde{n},1,j}\right\|_{L^1} \lesssim \Gamma_{q+1}^{-\mathsf{C_R}-1}\delta_{q+2}\lambda_{q+1}^k \mathcal{M}\left(k, \mathsf{N}_{\text{ind},t}, \Gamma_{q+1}^{i+1}\tau_q^{-1}, \Gamma_{q+1}^{-1}\tilde{\tau}_q^{-1}\right)$$

for all $k, m \leq 3\mathsf{N}_{\text{ind},v}$.

Proof of Lemma 8.13. The divergence corrector errors are given in (8.32), (8.50), and (8.68). The estimates for $j = \{2,3\}$ are each similar, and so we shall only prove the case $j = 2$. Thus we estimate

$$\left\| \psi_{i,q} D^k D_{t,q}^m \sum_{\xi, i', j, k, \widetilde{p}, \vec{l}} \left(\left(a_{(\xi)} \nabla \Phi_{(i',k)}^{-1} \mathbb{W}_{\xi, q+1, \widetilde{n}} \circ \Phi_{(i,k)} \right) \right. \right.$$

$$\left. \left. \otimes \left(\nabla a_{(\xi)} \times \left(\nabla \Phi_{(i',k)}^T \mathbb{U}_{\xi, q+1, \widetilde{n}} \circ \Phi_{(i,k)} \right) \right) \right) \right\|_{L^1}. \qquad (8.135)$$

Recall that ξ takes only six distinct values and that $j \leq j_{\max}$, $\widetilde{p} \leq p_{\max}$ are bounded independently of q. Furthermore, on the support of $\psi_{i,q}$, only $\psi_{i-1,q}$, $\psi_{i,q}$, and $\psi_{i+1,q}$ are non-zero from (6.19). As a result, only time cutoffs $\chi_{i-1,k,q}$, $\chi_{i,k,q}$, and $\chi_{i+1,k,q}$ may be non-zero. Since for each i the $\chi_{i,k,q}$'s form a partition of unity in time for which only two cutoff functions are non-zero at any fixed time, for every time, the sum in (8.135) is a finite sum for which the number of non-zero terms in the summand is bounded independently of q. Similarly, the sum over $\vec{l}$ forms a partition of unity which only finitely many cutoff functions overlap at any fixed point in space and time. Therefore we may absorb the effects of ξ, j, k, $\widetilde{p}$, and $\vec{l}$ in the implicit constant in the inequality.

Using Hölder's inequality and estimates (8.17) and (8.18) from Corollary 8.2 with $r = 2$, $r_2 = 1$, and $r_1 = \infty$, we have that

$$\sum_{\xi, i', j, k, \widetilde{p}, \vec{l}} \left\| \psi_{i,q} D^k D_{t,q}^m \left(\left(a_{(\xi)} \nabla \Phi_{(i',k)}^{-1} \mathbb{W}_{\xi, q+1, \widetilde{n}} \circ \Phi_{(i',k)} \right) \right. \right.$$

$$\left. \left. \otimes \left(\nabla a_{(\xi)} \times \left(\nabla \Phi_{(i',k)}^T \mathbb{U}_{\xi, q+1, \widetilde{n}} \circ \Phi_{(i',k)} \right) \right) \right) \right\|_{L^1}$$

$$\lesssim \Gamma_{q+1}^{8+\mathsf{C}_b} \delta_{q+1, \widetilde{n}, \widetilde{p}} \lambda_{q+1}^k \mathcal{M} \left(m, \mathsf{N}_{\mathrm{ind}, t}, \tau_q^{-1} \Gamma_{q+1}^{i-\mathsf{c}_{\widetilde{n}}+4}, \widetilde{\tau}_q^{-1} \Gamma_{q+1}^{-1} \right) \frac{\lambda_{q, \widetilde{n}, \widetilde{p}}}{\lambda_{q+1}}$$

$$\lesssim \Gamma_{q+1}^{-\mathsf{C}_R-1} \delta_{q+2} \lambda_{q+1}^k \mathcal{M} \left(m, \mathsf{N}_{\mathrm{ind}, t}, \tau_q^{-1} \Gamma_{q+1}^{i+1}, \widetilde{\tau}_q^{-1} \Gamma_{q+1}^{-1} \right),$$

for $N, M \leq \lfloor 1/2 \left(\mathsf{N}_{\mathrm{fin}, \widetilde{n}} - \mathsf{N}_{\mathrm{cut}, t} - \mathsf{N}_{\mathrm{cut}, x} - 2\mathsf{N}_{\mathrm{dec}} - 9 \right) \rfloor$, which proves the desired estimate after recalling that for all $\widetilde{n}$,

$$\lfloor 1/2 \left(\mathsf{N}_{\mathrm{fin}, \widetilde{n}} - \mathsf{N}_{\mathrm{cut}, t} - \mathsf{N}_{\mathrm{cut}, x} - 2\mathsf{N}_{\mathrm{dec}} - 9 \right) \rfloor \geq 3\mathsf{N}_{\mathrm{ind}, v},$$

$$\Gamma_{q+1}^{8+\mathsf{C}_b} \frac{\delta_{q+1, \widetilde{n}, \widetilde{p}} \lambda_{q, \widetilde{n}, \widetilde{p}}}{\lambda_{q+1}} \leq \delta_{q+2} \Gamma_{q+1}^{-\mathsf{C}_R-1},$$

$$-\mathsf{c}_{\widetilde{n}} + 4 \leq 1,$$

which follow from (9.60b), (9.34) and (9.54), and (9.42), respectively. $\qquad \square$

8.9 TIME SUPPORT OF PERTURBATIONS AND STRESSES

First, we prove (7.12). Indeed, appealing to (5.1), which defines $\mathring{R}_{\ell_q}$ in terms of a mollifier applied to $\mathring{R}_q$, (9.20), which defines the scale at which $\mathring{R}_q$ is mollified, and (6.104), which ensures that the time support of $w_{q+1,0}$ is only enlarged relative to the time support of $\mathring{R}_{\ell_q}$ by $2\left(\delta_q^{1/2}\lambda_q\Gamma_{q+1}^2\right)^{-1}$, we achieve (7.12). To prove (7.14) and (7.16), first note that application of the inverse divergence operators $\mathcal{H}$ and $\mathcal{R}^*$ *commutes* with multiplication by $\overline{\chi}_{q,n,p}$.[8] Then by the definition of $\mathring{R}_{q+1}^0$ and $\mathring{H}_{q,n,p}^0$ in Section 8.3, we achieve (7.14) and (7.16). Proving the inclusions in (7.19), (7.21), (7.23), (7.26), (7.28), and (7.30) follows similarly from (6.104), the properties of $\mathcal{H}$ and $\mathcal{R}^*$, and the definitions of $\mathring{R}_{q+1}^{\tilde{n}}$ and $\mathring{H}_{q,n,p}^{\tilde{n}}$ in Section 8.3. Finally, to see that (7.4) follows from the inclusions already demonstrated, notice that the threshold in (7.4) is weaker than any of the previous inclusions by a factor of Γ_{q+1}, and so we may allow the time support of $\mathring{R}_{q+1}^{\tilde{n}}$ to expand slightly as $\tilde{n}$ increases from 0 to $n_{\max}$ while still meeting the desired inclusion.

---

[8]This is simple to check from the formula given in Proposition A.17 and the formula for the standard nonlocal inverse divergence operator given in (A.100), both of which involve operations which are purely spatial, such as differentiation and application of Fourier multipliers.

Chapter Nine

Parameters

The purpose of this section is to provide an exhaustive delineation of the many parameters, inequalities, and notations which arise throughout the bulk of the book. In Section 9.1, we define the q-independent parameters *in order*, beginning with the regularity index β, and ending with the number a_*, which will be used to absorb every implicit constant throughout the book. Then in Section 9.2, we define the parameters which depend on q, as well as the parameters which depend in addition on n and p. The definitions of both the q-independent and q-dependent parameters will appear rather arbitrary, but are justified in Section 9.3. This section contains, in no particular order, consequences of the definitions made in the previous two sections which are necessary to close the estimates in the proof. Finally, Sections 9.4 and 9.5 contain the definitions of a few operators and some notations that are used throughout the book.

9.1 DEFINITIONS AND HIERARCHY OF THE PARAMETERS

The parameters in our construction are chosen as follows:

1. Choose an arbitrary regularity parameter $\beta \in [1/3, 1/2)$. In light of [11, 43], there is no reason to consider the regime $\beta < 1/3$.
2. Choose $b \in (1, 3/2)$ sufficiently small such that

$$2\beta b < 1. \tag{9.1}$$

   The heuristic reason for (9.1) is given by (2.8). Note that (9.1) and the inequality $\beta < 1/2$ imply that $\beta(2b + 1) < 3/2$, which is a required inequality for the heuristic estimate (2.22).
3. With β and b chosen, we may now designate a number of parameters:

   a) The parameter $n_{\max}$, which per Section 2.4.2 denotes the total number of higher order stresses $\mathring{R}_{q,n}$ and thus primary frequency divisions in between λ_q and λ_{q+1}, is defined as the smallest integer for which

$$1 - 2\beta b > \frac{5}{6}\left(\frac{4}{5}\right)^{n_{\max}-1}. \tag{9.2}$$

   b) The parameter $p_{\max}$, which per Section 2.4.2 denotes the total number of subdivided components $\mathring{R}_{q,n,p}$ of a higher order stress $\mathring{R}_{q,n}$ and thus secondary frequency divisions in between λ_q and λ_{q+1}, is defined as the smallest integer for which

$$\frac{1}{p_{\max}} < \frac{1-2\beta b}{10}. \tag{9.3}$$

c) The parameter C_b appearing in (3.21) is use to quantify the L^1 norm of the velocity cutoff functions $\psi_{i,q}$. It is defined as

$$C_b = \frac{b+4}{b-1}. \tag{9.4}$$

d) The exponent C_R is used in order to define a small parameter in the estimate for the Reynolds stress; cf. (3.15). This parameter is then used in the proof to absorb geometric constants in the construction. It is defined as

$$C_R = 4b+1. \tag{9.5}$$

4. The parameter c_0, which is first introduced in (3.20) and utilized in Sections 7 and 8 to control small losses in the sharp material derivative estimates, is defined in terms of $n_{\max}$ as

$$c_0 = 4n_{\max} + 5. \tag{9.6}$$

5. The parameter $\varepsilon_\Gamma > 0$, which is used in (9.18) to quantify the *finest* frequency scale between λ_q and λ_{q+1} utilized throughout the scheme, is defined as the greatest real number for which the following inequalities hold:

$$\varepsilon_\Gamma \left(7 + C_R + n_{\max}(8 + C_b)) \right) < \frac{1-2\beta}{10} \tag{9.7a}$$

$$\varepsilon_\Gamma < \frac{1}{100} \left(\frac{4}{5} \right)^{n_{\max}-1} \tag{9.7b}$$

$$\varepsilon_\Gamma < \frac{b}{9(b-1)} \tag{9.7c}$$

$$2b\varepsilon_\Gamma(c_0 + 7) < 1 - \beta. \tag{9.7d}$$

6. The parameter $\alpha_R > 0$ from the L^1 loss of the inverse divergence operator is now defined as

$$\alpha_R = \frac{\varepsilon_\Gamma(b-1)}{2b}. \tag{9.8}$$

7. The parameters $N_{cut,t}$ and $N_{cut,x}$ are used in Chapter 6 in order to define the velocity and stress cutoff functions. $N_{cut,x}$ is the number of space

derivatives which are embedded into the definitions of these cutoff functions, while $N_{cut,t}$ is the number of material derivatives. See (6.6), (6.14), and (6.119). These large parameters are chosen solely in terms of b and ε_Γ as

$$\frac{1}{2}N_{cut,x} = N_{cut,t} = \left\lceil \frac{3b}{\varepsilon_\Gamma(b-1)} + \frac{15b}{2} \right\rceil. \tag{9.9}$$

8. The parameter $N_{ind,t}$, which is the number of sharp material derivatives propagated on stresses and velocities in Chapters 3 through 8, is chosen as the smallest integer for which we have

$$N_{ind,t} = \left\lceil \frac{4}{\varepsilon_\Gamma(b-1)} \right\rceil N_{cut,t}. \tag{9.10}$$

9. The parameter $N_{ind,v}$, whose primary role is to quantify the number of sharp space derivatives propagated on the velocity increments and stresses—cf. (3.12) and (3.15)—is chosen as the smallest integer for which we have the bounds

$$4bN_{ind,t} + 8 + b(C_R + 3)\varepsilon_\Gamma(b-1) + 2\beta(b^3 - 1) < \varepsilon_\Gamma(b-1)N_{ind,v}. \tag{9.11}$$

10. The value of the decoupling parameter N_{dec}, which is used in the L^p decorrelation Lemma A.2, is chosen as the smallest integer for which we have

$$N_{dec}\left(\frac{1}{30} \left(\frac{4}{5} \right)^{n_{max}} - \varepsilon_\Gamma \right) > \frac{4b}{b-1}. \tag{9.12}$$

11. The value of the parameter d, which in essence is used in the inverse divergence operator of Proposition A.18 to count the order of a parametric expansion, is chosen as the smallest integer for which we have

$$(d-1)\left(\frac{1}{30} \left(\frac{4}{5} \right)^{n_{max}} - \varepsilon_\Gamma \right) > \frac{(12N_{ind,v} + 7)b}{b-1}. \tag{9.13}$$

12. The value of N_{fin}, which is introduced in Chapter 3 and used to quantify the highest order derivative estimates utilized throughout the scheme, is chosen as the smallest integer such that

$$\frac{3}{2}N_{fin} > (2N_{cut,t} + N_{cut,x} + 14N_{ind,v} + 2d + 2N_{dec} + 12)2^{n_{max}+1}. \tag{9.14}$$

13. Having chosen all the previous parameters in items (1)–(12), there exists a *sufficiently large* parameter $a_* \geq 1$ which depends on all the parameters listed above (which recursively means that $a_* = a_*(\beta, b)$), and which

allows us to choose a an *arbitrary number* in the interval $[a_*, \infty)$. While we do not give a formula for a_* explicitly, it is chosen so that $a_*^{(b-1)\varepsilon_\Gamma}$ is at least twice larger than *all the implicit constants in the $\lesssim$ symbols throughout the book*; note that these constants depend only on the parameters in items (1)–(12) —never on q— which justifies the existence of a_*.

Having made the choices in items (1)–(13) above, we are now ready to define the q-dependent parameters which appear in the proof.

9.2   DEFINITIONS OF THE Q-DEPENDENT PARAMETERS

9.2.1   Parameters which depend on q

For $q \geq 0$, we define the fundamental frequency parameter used in this book as

$$\lambda_q = 2^{\left\lceil (b^q)\log_2 a \right\rceil}. \tag{9.15}$$

Definition (9.15) gives that λ_q is an integer power of 2, and that we have the bounds

$$a^{(b^q)} \leq \lambda_q \leq 2a^{(b^q)} \quad \text{and} \quad \frac{1}{3}\lambda_q^b \leq \lambda_{q+1} \leq 2\lambda_q^b \tag{9.16}$$

for all $q \geq 0$. Throughout the book the above two inequalities are used by putting the factors of $1/3$ and 2 into the implicit constants of $\lesssim$ symbols. In terms of λ_q, the fundamental amplitude parameter used in the book is

$$\delta_q = \lambda_1^{(b+1)\beta}\lambda_q^{-2\beta}. \tag{9.17}$$

In terms of the parameter ε_Γ from (9.7), we introduce a parameter which is used repeatedly throughout the book to mean "a tiny power of the frequency parameter":

$$\Gamma_{q+1} = \left(\frac{\lambda_{q+1}}{\lambda_q}\right)^{\varepsilon_\Gamma}. \tag{9.18}$$

In order to cap off our derivative losses, we need to mollify in space and time using the operators described in Section 9.4 below. This is done in terms of the following space and time parameters:

$$\widetilde{\lambda}_q = \lambda_q \Gamma_{q+1}^5 \tag{9.19}$$

$$\widetilde{\tau}_q^{-1} = \tau_q^{-1}\widetilde{\lambda}_q^3\widetilde{\lambda}_{q+1}. \tag{9.20}$$

While $\tilde{\tau}_q$ is used for mollification and thus for rough material derivative bounds, the fundamental time parameter used in the book for sharp material derivative bounds is

$$\tau_q = \left(\delta_q^{1/2} \tilde{\lambda}_q \Gamma_{q+1}^{c_0+6} \right)^{-1} . \tag{9.21}$$

Note that besides depending on the parameters introduced in (1)–(13), the parameters introduced above only depend on q, but are independent of n and p.

9.2.2 Parameters which depend also on n and p

The rest of the parameters depend on $n \in \{0, \ldots, n_{\max}\}$ and $p \in \{0, \ldots, p_{\max}\}$. We start by defining the frequency parameter $\lambda_{q,n}$ and the intermittency parameter $r_{q+1,n}$ by

$$\lambda_{q,n} = 2^{\left\lceil \left(\frac{4}{5}\right)^{n+1} \log_2 \lambda_q + \left(1 - \left(\frac{4}{5}\right)^{n+1}\right) \log_2 \lambda_{q+1} \right\rceil} \tag{9.22}$$

$$r_{q+1,n} = \frac{\lambda_{q,n}}{\lambda_{q+1}} \tag{9.23}$$

for $0 \le n \le n_{\max}$. In particular, (9.22) shows that $\lambda_{q+1} r_{q+1,n}$ is an integer power of 2, and we have the bound

$$\lambda_q^{\left(\frac{4}{5}\right)^{n+1}} \lambda_{q+1}^{1-\left(\frac{4}{5}\right)^{n+1}} \le \lambda_{q,n} \le 2\lambda_q^{\left(\frac{4}{5}\right)^{n+1}} \lambda_{q+1}^{1-\left(\frac{4}{5}\right)^{n+1}} , \tag{9.24}$$

while (9.23) implies that r_{q+1}^{-1} is an integer power of 2, and we have the estimates

$$\left(\frac{\lambda_q}{\lambda_{q+1}} \right)^{\left(\frac{4}{5}\right)^{n+1}} \le r_{q+1,n} \le 2 \left(\frac{\lambda_q}{\lambda_{q+1}} \right)^{\left(\frac{4}{5}\right)^{n+1}} . \tag{9.25}$$

As with (9.16) we absorb the factors of 2 in (9.24) and (9.25) into the implicit constants in $\lesssim$ symbols.

We also define the frequency parameters $\lambda_{q,n,p}$ by

$$\lambda_{q,0,p} = \Gamma_{q+1} \tilde{\lambda}_q \qquad\qquad n = 0, 0 \le p \le p_{\max} \tag{9.26}$$

$$\lambda_{q,1,0} = \lambda_q^{\frac{4}{5}} \lambda_{q+1}^{\frac{1}{5}} \qquad\qquad n = 1, p = 0 \tag{9.27}$$

$$\lambda_{q,n,0} = \lambda_q^{\left(\frac{4}{5}\right)^{n-1} \cdot \frac{5}{6}} \lambda_{q+1}^{1-\left(\frac{4}{5}\right)^{n-1} \cdot \frac{5}{6}} \qquad\qquad 2 \le n \le n_{\max} + 1 \tag{9.28}$$

$$\lambda_{q,n,p} = \lambda_{q,n,0}^{1-p/p_{\max}} \lambda_{q,n+1,0}^{p/p_{\max}} \qquad\qquad 1 \le n \le n_{\max}, 0 \le p \le p_{\max}. \tag{9.29}$$

For $0 \le n \le n_{\max}$, we define

$$f_{q,0} = 1, \qquad\qquad n = 0, \tag{9.30}$$

$$f_{q,n} = \left(\frac{\lambda_{q,n+1,0}}{\lambda_{q,n,0}} \right)^{1/p_{\max}}, \qquad 1 \le n \le n_{\max}. \qquad (9.31)$$

We define $\delta_{q+1,0,p}$ by

$$\delta_{q+1,0,1} = \Gamma_q^{-C_R} \delta_{q+1}, \qquad\qquad p = 1, \qquad (9.32)$$

$$\delta_{q+1,0,p} = 0, \qquad\qquad 2 \le p \le p_{\max}. \qquad (9.33)$$

When $1 \le n \le n_{\max}$ and $1 \le p \le p_{\max}$, we define $\delta_{q+1,n,p}$ by

$$\delta_{q+1,n,p} = \Gamma_q^{-C_R} \delta_{q+1} \cdot \left(\frac{\widetilde{\lambda}_q}{\lambda_{q,n,p-1}} \right) \cdot \prod_{n'<n} \left(f_{q,n'} \Gamma_{q+1}^{8+C_b} \right). \qquad (9.34)$$

We remark that by the definition of $\lambda_{q,1,0}$ given in (9.27), and more generally $\lambda_{q,n,p}$ in (9.29), the fact that $n \ge 1$, and a large choice of $p_{\max}$ which makes $f_{q,n}$ (defined in (9.31)) small, $\delta_{q+1,n,p}$ is significantly smaller than $\Gamma_q^{-C_R} \delta_{q+1}$.

For $1 \le n \le n_{\max}$, we define c_n in terms of c_0 by

$$c_n = c_0 - 4n. \qquad (9.35)$$

For $n = 0$, we set

$$N_{\text{fin},0} = \frac{3}{2} N_{\text{fin}}, \qquad (9.36)$$

while for $1 \le n \le n_{\max}$, we define $N_{\text{fin},n}$ inductively on n by using (9.36) and the formula

$$N_{\text{fin},n} = \left\lfloor \frac{1}{2} \left(N_{\text{fin},n-1} - N_{\text{cut},t} - N_{\text{cut},x} - 6 \right) - d \right\rfloor. \qquad (9.37)$$

9.3 INEQUALITIES AND CONSEQUENCES OF THE PARAMETER DEFINITIONS

The definitions made in the previous two sections have the following consequences, which will be used frequently throughout the book.

Due to (9.15) we have that $\Gamma_{q+1} \ge (1/2)^{b\varepsilon_\Gamma} \lambda_q^{(b-1)\varepsilon_\Gamma} \ge (1/2)^{b\varepsilon_\Gamma} \lambda_0^{(b-1)\varepsilon_\Gamma} \ge (1/2)a_*^{(b-1)\varepsilon_\Gamma}$. As was already mentioned in item (13), we have chosen a_* to be sufficiently large so that $a_*^{(b-1)\varepsilon_\Gamma}$ is at least twice larger than all the implicit constants appearing in all $\lesssim$ symbols throughout the book. Therefore, for any $q \ge 0$, we may use a single power of Γ_{q+1} to absorb any implicit constant in the book: an inequality of the type $A \lesssim B$ may be rewritten as $A \le \Gamma_{q+1} B$.

From (9.18), (9.19), and (9.7c), we have that

$$\Gamma_{q+1}^4 \widetilde{\lambda}_q \le \lambda_{q+1}. \qquad (9.38)$$

From the definition (9.21) of τ_q and from (9.35), which gives that c_n is decreasing with respect to n, we have that for all $0 \le n \le n_{\max}$

$$\Gamma_{q+1}^{c_n+6} \delta_q^{1/2} \widetilde{\lambda}_q \le \tau_q^{-1}. \tag{9.39}$$

Using the definitions (9.17), (9.18), (9.19), and (9.21), writing out everything in terms of λ_{q-1}, and appealing to (9.7d), we have that

$$\tau_{q-1}^{-1} \Gamma_{q+1}^{3+c_0} \le \tau_q^{-1} \tag{9.40}$$

$$\tau_{q-1}^{-1} \Gamma_{q+1} \le \delta_q^{1/2} \lambda_q. \tag{9.41}$$

From the definitions (9.6) of c_0 and (9.35) of c_n, we have that for all $0 \le n \le n_{\max}$,

$$-c_n + 4 \le -1. \tag{9.42}$$

From the definition of $\widetilde{\tau}_q$, it is immediate that

$$\tau_q^{-1} \widetilde{\lambda}_q^4 \le \widetilde{\tau}_q^{-1} \le \tau_q^{-1} \widetilde{\lambda}_q^3 \widetilde{\lambda}_{q+1}. \tag{9.43}$$

From (9.7d), the assumption that $\beta \ge 1/3$, and the assumption that $b \le 3/2$, we can write everything out in terms of λ_q to deduce that

$$\tau_q^{-1} \Gamma_{q+1}^9 \le \tau_{q+1}^{-1}. \tag{9.44}$$

From the definitions (9.22) and (9.26)–(9.29), for all $0 \le n \le n_{\max}$ and $0 \le p \le p_{\max}$ we have

$$\frac{\lambda_{q,n,p}}{\lambda_{q,n}} \ll 1.$$

More precisely, when $n = 0$ we have that

$$\frac{\Gamma_{q+1} \lambda_{q,n,p}}{\lambda_{q,n}} = \frac{\Gamma_{q+1}^2 \widetilde{\lambda}_q}{\lambda_{q,0}} = \frac{\Gamma_{q+1}^7 \lambda_q}{\lambda_{q,0}} = \left(\frac{\lambda_{q+1}}{\lambda_q} \right)^{-\frac{1}{5} + 7\varepsilon_\Gamma} \tag{9.45}$$

while for $n \ge 1$ it holds that

$$\frac{\Gamma_{q+1} \lambda_{q,n,p}}{\lambda_{q,n}} \le \frac{\Gamma_{q+1} \lambda_{q,n+1,0}}{\lambda_{q,n}} = \left(\frac{\lambda_{q+1}}{\lambda_q} \right)^{(\frac{4}{5})^n (\frac{4}{5} - \frac{5}{6}) + \varepsilon_\Gamma} \le \left(\frac{\lambda_{q+1}}{\lambda_q} \right)^{-\frac{1}{30}(\frac{4}{5})^{n_{\max}} + \varepsilon_\Gamma} \tag{9.46}$$

as it is clear that the quotient on the left-hand side is largest when $n = n_{\max}$. Note that due to (9.2) we have $\frac{1}{30} \left(\frac{4}{5} \right)^{n_{\max}} - \varepsilon_\Gamma < \frac{1-2\beta b}{30} - \varepsilon_\Gamma \le \frac{1}{5} - 7\varepsilon_\Gamma$; here we also used that $\varepsilon_\Gamma \le \frac{1}{36}$, which handily follows from (9.7b). Combining (9.45) and (9.46) we thus arrive at

$$\frac{\Gamma_{q+1} \lambda_{q,n,p}}{\lambda_{q,n}} \le \left(\frac{\lambda_{q+1}}{\lambda_q} \right)^{-\frac{1}{30}(\frac{4}{5})^{n_{\max}} + \varepsilon_\Gamma} \le \left(2\lambda_q^{b-1} \right)^{-\frac{1}{30}(\frac{4}{5})^{n_{\max}} + \varepsilon_\Gamma} \tag{9.47}$$

for all $0 \le n \le n_{\max}$ and $0 \le p \le p_{\max}$. Combining the above estimate with our choice of N_{dec} in (9.12), we thus arrive at

$$\lambda_{q+1}^4 \le \left(\frac{\lambda_{q,\tilde{n}}}{2\pi\sqrt{3}\Gamma_{q+1}\lambda_{q,\tilde{n},\tilde{p}}} \right)^{N_{\mathrm{dec}}} . \tag{9.48}$$

for all $0 \le \tilde{n} \le n_{\max}$ and $1 \le \tilde{p} \le p_{\max}$.

Next, we a list a few consequences of the fact that $N_{\mathrm{ind},v} \gg N_{\mathrm{ind},t}$, as specified in (9.11). First, we note from (9.43) that

$$\tilde{\tau}_{q-1}^{-1}\tau_{q-1} \le \tilde{\lambda}_{q-1}^3 \tilde{\lambda}_q \le \lambda_q^4 \tag{9.49}$$

where in the second inequality we have used the fact that $\varepsilon_\Gamma \le \frac{3}{20b}$. In turn, the above inequality combined with (9.11) implies the following estimates, all of which are used for the first time in Chapter 5:

$$\lambda_{q-1}^8 \Gamma_{q+1}^{1+C_{\mathrm{R}}\frac{\delta_{q-1}}{\delta_{q+2}}} \left(\tilde{\tau}_{q-1}^{-1}\tau_{q-1} \right)^{N_{\mathrm{ind},t}} \le \Gamma_q^{N_{\mathrm{ind},v}-2} \tag{9.50a}$$

$$\tilde{\lambda}_q^2 \left(\tilde{\tau}_{q-1}^{-1}\tau_{q-1} \right)^{N_{\mathrm{ind},t}} \le \Gamma_{q+1}^{5N_{\mathrm{ind},v}} \tag{9.50b}$$

$$\lambda_{q-1}^4 \delta_{q-1}^{1/2}\Gamma_q^2\delta_q^{-1/2}\left(\tilde{\tau}_{q-1}^{-1}\tau_{q-1} \right)^{N_{\mathrm{ind},t}} \le \Gamma_q^{N_{\mathrm{ind},v}} . \tag{9.50c}$$

Next, as a consequence of our choice of $N_{\mathrm{cut},t}$ and $N_{\mathrm{cut},x}$ in (9.9), we obtain the following bounds, which are used in Chapter 6:

$$\tilde{\lambda}_q^{3/2}\Gamma_q^{-N_{\mathrm{cut},t}} \le \lambda_q^3\Gamma_q^{-N_{\mathrm{cut},t}} \le 1 \tag{9.51}$$

for all $q \ge 0$. The fact that $N_{\mathrm{ind},t}$ is taken to be much larger than $N_{\mathrm{cut},t}$, as expressed in (9.10), implies when combined with (9.49) the following bound, which is also used in Chapter 6:

$$\left(\tau_q\tilde{\tau}_q^{-1} \right)^{N_{\mathrm{cut}}} \le \lambda_{q+1}^{4N_{\mathrm{cut}}} \le \Gamma_{q+1}^{N_{\mathrm{ind},t}} \tag{9.52}$$

for all $q \ge 1$.

The parameter α_{R} is chosen in (9.8) in order to ensure the inequality

$$\lambda_{q+1}^{\alpha_{\mathrm{R}}} \le \Gamma_{q+1} \tag{9.53}$$

for all $q \ge 0$. This fact is used in Chapter 8. Several other, much more hideous, parameter inequalities are used in Chapter 8, and for the reader's convenience we list them next. First, we claim that

$$\Gamma_{q+1}\Gamma_q^{-C_{\mathrm{R}}}\delta_{q+1}\tilde{\lambda}_q \prod_{n' \le n_{\max}} \left(f_{q,n'}\Gamma_{q+1}^{8+C_b} \right) \lambda_{q,n_{\max}+1,0}^{-1} \le \Gamma_{q+1}^{-C_{\mathrm{R}}}\Gamma_{q+1}^{-1}\delta_{q+2} . \tag{9.54}$$

In order to verify the above bound, we appeal to the choices made in (9.1), (9.2), and (9.3), to the definitions (9.19), (9.27), (9.28), and (9.31), and to the fact that $\tilde{n} \leq n_{\max}$ to deduce that the left side of (9.54) is bounded from above by

$$\delta_{q+1} \Gamma_{q+1}^{6+n_{\max}(8+C_b)} \frac{\lambda_q}{\lambda_{q,n_{\max}+1,0}} \left(\frac{\lambda_{q,n_{\max}+1,0}}{\lambda_{q,1,0}} \right)^{\frac{1}{p_{\max}}}$$

$$= \delta_{q+1} \Gamma_{q+1}^{6+n_{\max}(8+C_b)} \left(\frac{\lambda_q}{\lambda_{q+1}} \right)^{(1-(\frac{4}{5})^{n_{\max}}\frac{5}{6})} \left(\frac{\lambda_{q+1}}{\lambda_q} \right)^{\frac{1}{p_{\max}}\left(\frac{4}{5}-(\frac{4}{5})^{n_{\max}}\frac{5}{6}\right)}$$

$$\leq \frac{\lambda_q \delta_{q+1}}{\lambda_{q+1}} \Gamma_{q+1}^{6+n_{\max}(8+C_b)} \left(\frac{\lambda_{q+1}}{\lambda_q} \right)^{(1-2\beta b)\frac{4}{5}} \left(\frac{\lambda_{q+1}}{\lambda_q} \right)^{\frac{1-2\beta b}{10}\frac{4}{5}}$$

$$\leq \left(\Gamma_{q+1}^{-C_R} \Gamma_{q+1}^{-1} \delta_{q+2} \right) \frac{\lambda_q \delta_{q+1}}{\lambda_{q+1} \delta_{q+2}} \Gamma_{q+1}^{7+C_R+n_{\max}(8+C_b)} \left(\frac{\lambda_{q+1}}{\lambda_q} \right)^{(1-2\beta b)\frac{22}{25}}$$

$$\leq \left(\Gamma_{q+1}^{-C_R} \Gamma_{q+1}^{-1} \delta_{q+2} \right) \Gamma_{q+1}^{7+C_R+n_{\max}(8+C_b)} \left(\frac{\lambda_{q+1}}{\lambda_q} \right)^{-(1-2\beta b)\frac{3}{25}}.$$

The proof of (9.54) is now completed by appealing to (9.7a), which ensures that Γ_{q+1} represents a sufficiently small power of λ_{q+1}/λ_q.

Next, we claim that due to our choice of d, we have

$$\Gamma_q^{-C_R} \delta_{q+1} \tilde{\lambda}_q \prod_{n' \leq n_{\max}} \left(f_{q,n'} \Gamma_{q+1}^{8+C_b} \right) \lambda_{q+1} \left(\frac{\Gamma_{q+1} \lambda_{q,\tilde{n},p_{\max}}}{\lambda_{q,\tilde{n}}} \right)^{d-1} \left(\lambda_{q+1}^4 \right)^{3N_{ind,v}}$$

$$\leq \frac{\delta_{q+2}}{\lambda_{q+1}^5}. \tag{9.55}$$

In order to verify the above bound we use the previously established estimate (9.54) in conjunction with (9.47); after dropping the helpful factor of $\Gamma_{q+1}^{-2-C_R}$, we deduce that the left side of (9.55) is bounded from above by

$$\delta_{q+2} \lambda_{q,n_{\max}+1,0} \lambda_{q+1} \left(\frac{\Gamma_{q+1} \lambda_{q,\tilde{n},p_{\max}}}{\lambda_{q,\tilde{n}}} \right)^{d-1} \left(\lambda_{q+1}^4 \right)^{3N_{ind,v}}$$

$$\leq \frac{\delta_{q+2}}{\lambda_{q+1}^5} \lambda_{q+1}^3 \left(2\lambda_q^{b-1} \right)^{-(d-1)\left(\frac{1}{30}(\frac{4}{5})^{n_{\max}}-\varepsilon_\Gamma\right)} \lambda_{q+1}^{12N_{ind,v}}.$$

The choice of d in (9.13) shows that the above estimate directly implies (9.55).

The amplitudes of the higher order corrections $w_{q+1,n,p}$ must meet the inductive assumptions stated in (3.13). In order to meet the satisfactory bound in Remark 8.3, from (9.32)–(9.34) we deduce the bound

$$\delta_{q+1,\tilde{n},\tilde{p}}^{1/2} \leq \Gamma_{q+1}^{-2} \delta_{q+1}^{1/2}. \tag{9.56}$$

Indeed, the case $\tilde{n} = 0$ follows from the definition of C_R in (9.5), while the case $\tilde{n} \geq 1$ is a consequence of the definition (9.34), which implies that $\delta_{q,\tilde{n},\tilde{p}} \leq \delta_{q,0,1}$, for any $\widetilde{n} \geq 1$ and any $\widetilde{p} \geq 1$.

Another parameter inequality which is necessary to estimate the transport and Nash errors in Sections 8.4 and 8.5 is

$$\Gamma_{q+1}^{4+\frac{c_b}{2}} \delta_{q+1,\widetilde{n},1}^{\frac{1}{2}} \tau_q^{-1} r_{q+1,\widetilde{n}} \lambda_{q+1}^{-1} \leq \Gamma_{q+1}^{-c_R-1} \delta_{q+2} \tag{9.57}$$

for all $0 \leq \widetilde{n} \leq n_{\max}$. When $\widetilde{n} = 0$, this inequality may be deduced by writing everything out in terms of λ_q, appealing to the appropriate definitions, and then using the fact that $\beta < 1/2$ from item 1, (9.1), (9.4), (9.5), (9.6), and (9.7b), after which one arrives at

$$\varepsilon_\Gamma \left(4 + \frac{b-4}{2} + \frac{1}{2} - c_R + c_0 + 12 \right) + \beta(2b+1) < \frac{1}{100} + \frac{3}{2} < \frac{9}{5}.$$

It is clear there is quite a bit of room in the above inequality, and similarly, (9.57) becomes *most* restrictive when $\widetilde{n} = n_{\max}$. In this case, one may again write everything out in terms of λ_q, move everything to the left-hand side, and appeal to most of the same referenced inequalities as before to see that

$$\varepsilon_\Gamma \left(22 + 4n_{\max} \right) + \beta(2b+1) - \frac{3}{2} \leq \varepsilon_\Gamma \left(22 + 4n_{\max} \right) + \beta - \frac{1}{2} < 0,$$

where in the last inequality we have instead appealed to (9.7a) rather than (9.7b), proving (9.57) in the remaining cases $1 \leq \widetilde{n} \leq n_{\max}$.

Parameter inequalities which play a crucial role in showing that the Oscillation 2 type errors vanish, see—Section 8.7—are:

$$16 C_* \mathcal{C}_{\text{pipe}} \mathcal{C}_\eta \Gamma_{q+1} \left(\frac{\lambda_q}{\lambda_{q+1}} \right)^{\left(\frac{4}{5}\right)^{\widetilde{n}+1}\cdot 4} < \left(\frac{\lambda_q}{\lambda_{q+1}} \right)^{\left(\frac{4}{5}\right)^{\widetilde{n}-1}\cdot\frac{5}{6}\cdot 3} \Gamma_{q+1}^{-3}, \quad \text{for} \quad \widetilde{n} \geq 2, \tag{9.58a}$$

$$16 C_* \mathcal{C}_{\text{pipe}} \mathcal{C}_\eta \Gamma_{q+1} \left(\frac{\lambda_q}{\lambda_{q+1}} \right)^{\frac{4}{5}\cdot 4} < \left(\frac{\widetilde{\lambda}_q}{\lambda_{q+1}} \right)^3, \tag{9.58b}$$

$$16 C_* \mathcal{C}_{\text{pipe}} \mathcal{C}_\eta \Gamma_{q+1}^4 \left(\frac{\lambda_q}{\lambda_{q+1}} \right)^{\left(\frac{4}{5}\right)^2\cdot 4} < \left(\frac{\lambda_q}{\lambda_{q+1}} \right)^{\frac{4}{5}\cdot 3}, \tag{9.58c}$$

where C_* is the geometric constant from Lemma 4.8, estimate (4.31), $\mathcal{C}_{\text{pipe}}$ is a geometric constant which appears in Lemma 8.11, estimate (8.121), and $\mathcal{C}_\eta$ is the constant from Lemma 8.8. In order to verify (9.58), we first note that $C_* \mathcal{C}_{\text{pipe}} \mathcal{C}_\eta \leq \Gamma_{q+1}$, since a_* was chosen to be sufficiently large. Inequality (9.58b) is then an immediate consequence of the fact that $16/5 > 3$. The bound (9.58a) follows from

$$\Gamma_{q+1}^5 < \left(\frac{\lambda_{q+1}}{\lambda_q} \right)^{\left(\frac{4}{5}\right)^{n_{\max}-1}\left(\frac{64}{25}-\frac{5}{2}\right)} \leq \left(\frac{\lambda_{q+1}}{\lambda_q} \right)^{\left(\frac{4}{5}\right)^{\widetilde{n}+1}\cdot 4 - \left(\frac{4}{5}\right)^{\widetilde{n}-1}\cdot\frac{5}{6}\cdot 3}. \tag{9.59}$$

The second inequality in the above display is a consequence of $\tilde{n} \leq n_{\max}$, while the first one follows from (9.7b). Finally, inequality (9.58c) is a consequence of the fact that $64/25 - 12/5 > 64/25 - 5/2$ and the first inequality in (9.59), which bounds Γ_{q+1}^5.

We conclude this section by verifying a few inequalities concerning the parameter $N_{\mathrm{fin},n}$, which counts the number of available space-plus-material derivatives for the residual stress $\mathring{R}_{q,n}$. For all $0 \leq n \leq n_{\max}$ we require that

$$N_{\mathrm{ind},t}, 2N_{\mathrm{dec}} + 4 \leq \lfloor 1/2 \left(N_{\mathrm{fin},n} - N_{\mathrm{cut},t} - N_{\mathrm{cut},x} - 5\right)\rfloor - d, \tag{9.60a}$$

$$14N_{\mathrm{ind},v} \leq N_{\mathrm{fin},n} - N_{\mathrm{cut},t} - N_{\mathrm{cut},x} - 2N_{\mathrm{dec}} - 9, \tag{9.60b}$$

$$6N_{\mathrm{ind},v} \leq \lfloor 1/2 \left(N_{\mathrm{fin},n} - N_{\mathrm{cut},t} - N_{\mathrm{cut},x} - 6\right)\rfloor - d, \tag{9.60c}$$

$$6N_{\mathrm{ind},v} \leq \lfloor 1/4 \left(N_{\mathrm{fin},n} - N_{\mathrm{cut},t} - N_{\mathrm{cut},x} - 7\right)\rfloor \tag{9.60d}$$

for all $0 \leq n \leq n_{\max}$. Additionally for $0 \leq \tilde{n} < n \leq n_{\max}$, we require that

$$\lfloor 1/2 \left(N_{\mathrm{fin},\tilde{n}} - N_{\mathrm{cut},t} - N_{\mathrm{cut},x} - 6\right)\rfloor - d \geq N_{\mathrm{fin},n} \tag{9.61}$$

holds. The inequality (9.61) is a direct consequence of the formula (9.37) and of the fact that the sequence $N_{\mathrm{fin},n}$ is monotone decreasing with respect to n. Using (9.36) and (9.37) one may show that

$$N_{\mathrm{fin},n} \geq 2^{-n} N_{\mathrm{fin},0} - (2d + N_{\mathrm{cut},t} + N_{\mathrm{cut},x} + 8).$$

Noting that the bounds (9.60) are most restrictive for $n = n_{\max}$, they now readily follow from our choice (9.14).

9.4 MOLLIFIERS AND FOURIER PROJECTORS

Let $\phi(\zeta) : \mathbb{R} \to \mathbb{R}$ be a smooth, C^∞ function compactly supported in the set $\{\zeta : |\zeta| \leq 1\}$, which in addition satisfies

$$\int \phi(\zeta)\, d\zeta = 1, \qquad \int \phi(\zeta)\zeta^n = 0 \quad \forall n = 1, 2, ..., N_{\mathrm{ind},v}. \tag{9.62}$$

Let $\tilde{\phi}(x) : \mathbb{R}^3 \to \mathbb{R}$ be defined by $\tilde{\phi}(x) = \phi(|x|)$. For $\lambda, \mu \in \mathbb{R}$, define

$$\phi_\lambda^{(x)}(x) = \lambda^3 \tilde{\phi}(\lambda x), \qquad \phi_\mu^{(t)}(t) = \mu\phi(\mu t). \tag{9.63}$$

For $q \in \mathbb{N}$, we will define the spatial and temporal convolution operators

$$\mathcal{P}_{q,x} := \phi_{\lambda_q}^{(x)} *, \qquad \mathcal{P}_{q,t} := \phi_{\tilde{\tau}_{q-1}^{-1}}^{(t)} *, \qquad \mathcal{P}_{q,x,t} := \mathcal{P}_{q,x} \circ \mathcal{P}_{q,t}. \tag{9.64}$$

We will use the notation $\mathbb{P}_{\leq\lambda}$ to denote the standard (Littlewood-Paley)

Fourier projection operators onto spatial frequencies which are less than or equal to λ, $\mathbb{P}_{\geq\lambda}$ to denote the standard Littlewood-Paley projection operators onto spatial frequencies which are greater than or equal to λ, and the notation

$$\mathbb{P}_{[\lambda_1,\lambda_2)}$$

to denote the Fourier projection operator onto spatial frequencies ξ such that $\lambda_1 \leq |\xi| < \lambda_2$. If $\lambda_1 = \lambda_2$, we adopt the convention that $\mathbb{P}_{[\lambda_1,\lambda_2)}f = 0$ for any f.

9.5   NOTATIONS

$$\mathcal{M}(n,N,\lambda,\Lambda) = \lambda^{\min\{n,N\}}\Lambda^{\max\{n-N,0\}}$$

$$a \otimes_{\mathrm{s}} b = \frac{1}{2}(a \otimes b + b \otimes a) \tag{9.65}$$

$$a \mathring{\otimes}_{\mathrm{s}} b = \frac{1}{2}(a \mathring{\otimes} b + b \mathring{\otimes} a) \tag{9.66}$$

$$\mathrm{supp}_t f = \overline{\{t : f|_{\mathbb{T}^3 \times \{t\}} \not\equiv 0\}} \tag{9.67}$$

We will use repeatedly the notation (noted in the introduction in (2.3) and (2.4) and in Remark 3.2)

$$\|f\|_{L^p} := \|f\|_{L_t^\infty(L^p(\mathbb{T}^3))} . \tag{9.68}$$

That is, all L^p norms stand for L^p *norms in space, uniformly in time*. Similarly, when we wish to emphasize a set dependence on $\Omega \subset \mathbb{R} \times \mathbb{T}^3$ of an L^p norm, we write

$$\|f\|_{L^p(\Omega)} := \|\mathbf{1}_\Omega\, f\|_{L_t^\infty(L^p(\mathbb{T}^3))} . \tag{9.69}$$

Appendix A

Useful Lemmas

This appendix contains a collection of auxiliary lemmas which are used throughout the book:

- Section A.1 recalls the classical C^N estimates for solutions of the transport equation. This is, for instance, used in Section 6.4.
- Section A.2 gives the detailed construction of the basic cutoff functions $\widetilde{\psi}_{m,q}$ and $\psi_{m,q}$, which are used in Chapter 6 to construct the velocity and the stress cutoff functions.
- Section A.3 recalls the fundamental fact that the L^p norm of the product of a slowly oscillating function and a fast periodic function is essentially bounded by the product of their L^p norms.
- Section A.4 contains a version of the Sobolev inequality which takes into account the support of the velocity cutoff functions.
- Section A.5 contains a number of consequences of the multivariate Faà di Bruno formula. Most of the results here are used for bounding the space and material derivatives of the cutoff functions in Chapter 6. We also present here —cf. Lemma A.7— a version of the L^p decorrelation lemma from Section A.3 in which the fast periodic function is composed with a volume-preserving flow map. Lemma A.7 plays a crucial role in estimating the L^2 norms of the velocity increments in Section 8.2.
- Sections A.6 and A.7 contain a number of lemmas which allow us to go back and forth between information for (arbitrarily) high order derivative bounds in Eulerian and Lagrangian variables. These lemmas concerning sums of operators and commutators with material derivatives are frequently used throughout the book to overcome the fact that material derivatives and spatial/temporal derivatives do not commute.
- Section A.8 introduces in Proposition A.18 the inverse divergence operator used in this book. We call this operator "intermittency friendly" because it is composed of a principal part which precisely maintains the spatial support of the vector field it is applied to, plus a secondary part which is nonlocal, but whose amplitude is incredibly small. It is here that the definition (4.10) for the density of our pipe flows plays an important role, as the high order d of the Laplacian present in (4.10) allows us to perform a parametric expansion which maintains (to leading order) the support of pipes, and also takes into account deformations due to the flow map.

A.1   TRANSPORT ESTIMATES

We shall require the following estimates for smooth solutions of transport equations. For proofs we refer the reader to [8, Appendix D].

Lemma A.1 (Transport estimates). *Consider the transport equation*

$$\partial_t f + u \cdot \nabla f = g, \qquad\qquad f|_{t_0} = f_0,$$

where $f, g : \mathbb{T}^n \to \mathbb{R}$ and $u : \mathbb{T}^n \to \mathbb{R}^n$ are smooth functions. Let X be the flow of u, defined by

$$\frac{d}{dt} X = u(X, t), \qquad X(x, t_0) = x,$$

and let Φ be the inverse of the flow of X, which is the identity at time t_0. Then the following hold:

1. $\|f(t)\|_{C^0} \leq \|f_0\|_{C^0} + \displaystyle\int_{t_0}^t \|g(s)\|_{C^0}\, ds.$

2. $\|Df(t)\|_{C^0} \leq \|Df_0\|_{C^0} e^{(t-t_0)\|Du\|_{C^0}} + \displaystyle\int_{t_0}^t e^{(t-s)\|Du\|_{C^0}} \|Dg(s)\|_{C^0}\, ds.$

3. *For any $N \geq 2$, there exists a constant $C = C(N)$ such that*

$$\|D^N f(t)\|_{C^0}$$
$$\leq \left(\|D^N f_0\|_{C^0} + C(t-t_0)\|D^n u\|_{C^0}\|Df\|_{C^0}\right) e^{C(t-t_0)\|Du\|_{C^0}}$$
$$+ \int_{t_0}^t e^{C(t-s)\|Du\|_{C^0}} \left(\|D^N g(s)\|_{C^0} + C(t-s)\|D^N u\|_{C^0}\|Dg(s)\|_{C^0}\right)\, ds.$$

4. $\|D\Phi(t) - \mathrm{Id}\|_{C^0} \leq e^{(t-t_0)\|Du\|_{C^0}} - 1 \leq (t-t_0)\|Du\|_{C^0} e^{(t-t_0)\|Du\|_{C^0}}.$
5. *For $N \geq 2$ and a constant $C = C(N)$,*

$$\|D^N \Phi(t)\|_{C^0} \leq C(t-t_0)\|D^N u\|_{C^0} e^{C(t-t_0)\|Du\|_{C^0}}.$$

A.2   PROOF OF LEMMA 6.2

We first consider the function

$$f(x) = \begin{cases} 0 & \text{if } x \leq 0 \\ e^{-\frac{1}{x^2}} & \text{if } x > 0. \end{cases} \qquad\qquad (A.1)$$

We claim that for all $0 \leq N \leq \mathsf{N}_{\text{fin}}$ and $x > 0$,

$$\frac{|D^N f(x)|}{(f(x))^{1-\frac{N}{\mathsf{N}_{\text{fin}}}}} \lesssim 1. \qquad\qquad (A.2)$$

The proof of this is achieved in two steps; first, one can show by induction that for all $0 \leq N \leq N_{\text{fin}}$, there exist constants K_N and c_k for $0 \leq k \leq K_N$ such that

$$D^N\left(e^{-\frac{1}{x^2}}\right) = \sum_{k=0}^{K_N} \frac{c_k}{x^k} e^{-\frac{1}{x^2}}. \tag{A.3}$$

Next, one may also check that for any powers $p, q > 0$,

$$\lim_{x \to 0^+} e^{-\frac{q}{x^2}} \frac{1}{x^p} = 0. \tag{A.4}$$

Then for $1 \leq N \leq N_{\text{fin}}$, we see that $0 \leq 1 - \frac{N}{N_{\text{fin}}} < 1$, and so using (A.3), we have that the left-hand side of (A.2) may be split into a finite linear combination of terms of the form in (A.4), showing that (A.2) is valid.

We now glue together two versions of f as follows with the goal of forming a prototypical cutoff function ψ. First, let $x_0 = \sqrt{\frac{1}{\ln(2)}}$ so that $f(x_0) = \frac{1}{2}$. Now consider the function $\widetilde{f}(x) = f(2x_0 - x)$, and set

$$F(x) = \begin{cases} f(x) & \text{if } x \leq x_0 \\ 1 - f(2x_0 - x) & \text{if } x > x_0. \end{cases} \tag{A.5}$$

Then $F(x)$ is continuous everywhere, and C^∞ everywhere except x_0, where it is not necessarily differentiable. Furthermore, one can check that by the definition of F and (A.2), for all $0 \leq N \leq N_{\text{fin}}$,

$$\frac{|D^N F(x)|}{(F(x))^{1-\frac{N}{N_{\text{fin}}}}} \lesssim 1 \text{ for all } 0 < x < x_0, \tag{A.6a}$$

$$\frac{\left|D^N\left(1 - (F(x))^2\right)^{\frac{1}{2}}\right|}{\left(1 - (F(x))^2\right)^{\frac{1}{2}\left(1-\frac{N}{N_{\text{fin}}}\right)}} \lesssim 1 \text{ for all } x_0 < x < 2x_0. \tag{A.6b}$$

The latter inequality follows from noticing that for x close to $2x_0$,

$$\left(1 - (F(x))^2\right)^{\frac{1}{2}} = ((1 + F(x))(1 - F(x)))^{\frac{1}{2}} = (1 + F(x))^{\frac{1}{2}} (f(2x_0 - x))^{\frac{1}{2}}.$$

Since multiplying by a smooth function strictly larger than 1, rescaling f by a fixed parameter, and raising f to a positive power preserves the estimate (A.2) up to implicit constants (in fact, raising f to a power is equivalent to rescaling), (A.6) is verified.

Towards the goal of adjusting F to be differentiable at x_0, let E be the set $\left(\frac{x_0}{2}, \frac{3x_0}{2}\right)$, and let ϕ be a compactly supported, C^∞ mollifier such that the support of the mollified characteristic function $\mathcal{X}_E * \phi(x)$ is contained in $\left(\frac{x_0}{4}, \frac{7x_0}{4}\right)$. Setting

$$\psi(x) = (\mathcal{X}_E * \phi(x)) \, \phi * F(x) + (1 - \mathcal{X}_E * \phi(x)) \, F(x), \tag{A.7}$$

one may check that ψ is C^∞ and has the following properties:

$$\psi(x) = 0 \text{ for } x \leq 0 \tag{A.8}$$

$$0 < \psi(x) < 1 \text{ for } 0 < x < 2x_0 \tag{A.9}$$

$$\psi(x) = 1 \text{ for } x \geq 2x_0 \tag{A.10}$$

$$\frac{|D^N \psi(x)|}{(\psi(x))^{1-\frac{N}{N_{\text{fin}}}}} \lesssim 1 \text{ for all } 0 < x \tag{A.11}$$

$$\frac{|D^N \left(1 - (\psi(x))^2\right)^{\frac{1}{2}}|}{(1 - (\psi(x))^2)^{\frac{1}{2}\left(1 - \frac{N}{N_{\text{fin}}}\right)}} \lesssim 1 \text{ for all } 0 < x < 2x_0. \tag{A.12}$$

We can now build $\widetilde{\psi}_{m,q}$. By rescaling and translating ψ and using (A.8)–(A.10), one can check that

$$\widetilde{\psi}_{m,q}(x) = \psi\left(\frac{x - \Gamma_q^{2(m+1)}}{\frac{1}{2x_0}\left(\frac{1}{4} - 1\right)\Gamma_q^{2(m+1)}}\right) \tag{A.13}$$

satisfies all components of (1). Notice that this rescaling involves a factor proportional to $\Gamma_q^{-2(m+1)}$. Then using (A.11) and the fact that every derivative $\psi_{m,q}$ introduces another factor of $\Gamma_q^{-2(m+1)}$, we have that (6.3) is satisfied.

We now outline how to construct $\psi_{m,q}(\Gamma_q^{-2(m+1)}y)$, which is the first term in the series in (6.1), and will define $\psi_{m,q}(y)$. The basic idea is that the region $(\frac{1}{4}\Gamma_{q+1}^{2(m+1)}, \Gamma_{q+1}^{2(m+1)})$ where $\widetilde{\psi}_{m,q}$ decreases from 1 to 0 will be the region where $\psi_{m,q}(\Gamma_{q+1}^{-2(m+1)}y)$ increases from 0 to 1, and furthermore in order to satisfy (6.1), we have a formula for $\psi_{m,q}(\Gamma_{q+1}^{-2(m+1)}y)$ for these y-values. Specifically, in order to ensure (6.1) for $y \in (\frac{1}{4}\Gamma_{q+1}^{2(m+1)}, \Gamma_{q+1}^{2(m+1)})$, we define

$$\psi_{m,q}^2\left(\Gamma_{q+1}^{-2(m+1)}y\right) = 1 - \widetilde{\psi}_{m,q}^2(y)$$

in this range of y-values. Then by adjusting (A.12) to reflect the rescalings present in the definition of $\widetilde{\psi}_{m,q}$ and $\psi_{m,q}(\Gamma_q^{-2(m+1)}y)$, we have that for $y \in \left(\frac{1}{4}, 1\right)$, $\psi_{m,q}$ is well-defined and (6.4) holds. To define $\psi_{m,q}(\Gamma_q^{-2(m+1)}y)$ for $y \in [\frac{1}{4}\Gamma_q^{4(m+1)}, \Gamma_q^{4(m+1)}]$ and thus $\psi_{m,q}(y)$ for $y \in [\frac{1}{4}\Gamma_q^{2(m+1)}, \Gamma_q^{2(m+1)}]$, we can use that for $y \in [\frac{1}{4}\Gamma_q^{4(m+1)}, \Gamma_q^{4(m+1)}]$, the rescaled function $\psi_{m,q}(\Gamma_{q+1}^{-4(m+1)}y)$ (i.e., the term in (6.1) with $i = 2$) is now well-defined. Then we can set

$$\psi_{m,q}^2\left(\Gamma_{q+1}^{-2(m+1)}y\right) = 1 - \psi_{m,q}^2\left(\Gamma_{q+1}^{-4(m+1)}y\right)$$

so that $\psi_{m,q}$ is well-defined for $y \in [\frac{1}{4}\Gamma_q^{2(m+1)}, \Gamma_q^{2(m+1)}]$ and (6.1) holds in this range of y-values. Appealing again to (A.11) and (A.12), we have that (6.5) is satisfied in the claimed range of y-values. Finally, in the missing interval $[1, \frac{1}{4}\Gamma_q^{2(m+1)}]$, we set $\psi_{m,q} \equiv 1$. One can now check that (6.1) holds for all $y \geq 0$, and that (6.2) follows from (1), (2), and (6.1), concluding the proof.

A.3 L^p DECORRELATION

The following lemma may be found in [13, Lemma 3.7].

Lemma A.2 (L^p **de-correlation estimate**). *Fix integers* $N_{\text{dec}} \geq 1$ *and* $\mu \geq \lambda \geq 1$ *and assume that they obey*

$$\lambda^{N_{\text{dec}}+4} \leq \left(\frac{\mu}{2\pi\sqrt{3}}\right)^{N_{\text{dec}}}. \tag{A.14}$$

Let $p \in \{1, 2\}$, *and let* f *be a* $\mathbb{T}^3$-*periodic function such that*

$$\max_{0 \leq N \leq N_{\text{dec}}+4} \lambda^{-N} \|D^N f\|_{L^p} \leq \mathcal{C}_f \tag{A.15}$$

for a constant $\mathcal{C}_f > 0$.[1] *Then, for any* $(\mathbb{T}/\mu)^3$-*periodic function* g, *we have that*

$$\|fg\|_{L^p} \lesssim \mathcal{C}_f \|g\|_{L^p},$$

where the implicit constant is universal (in particular, independent of μ *and* λ*).*

A.4 SOBOLEV INEQUALITY WITH CUTOFFS

Lemma A.3. *Let* $0 \leq \psi_i \leq 1$ *be cutoff functions such that* $\psi_{i\pm} = (\psi_{i-1}^2 + \psi_i^2 + \psi_{i+1}^2)^{1/2} = 1$ *on* $\text{supp}(\psi_i)$, *and such that for some* $\rho > 0$ *we have*

$$|D^K \psi_i(x)| \lesssim \psi_i^{1-K/N_{\text{fin}}}(x)\rho^K \tag{A.16}$$

for all $K \leq 4$. *Fix parameters* $p \in [1, \infty]$, $0 < \lambda \leq \widetilde{\lambda}$, $0 < \mu_i \leq \widetilde{\mu}_i$, *and* $N_x, N_t \geq 0$, *and assume that the sequences* $\{\mu_i\}_{i \geq 0}$ *and* $\{\widetilde{\mu}_i\}_{i \geq 0}$ *are nondecreasing. Assume that there exist* $N_*, M_* \geq 0$ *such that the function* $f \colon \mathbb{T}^3 \to \mathbb{R}$ *obeys the estimate*

$$\left\|\psi_i D^N D_t^M f\right\|_{L^p} \lesssim \mathcal{C}_f \mathcal{M}\left(N, N_x, \lambda, \widetilde{\lambda}\right) \mathcal{M}\left(M, N_t, \mu_i, \widetilde{\mu}_i\right) \tag{A.17}$$

---

[1]For instance, if f has frequency support in the ball of radius λ around the origin, we have that $\mathcal{C}_f \approx \|f\|_{L^p}$.

for all $N \leq N_*$ and $M \leq M_*$. Then, we have that

$$\left\| \psi_i^2 D^N D_t^M f \right\|_{L^\infty} \lesssim C_f (\max\{1, \rho, \widetilde{\lambda}\})^{3/p}$$
$$\times \mathcal{M}\left(N, N_x, \lambda, \widetilde{\lambda}\right) \mathcal{M}\left(M, N_t, \mu_i, \widetilde{\mu}_i\right) \qquad \text{(A.18a)}$$

$$\left\| D^N D_t^M f \right\|_{L^\infty(\text{supp }\psi_i)} \lesssim C_f (\max\{1, \rho, \widetilde{\lambda}\})^{3/p}$$
$$\times \mathcal{M}\left(N, N_x, \lambda, \widetilde{\lambda}\right) \mathcal{M}\left(M, N_t, \mu_{i+1}, \widetilde{\mu}_{i+1}\right) \qquad \text{(A.18b)}$$

for all $N \leq N_* - \lfloor 3/p \rfloor - 1$ and $M \leq M_*$.

Lastly, if the inequality (A.17) holds for all $N + M \leq N_\circ$ for some $N_\circ \geq 0$ (instead of $N \leq N_*$ and $M \leq M_*$), then the bounds (A.18a) and (A.18b) hold for $N + M \leq N_\circ - \lfloor 3/p \rfloor - 1$.

Proof of Lemma A.3. The proof uses that $\lfloor 3/p \rfloor + 1 > 3/p$ for all $p \in [1, \infty]$, and that $W^{s,p} \subset L^\infty$ for $s > 3/p$. Moreover, the proof of (A.18a) is nearly identical to that of (A.18b), and thus we only give the proof of (A.18b); moreover, for simplicity we only give the proof for $p = 2$, as all the other Lebesgue exponents are treated in the same way. By Gagliardo-Nirenberg-Sobolev interpolation we have

$$\left\| D^N D_t^M f \right\|_{L^\infty(\text{supp }\psi_i)} \leq \left\| \psi_{i\pm}^2 D^N D_t^M f \right\|_{L^\infty(\mathbb{T}^3)}$$
$$\lesssim \left\| \psi_{i\pm}^2 D^N D_t^M f \right\|_{L^2(\mathbb{T}^3)}^{1/4} \left\| \psi_{i\pm}^2 D^N D_t^M f \right\|_{\dot{H}^2(\mathbb{T}^3)}^{3/4}$$
$$+ \left\| \psi_{i\pm}^2 D^N D_t^M f \right\|_{L^2(\mathbb{T}^3)} .$$

Using (A.16), (A.17), and the monotonicity of the μ_i and $\widetilde{\mu}_i$, we obtain

$$\left\| \psi_{i\pm}^2 D^N D_t^M f \right\|_{\dot{H}^2(\mathbb{T}^3)}$$
$$\lesssim \left\| \psi_{i\pm} D^{N+2} D_t^M f \right\|_{L^2} + \left\| D\psi_{i\pm} \right\|_{L^\infty} \left\| \psi_{i\pm} D^{N+1} D_t^M f \right\|_{L^2}$$
$$+ \left\| \frac{D^2(\psi_{i\pm}^2)}{\psi_{i\pm}} \right\|_{L^\infty} \left\| \psi_{i\pm} D^N D_t^M f \right\|_{L^2}$$
$$\lesssim \left\| \psi_{i\pm} D^{N+2} D_t^M f \right\|_{L^2} + \rho \left\| \psi_{i\pm} D^{N+1} D_t^M f \right\|_{L^2} + \rho^2 \left\| \psi_{i\pm} D^N D_t^M f \right\|_{L^2}$$
$$\lesssim (\max\{\widetilde{\lambda}, \rho\})^2 C_f \mathcal{M}\left(N, N_x, \lambda, \widetilde{\lambda}\right) \mathcal{M}\left(M, N_t, \mu_{i+1}, \widetilde{\mu}_{i+1}\right) ,$$

for all $N \leq N_* - 2$ and $M \leq M_*$. In the second inequality above we have used that $|D^2(\psi_{i\pm}^2)| \lesssim \rho^2 \psi_{i\pm}(x)$, which follows from (A.16). Combining the above two displays proves (A.18b).

Note that for $p = 1$ we require that $|D^4(\psi_{i\pm}^2)| \lesssim \rho^4 \psi_{i\pm}(x)$, which also follows from (A.16) since $N_{\text{fin}} \geq 4$, and this is why we have assumed this inequality to hold for all $K \leq 4$.

Lastly, assume that (A.17) holds for all $N + M \leq N_\circ$, and fix any $N', M' \geq 0$ such that $N' + M' \leq N_\circ - \lfloor 3/p \rfloor - 1$. Let $N_* = N' + \lfloor 3/p \rfloor + 1$ and $M_* = M'$. Then

(A.17) gives a bound for $\|\psi_i D^{N''} D_t^{M''} f\|_{L^p}$ for all $N'' \le N_*$ and $M'' \le M_*$. The bounds (A.18a) and (A.18b) thus give an estimate for $\|\psi_i D^{N'} D_t^{M'} f\|_{L^p}$, which concludes the proof. $\qquad\square$

A.5 CONSEQUENCES OF THE FAÀ DI BRUNO FORMULA

We are using the following version of the multivariable Faà di Bruno formula [25, Theorem 2.1]. Let $g = g(x_1, \ldots, x_d) = f(h(x_1, \ldots, x_d))$, where $f \colon \mathbb{R}^m \to \mathbb{R}$ and $h \colon \mathbb{R}^d \to \mathbb{R}^m$ are C^n smooth functions of their respective variables. Let $\alpha \in \mathbb{N}_0^d$ be s.t. $|\alpha| = n$, and let $\beta \in \mathbb{N}_0^m$ be such that $1 \le |\beta| \le n$. We then define

$$
p(\alpha, \beta) = \Bigg\{ (k_1, \ldots, k_n; \ell_1, \ldots, \ell_n) \in (\mathbb{N}_0^m)^n \times (\mathbb{N}_0^d)^n : \exists s \text{ with } 1 \le s \le n \text{ s.t.}
$$

$$
|k_j|, |\ell_j| > 0 \Leftrightarrow 1 \le j \le s, \; 0 \prec \ell_1 \prec \ldots \prec \ell_s,
$$

$$
\sum_{j=1}^{s} k_j = \beta, \sum_{j=1}^{s} |k_j| \ell_j = \alpha \Bigg\}. \tag{A.19}
$$

Then the multivariable Faà di Bruno formula states that we have the equality

$$
\partial^\alpha g(x) = \alpha! \sum_{|\beta|=1}^{n} (\partial^\beta f)(h(x)) \sum_{p(\alpha,\beta)} \prod_{j=1}^{n} \frac{(\partial^{\ell_j} h(x))^{k_j}}{k_j! (\ell_j!)^{k_j}}. \tag{A.20}
$$

Note that in (A.19) we have that $k_j = 0 \in \mathbb{N}_0^m$ and $\ell_j = 0 \in \mathbb{N}_0^d$ for $j \ge s+1$. Therefore, we could write the sums and products with $j \in \{1, \ldots, s\}$ as sums for $j \in \{1, \ldots, n\}$. Keeping in mind this convention, we importantly note that in (A.20) we can have $|\ell_j| = 0$ only if $|k_j| = 0$, and in this case the entire term in the product is equal to 1. That is, the product in (A.20) only goes from 1 to s, and in this case $|\ell_j| \ge 1$ for $j \in \{1, \ldots, s\}$. This fact will be used frequently.

For applications to cutoff functions we apply this formula for scalar functions h, i.e., $m = 1$, while for applications to the perturbation or Reynolds stress sections we apply this formula for vector fields h, i.e., $m = 3$.

Since throughout this manuscript the number of derivatives that we need to estimate is uniformly bounded (say by N_{fin}), we may ignore the factorial terms in (A.20) and include them in the implicit constant of $\lesssim$. Using this convention, we summarize in the following lemma a useful consequence of the Faà di Bruno formula above.

Lemma A.4 (Faà di Bruno). *Fix $N \le \mathsf{N}_{\text{fin}}$. Let $\psi \colon [0, \infty) \to [0, 1]$ be a smooth function obeying*

$$
|D^B \psi| \lesssim \Gamma_\psi^{-2B} \psi^{1 - B/\mathsf{N}_{\text{fin}}} \tag{A.21}
$$

for all $B \leq N$, and some $\Gamma_\psi > 0$. Let $\Gamma, \lambda, \Lambda > 0$ and $N_* \leq N$. Furthermore, let $h\colon \mathbb{T}^3 \times \mathbb{R} \to \mathbb{R}$ and denote

$$g(x) = \psi(\Gamma^{-2}h(x)).$$

Assume the function h obeys

$$\left\|D^B h\right\|_{L^\infty(\mathrm{supp}\,g)} \lesssim \mathcal{C}_h \mathcal{M}(B, N_*, \lambda, \Lambda) \tag{A.22}$$

for all $B \leq N$, where the implicit constant is independent of $\lambda, \Lambda, \Gamma, \mathcal{C}_h > 0$. Then, we have that for all points $(x,t) \in \mathrm{supp}\,h$, the bound

$$\frac{|D^N g|}{g^{1-N/\mathsf{N}_\mathrm{fin}}} \lesssim \mathcal{M}(N, N_*, \lambda, \Lambda) \max\{(\Gamma_\psi\Gamma)^{-2}\mathcal{C}_h, (\Gamma_\psi\Gamma)^{-2N}\mathcal{C}_h^N\} \tag{A.23}$$

holds. If the $\psi^{1-B/\mathsf{N}_\mathrm{fin}}$ factor on the right side of (A.21) is replaced by 1, then the $g^{1-N/\mathsf{N}_\mathrm{fin}}$ factor on the left side of (A.23) also has to be replaced by 1.

Proof of Lemma A.4. The goal is to apply (A.19)–(A.20) with $f(x) = \psi(\Gamma^{-2}x)$. For $(x,t) \in \mathrm{supp}\,(g)$ we obtain from (3.9), (A.21), and (A.23) that

$$\frac{|D^N g|}{g^{1-N/\mathsf{N}_\mathrm{fin}}} \lesssim \sum_{B=1}^{N} \frac{|D^B \psi|}{\psi^{1-B/\mathsf{N}_\mathrm{fin}}} \psi^{(N-B)/\mathsf{N}_\mathrm{fin}} \Gamma^{-2B} \sum_{p(\alpha,B)} \prod_{j=1}^{n} \left\|\partial^{\ell_j} h\right\|_{L^\infty(\mathrm{supp}\,g)}^{k_j}$$

$$\lesssim \sum_{B=1}^{N} (\Gamma_\psi\Gamma)^{-2B} \sum_{p(\alpha,B)} \prod_{j=1}^{n} (\mathcal{C}_h \mathcal{M}(\ell_j, N_*, \lambda, \Lambda))^{k_j}$$

$$\lesssim \sum_{B=1}^{N} (\Gamma_\psi\Gamma)^{-2B} \mathcal{C}_h^B \mathcal{M}(N, N_*, \lambda, \Lambda)$$

for any $1 \leq B \leq N$. The conclusion of the lemma follows upon bounding the geometric sum. $\qquad\square$

Frequently in the book, we need a version of Lemma A.4 which also deals with mixed spatial and material derivatives. A convenient statement is:

Lemma A.5 (Mixed derivative Faà di Bruno). *Fix $N, M \in \mathbb{N}$ such that $N + M \leq \mathsf{N}_\mathrm{fin}$. Let $\psi\colon [0, \infty) \to [0, 1]$ be a smooth function obeying*

$$|D^B \psi| \lesssim \Gamma_\psi^{-2B} \psi^{1-B/\mathsf{N}_\mathrm{fin}} \tag{A.24}$$

for all $B \leq N$ and a constant $\Gamma_\psi > 0$. Let v be a fixed vector field, and denote $D_t = \partial_t + v \cdot \nabla$, which is a scalar differential operator. Let $\Gamma, \lambda, \tilde{\lambda}, \mu, \tilde{\mu} \geq 1$ and $N_x, N_t \leq N$. Furthermore, let $h\colon \mathbb{T}^3 \times \mathbb{R} \to \mathbb{R}$ and denote

$$g(x,t) = \psi(\Gamma^{-2}h(x,t)).$$

Assume the function h obeys

$$\left\| D^{N'} D_t^{M'} h \right\|_{L^\infty(\operatorname{supp} g)} \lesssim \mathcal{C}_h \mathcal{M}\left(N', N_x, \lambda, \widetilde{\lambda}\right) \mathcal{M}\left(M', N_t, \mu, \widetilde{\mu}\right) \qquad (\text{A.25})$$

for all $N' \leq N$ and $M' \leq M$, where the implicit constant is independent of $\lambda, \widetilde{\lambda}, \mu, \widetilde{\mu}, \Gamma,$ and $\mathcal{C}_h$. Then, we have that for all points $(x,t) \in \operatorname{supp} h$, the bound

$$\frac{|D^N D_t^M g|}{g^{1-(N+M)/\mathsf{N}_{\mathrm{fin}}}} \lesssim \mathcal{M}\left(N, N_x, \lambda, \widetilde{\lambda}\right) \mathcal{M}\left(M, N_t, \mu, \widetilde{\mu}\right)$$
$$\times \max\left\{ (\Gamma_\psi \Gamma)^{-2} \mathcal{C}_h, ((\Gamma_\psi \Gamma)^{-2} \mathcal{C}_h)^{N+M} \right\} \qquad (\text{A.26})$$

holds. If the $\psi^{1-B/\mathsf{N}_{\mathrm{fin}}}$ factor on the right side of (A.24) is replaced by 1, then the $g^{1-(N+M)/\mathsf{N}_{\mathrm{fin}}}$ factor on the left side of (A.26) also has to be replaced by 1.

Proof of Lemma A.5. Let $X(a,t)$ be the flow induced by the vector field v, with initial condition $X(a,t) = x$. Denote by $a = X^{-1}(x,t)$ the inverse of the map X. We then note that

$$D_t^M g(x,t) = \left(\partial_t^M((g \circ X)(a,t))\right)|_{a=X^{-1}(x,t)}.$$

We wish to apply the above with the function $g(x,t) = \psi(\Gamma^{-2}h(x,t))$. We apply the Faà di Bruno formula (A.19)–(A.20) with the one-dimensional differential operator ∂_t^M to the composition $g \circ X$, note that $\partial_t^{\beta_i}(h(X(a,t),t)) = (D_t^{\beta_i}h)(X(a,t),t)$, and then evaluate the resulting expression at $a = X^{-1}(x,t)$, to obtain

$$D_t^M g(x,t) = M! \sum_{B=1}^{M} \Gamma^{-2B} \psi^{(B)}(\Gamma^{-2}h(x,t)) \sum_{\substack{\{\kappa,\beta \in \mathbb{N}^M: \\ |\kappa|=B, \kappa \cdot \beta = M\}}} \prod_{i=1}^{M} \frac{\left((D_t^{\beta_i}h)(x,t)\right)^{\kappa_i}}{\kappa_i!(\beta_i!)^{\kappa_i}}.$$

We now apply D^N to the above expression, use the Leibniz rule, and then appeal again to the Faà di Bruno formula (A.19)–(A.20), this time for spatial derivatives. We obtain

$$D^N D_t^M g(x,t) = M!N! \sum_{B=1}^{M} \sum_{K=0}^{N} \sum_{B'=0}^{K} \Gamma^{-2(B+B')} \psi^{(B+B')}(\Gamma^{-2}h(x,t))$$

$$\times \sum_{p(K,B')} \prod_{j=1}^{K} \frac{(D^{\ell_j}h(x,t))^{k_j}}{k_j!(\ell_j!)^{k_j}}$$

$$\times \sum_{\substack{\{\alpha \in \mathbb{N}^M: \\ |\alpha|=N-K\}}} \sum_{\substack{\{\kappa,\beta \in \mathbb{N}^M: \\ |\kappa|=B, \kappa \cdot \beta = M\}}} \prod_{i=1}^{M} \frac{D^{\alpha_i}(((D_t^{\beta_i}h)(x,t))^{\kappa_i})}{\alpha_i!\kappa_i!(\beta_i!)^{\kappa_i}}. \qquad (\text{A.27})$$

Upon dividing by $g^{1-(N+M)/N_{\text{fin}}}$ and noting that $B + B' \leq M + N$, from (A.24), identity (A.27), the Leibniz rule, and assumption (A.25), we obtain

$$\frac{|D^N D_t^M g|}{g^{1-(N+M)/N_{\text{fin}}}}$$

$$\lesssim \sum_{B=1}^{M} \sum_{K=0}^{N} \sum_{B'=0}^{K} (\Gamma_\psi \Gamma)^{-2(B+B')}$$

$$\times \mathcal{C}_h^{B'} \mathcal{M}\left(K, N_x, \lambda, \widetilde{\lambda}\right) \mathcal{C}_h^{B} \mathcal{M}\left(N - K, N_x, \lambda, \widetilde{\lambda}\right) \mathcal{M}\left(M, N_t, \mu, \widetilde{\mu}\right)$$

$$\lesssim \mathcal{M}\left(N, N_x, \lambda, \widetilde{\lambda}\right) \mathcal{M}\left(M, N_t, \mu, \widetilde{\mu}\right) \sum_{B=1}^{M} \sum_{B'=0}^{N} (\Gamma_\psi \Gamma)^{-2(B+B')} \mathcal{C}_h^{B'+B},$$

from which (A.26) follows by summing the geometric series. $\square$

Lemma A.6. *Given a smooth function $f \colon \mathbb{R}^3 \times \mathbb{R} \to \mathbb{R}$, suppose that for $\lambda \geq 1$ the vector field $\Phi \colon \mathbb{R}^3 \times \mathbb{R} \to \mathbb{R}^3$ satisfies the estimate*

$$\left\| D^{N+1} \Phi \right\|_{L^\infty(\mathrm{supp}\, f)} \lesssim \lambda^N \tag{A.28}$$

for $0 \leq N \leq N_*$. Then for any $1 \leq N \leq N_*$ we have

$$\left| D^N \left(f \circ \Phi \right)(x, t) \right| \lesssim \sum_{m=1}^{N} \lambda^{N-m} \left| (D^m f) \circ \Phi(x, t) \right| \tag{A.29}$$

and thus trivially we obtain

$$\left| D^N \left(f \circ \Phi \right)(x, t) \right| \lesssim \sum_{m=0}^{N} \lambda^{N-m} \left| (D^m f) \circ \Phi(x, t) \right|$$

for any $0 \leq N \leq N_*$.

Proof of Lemma A.6. Applying (A.20), noting that $|\ell_j| = 0$ implies $|k_j| = 0$, and employing assumption (A.28), we have that for any multi-index $\alpha \in \mathbb{N}_0^3$ with $|\alpha| = N$,

$$\left| \partial^\alpha \left(f \circ \Phi \right)(x, t) \right| \lesssim \sum_{|\beta|=1}^{N} \left| ((\partial^\beta f) \circ \Phi)(x, t) \right| \prod_{j=1}^{N} \sum_{p(\alpha, \beta)} \left| \left(\partial^{\ell_j} \Phi(x, t) \right)^{k_j} \right|$$

$$\lesssim \sum_{|\beta|=1}^{N} \left| (\partial^\beta f) \circ \Phi \right| \prod_{j=1}^{N} \sum_{p(\alpha, \beta)} \lambda^{(|\ell_j|-1)|k_j|}$$

$$\lesssim \sum_{m=1}^{N} \lambda^{N-m} \left| (D^m f) \circ \Phi \right|$$

by the definition (A.19). Thus we obtain (A.29).   □

In order to estimate the perturbation in L^p spaces as well as terms appearing in the Reynolds stress, we will need the following abstract lemma, which follows from Lemmas A.2 and A.6.

Lemma A.7. *Let $p \in \{1,2\}$, and fix integers $N_* \geq M_* \geq \mathsf{N}_{\mathrm{dec}} \geq 1$. Suppose $f \colon \mathbb{R}^3 \times \mathbb{R} \to \mathbb{R}$ and let $\Phi \colon \mathbb{R}^3 \times \mathbb{R} \to \mathbb{R}^3$ be a vector field advected by an incompressible velocity field v, i.e., $D_t \Phi = (\partial_t + v \cdot \nabla)\Phi = 0$. Denote by Φ^{-1} the inverse of the flow Φ, which is the identity at a time slice which intersects the support of f. Assume that for some $\lambda, \nu, \widetilde{\nu} \geq 1$ and $\mathcal{C}_f > 0$ the function f satisfies the estimates*

$$\left\| D^N D_t^M f \right\|_{L^p} \lesssim \mathcal{C}_f \lambda^N \mathcal{M}\left(M, N_t, \nu, \widetilde{\nu}\right) \tag{A.30}$$

for all $N \leq N_*$ and $M \leq M_*$, and that Φ and Φ^{-1} are bounded as

$$\left\| D^{N+1} \Phi \right\|_{L^\infty(\mathrm{supp}\, f)} \lesssim \lambda^N \tag{A.31}$$

$$\left\| D^{N+1} \Phi^{-1} \right\|_{L^\infty(\mathrm{supp}\, f)} \lesssim \lambda^N \tag{A.32}$$

for all $N \leq N_*$. Lastly, suppose that φ is $(\mathbb{T}/\mu)^3$-periodic, and that there exist parameters $\widetilde{\zeta} \geq \zeta \geq \mu$ and $\mathcal{C}_\varphi > 0$ such that

$$\left\| D^N \varphi \right\|_{L^p} \lesssim \mathcal{C}_\varphi \mathcal{M}\left(N, N_x, \zeta, \widetilde{\zeta}\right) \tag{A.33}$$

for all $0 \leq N \leq N_*$. If the parameters

$$\lambda \leq \mu \leq \zeta \leq \widetilde{\zeta}$$

satisfy

$$\widetilde{\zeta}^4 \leq \left(\frac{\mu}{2\pi\sqrt{3}\lambda}\right)^{\mathsf{N}_{\mathrm{dec}}}, \tag{A.34}$$

and we have

$$2\mathsf{N}_{\mathrm{dec}} + 4 \leq N_*, \tag{A.35}$$

then the bound

$$\left\| D^N D_t^M \left(f\, \varphi \circ \Phi \right) \right\|_{L^p} \lesssim \mathcal{C}_f \mathcal{C}_\varphi \mathcal{M}\left(N, N_x, \zeta, \widetilde{\zeta}\right) \mathcal{M}\left(M, M_t, \nu, \widetilde{\nu}\right) \tag{A.36}$$

holds for $N \leq N_*$ and $M \leq M_*$.

Remark A.8. We emphasize that (A.36) holds for the same range of N and M for which (A.30) holds, as soon as N_* is sufficiently large compared to $\mathsf{N}_{\mathrm{dec}}$ so that (A.35) holds.

Remark A.9. We note that if estimate (A.30) is known to hold for $N + M \leq N_\circ$ for some $N_\circ \geq 2N_{\text{dec}} + 4$ (instead of for $N \leq N_*$ and $M \leq M_*$), and if the bounds (A.31)–(A.32) hold for all $N \leq N_\circ$, then it follows from the following proof that the bound (A.36) holds for $N + M \leq N_\circ$ and $M \leq N_\circ - 2N_{\text{dec}} - 4$. The only modification required to the proof is that instead of considering the cases $N' \leq N_* - N_{\text{dec}} - 4$ and $N' > N_* - N_{\text{dec}} - 4$, we now have to split according to $N' + M \leq N_\circ - N_{\text{dec}} - 4$ and $N' + M > N_\circ - N_{\text{dec}} - 4$. In the second case we use the fact that $N - N'' \geq N_\circ - M - N_{\text{dec}} - 4 \geq N_{\text{dec}}$, which holds exactly because $M \leq N_\circ - 2N_{\text{dec}} - 4$.

Proof of Lemma A.7. Since $D_t \Phi = 0$ we have $D_t^M (\varphi \circ \Phi) = 0$. Using the fact that $\operatorname{div} v \equiv 0$, so that Φ and Φ^{-1} preserve volume, and Lemma A.6, which we may apply due to (A.31), we have

$$
\begin{aligned}
\left\| D^N D_t^M (f \, \varphi \circ \Phi) \right\|_{L^p} &\lesssim \sum_{N'=0}^{N} \left\| D^{N'} D_t^M f \, D^{N-N'} (\varphi \circ \Phi) \right\|_{L^p} \\
&\lesssim \sum_{N'=0}^{N} \sum_{N''=0}^{N-N'} \lambda^{N-N'-N''} \left\| D^{N'} D_t^M f \, (D^{N''} \varphi) \circ \Phi \right\|_{L^p} \\
&\lesssim \sum_{N'=0}^{N} \sum_{N''=0}^{N-N'} \lambda^{N-N'-N''} \left\| \left(D^{N'} D_t^M f \right) \circ \Phi^{-1} D^{N''} \varphi \right\|_{L^p} .
\end{aligned}
$$

$$(A.37)$$

In (A.37) let us first consider the case $N' \leq N_* - N_{\text{dec}} - 4$, so that $N' + M \leq N_* + M_* - N_{\text{dec}} - 4$. Under assumption (A.32) we may apply Lemma A.6, and using (A.30) we have

$$
\begin{aligned}
\left\| D^n \left((D^{N'} D_t^M f) \circ (\Phi^{-1}, t) \right) \right\|_{L^p} &\lesssim \sum_{n'=0}^{n} \lambda^{n-n'} \left\| (D^{n'+N'} D_t^M f) \circ \Phi^{-1} \right\|_{L^p} \\
&\lesssim \mathcal{C}_f \sum_{n'=0}^{n} \lambda^{n-n'} \lambda^{n'+N'} \mathcal{M}(M, N_t, \nu, \tilde{\nu}) \\
&\lesssim \left(\mathcal{C}_f \lambda^{N'} \mathcal{M}(M, N_t, \nu, \tilde{\nu}) \right) \lambda^n , \qquad (A.38)
\end{aligned}
$$

for all $n \leq N_{\text{dec}} + 4$. This bound matches (A.15), with the constant $\mathcal{C}_f$ replaced by $\mathcal{C}_f \lambda^{N'} \mathcal{M}(M, N_t, \nu, \tilde{\nu})$. Since, like φ, the function $D^{N''} \varphi$ is $(\mathbb{T}/\mu)^3$-periodic, due to (A.38), the fact that $\lambda \leq \tilde{\zeta}$, and assumption (A.34), we may apply Lemma A.2 to conclude

$$
\left\| \left(D^{N'} D_t^M f \right) \circ \Phi^{-1} D^{N''} \varphi \right\|_{L^p} \lesssim \mathcal{C}_f \lambda^{N'} \mathcal{M}(M, N_t, \nu, \tilde{\nu}) \left\| D^{N''} \varphi \right\|_{L^p} .
$$

Inserting this bound back into (A.37) and using (A.33) concludes the proof of

(A.36) for the values of N' considered in this case.

Next, let us consider the case $N' > N_* - \mathsf{N}_{\mathrm{dec}} - 4$. Since $0 \le N' \le N$, in particular, this means that $N > N_* - \mathsf{N}_{\mathrm{dec}} - 4$, and since $N'' \le N - N'$ we also obtain that $N - N'' \ge N' > N_* - \mathsf{N}_{\mathrm{dec}} - 4 \ge \mathsf{N}_{\mathrm{dec}}$. Here we have used (A.35). Then the Hölder inequality, the fact that Φ^{-1} is volume preserving, the Sobolev embedding $W^{4,p} \subset L^\infty$, the ordering $\widetilde{\zeta} \ge \zeta \ge \mu \ge 1$, and assumption (A.34), imply that

$$
\begin{aligned}
&\lambda^{N-N'-N''} \left\| \left(D^{N'} D_t^M f \right) \circ \Phi^{-1} D^{N''} \varphi \right\|_{L^p} \\
&\lesssim \lambda^{N-N'-N''} \left\| D^{N'} D_t^M f \right\|_{L^p} \left\| D^{N''} \varphi \right\|_{L^\infty} \\
&\lesssim \lambda^{N-N'-N''} C_f \lambda^{N'} \mathcal{M}\left(M, N_t, \nu, \widetilde{\nu} \right) C_\varphi \mathcal{M}\left(N'' + 4, N_x, \zeta, \widetilde{\zeta} \right) \\
&\lesssim C_f C_\varphi \mathcal{M}\left(N, N_x, \zeta, \widetilde{\zeta} \right) \mathcal{M}\left(M, N_t, \nu, \widetilde{\nu} \right) \widetilde{\zeta}^4 \left(\frac{\lambda}{\zeta} \right)^{N-N''} \\
&\lesssim C_f C_\varphi \mathcal{M}\left(N, N_x, \zeta, \widetilde{\zeta} \right) \mathcal{M}\left(M, N_t, \nu, \widetilde{\nu} \right) \widetilde{\zeta}^4 \left(\frac{\lambda}{\mu} \right)^{\mathsf{N}_{\mathrm{dec}}} \\
&\lesssim C_f C_\varphi \mathcal{M}\left(N, N_x, \zeta, \widetilde{\zeta} \right) \mathcal{M}\left(M, N_t, \nu, \widetilde{\nu} \right).
\end{aligned}
$$

Combining the above estimate with (A.37), we deduce that the bound (A.36) holds also for $N' > N_* - \mathsf{N}_{\mathrm{dec}} - 4$, concluding the proof of the lemma.   □

A.6   BOUNDS FOR SUMS AND ITERATES OF OPERATORS

For two differential operators A and B we have the expansion

$$
(A+B)^m = \sum_{k=1}^m \sum_{\substack{\alpha,\beta \in \mathbb{N}^k \\ |\alpha|+|\beta|=m}} \left(\prod_{i=1}^k A^{\alpha_i} B^{\beta_i} \right). \tag{A.39}
$$

Clearly (A.39) simplifies if $[A, B] = 0$. A lot of times we need to apply the above formula with

$$
A = v \cdot \nabla,
$$

for some vector field v. The question we would like to address in this section is the following: *Assume that we have already established estimates on* $(\prod_i D^{\alpha_i} B^{\beta_i})v$, *for* $|\alpha| + |\beta| \le m$. *Can we deduce estimates for the operator* $(A+B)^m = (v \cdot \nabla + B)^m$? The answer is *yes*, and is summarized in the following lemma:

Lemma A.10. *Fix* $N_x, N_t, N_* \in \mathbb{N}$ *and* $\Omega \in \mathbb{T}^3 \times \mathbb{R}$ *a space-time domain, and let* v *be a vector field. For* $k \ge 1$ *and* $\alpha, \beta \in \mathbb{N}^k$ *such that* $|\alpha| + |\beta| \le N_*$, *we* *assume that we have the bounds*

$$\left\|\left(\prod_{i=1}^{k} D^{\alpha_i} B^{\beta_i}\right) v\right\|_{L^{\infty}(\Omega)} \lesssim C_v \mathcal{M}\left(|\alpha|, N_x, \lambda_v, \widetilde{\lambda}_v\right) \mathcal{M}\left(|\beta|, N_t, \mu_v, \widetilde{\mu}_v\right) \quad (A.40)$$

for some $C_v \geq 0$, $1 \leq \lambda_v \leq \widetilde{\lambda}_v$, and $1 \leq \mu_v \leq \widetilde{\mu}_v$. With the same notation and restrictions on $|\alpha|, |\beta|$, let f be a function which for some $p \in [1, \infty]$ obeys

$$\left\|\left(\prod_{i=1}^{k} D^{\alpha_i} B^{\beta_i}\right) f\right\|_{L^{p}(\Omega)} \lesssim C_f \mathcal{M}\left(|\alpha|, N_x, \lambda_f, \widetilde{\lambda}_f\right) \mathcal{M}\left(|\beta|, N_t, \mu_f, \widetilde{\mu}_f\right) \quad (A.41)$$

for some $C_f \geq 0$, $1 \leq \lambda_f \leq \widetilde{\lambda}_f$, and $1 \leq \mu_f \leq \widetilde{\mu}_f$. Denote

$$\lambda = \max\{\lambda_f, \lambda_v\}, \quad \widetilde{\lambda} = \max\{\widetilde{\lambda}_f, \widetilde{\lambda}_v\}, \quad \mu = \max\{\mu_f, \mu_v\}, \quad \widetilde{\mu} = \max\{\widetilde{\mu}_f, \widetilde{\mu}_v\}.$$

Then, for

$$A = v \cdot \nabla$$

we have the bounds

$$\left\|D^n \left(\prod_{i=1}^{k} A^{\alpha_i} B^{\beta_i}\right) f\right\|_{L^{p}(\Omega)}$$

$$\lesssim C_f C_v^{|\alpha|} \mathcal{M}\left(n + |\alpha|, N_x, \lambda, \widetilde{\lambda}\right) \mathcal{M}(|\beta|, N_t, \mu, \widetilde{\mu}) \quad (A.42)$$

$$\lesssim C_f \mathcal{M}\left(n, N_x, \lambda, \widetilde{\lambda}\right) (C_v \widetilde{\lambda})^{|\alpha|} \mathcal{M}(|\beta|, N_t, \mu, \widetilde{\mu})$$

$$\lesssim C_f \mathcal{M}\left(n, N_x, \lambda, \widetilde{\lambda}\right) \mathcal{M}\left(|\alpha| + |\beta|, N_t, \max\{\mu, C_v \widetilde{\lambda}\}, \max\{\widetilde{\mu}, C_v \widetilde{\lambda}\}\right) \quad (A.43)$$

as long as $n + |\alpha| + |\beta| \leq N_*$. As a consequence, if $k = m$ then (A.39) and (A.43) imply the bound

$$\|D^n (A + B)^m f\|_{L^{p}(\Omega)}$$

$$\lesssim C_f \mathcal{M}\left(n, N_x, \lambda, \widetilde{\lambda}\right) \mathcal{M}\left(m, N_t, \max\{\mu, C_v \widetilde{\lambda}\}, \max\{\widetilde{\mu}, C_v \widetilde{\lambda}\}\right) \quad (A.44)$$

for $n + m \leq N_*$.

Remark A.11. The previous lemma is applied typically with $v = u_q$ and $B = D_{t,q-1}$ in order to obtain estimates for $D^n(\prod_i D_q^{\alpha_i} D_{t,q-1}^{\beta_i}) f$, and hence for $D^n D_t^m f$. A more non-standard application of this lemma uses $v = -v_{q-1}$ and $B = D_{t,q-1}$ in order to obtain estimates for time derivatives via $D^n \partial_t^m f = D^n(-v_{q-1} \cdot \nabla + D_{t,q-1})^m f$.

Proof of Lemma A.10. We recall (6.54)–(6.55) and note that we may write (ignoring the way in which tensors are contracted)

$$A^n = (v \cdot \nabla)^n = \sum_{j=1}^{n} f_{j,n} D^j \quad \text{where} \quad f_{j,n} = \sum_{\substack{\zeta \in \mathbb{N}^n \\ |\zeta|=n-j}} c_{n,j,\zeta} \prod_{\ell=1}^{n} (D^{\zeta_\ell} v), \quad (A.45)$$

where the $c_{n,j,\zeta}$ are certain combinatorial coefficients (tensors) with the dependence given in the subindex, and D^α represents ∂^α for some multi-index α with $|\alpha| = a$. Inserting (A.45) into the product of operators in (A.39), we see that

$$D^n \prod_{i=1}^{k} A^{\alpha_i} B^{\beta_i}$$

$$= \sum_{\substack{\gamma \in \mathbb{N}^k \\ 1^k \le \gamma \le \alpha}} D^n \prod_{i=1}^{k} (f_{\gamma_i,\alpha_i} D^{\gamma_i} B^{\beta_i})$$

$$= \sum_{\substack{\gamma \in \mathbb{N}^k \\ 1^k \le \gamma \le \alpha}} \sum_{\substack{0 \le n' \le n+|\gamma| \\ 0 \le m' \le |\beta|}} \sum_{\substack{\delta,\kappa \in \mathbb{N}^k \\ |\delta|=n+|\gamma|-n' \\ |\kappa|=|\beta|-m'}} \left(\prod_{i=1}^{k} \sum_{\substack{\delta_i',\kappa_i' \in \mathbb{N}^k \\ |\delta_i'|=\delta_i \\ |\kappa_i'|=\kappa_i}} \widetilde{c}_{(\ldots)} \left(\prod_{\ell_i=1}^{k} D^{\delta_{i,\ell_i}'} B^{\kappa_{i,\ell_i}'} \right) f_{\gamma_i,\alpha_i} \right)$$

$$\times \left(\sum_{\substack{\eta,\rho \in \mathbb{N}^k \\ |\eta|=n' \\ |\rho|=m'}} \bar{c}_{(\ldots)} \prod_{s=1}^{k} D^{\eta_s} B^{\rho_s} \right), \quad (A.46)$$

where the $\widetilde{c}_{(\ldots)}, \bar{c}_{(\ldots)} \ge 0$ are certain combinatorial coefficients (tensors). Combining (A.39)–(A.46), we obtain that

$$D^n \left(\prod_{i=1}^{k} A^{\alpha_i} B^{\beta_i} \right) f$$

$$= \sum_{\substack{\gamma \in \mathbb{N}^k \\ 1^k \le \gamma \le \alpha}} \sum_{\substack{0 \le n' \le n+|\gamma| \\ 0 \le m' \le |\beta|}} \sum_{\substack{\delta,\kappa \in \mathbb{N}^k \\ |\delta|=n+|\gamma|-n' \\ |\kappa|=|\beta|-m'}} \left(\sum_{\substack{\eta,\rho \in \mathbb{N}^k \\ |\eta|=n' \\ |\rho|=m'}} \bar{c}_{(\ldots)} \left(\prod_{s=1}^{k} D^{\eta_s} B^{\rho_s} \right) f \right)$$

$$\times \left(\prod_{i=1}^{k} \sum_{\substack{\delta_i', \kappa_i' \in \mathbb{N}^k \\ |\delta_i'| = \delta_i \\ |\kappa_i'| = \kappa_i}} \widetilde{c}_{(\dots)} \left(\prod_{\ell_i=1}^{k} D^{\delta_{i,\ell_i}'} B^{\kappa_{i,\ell_i}'} \right) \left(\sum_{\substack{\zeta_i \in \mathbb{N}^{\alpha_i} \\ |\zeta_i| = \alpha_i - \gamma_i}} c_{(\dots)} \prod_{r_i=1}^{\alpha_i} (D^{\zeta_{i,r_i}} v) \right) \right),$$

$$(A.47)$$

where the $c_{(\dots)}, \widetilde{c}_{(\dots)}, \overline{c}_{(\dots)} \geq 0$ are certain combinatorial coefficients (tensors) whose dependence is omitted for simplicity (they may depend on all the parameters in the sums and products). The above expansion combined with the Leibniz rule, the bound (3.9), and assumptions $(A.40)$–$(A.41)$, implies

$$\left\| D^n \left(\prod_{i=1}^{k} A^{\alpha_i} B^{\beta_i} \right) f \right\|_{L^p(\Omega)}$$

$$\lesssim \sum_{\substack{\gamma \in \mathbb{N}^k \\ 1^k \leq \gamma \leq \alpha}} \sum_{0 \leq n' \leq n+|\gamma|} \sum_{\substack{\delta, \kappa \in \mathbb{N}^k \\ 0 \leq m' \leq |\beta| \\ |\kappa| = |\beta| - m'}} \left(\sum_{\substack{\eta, \rho \in \mathbb{N}^k \\ |\eta| = n' \\ |\rho| = m'}} \left\| \left(\prod_{s=1}^{k} D^{\eta_s} B^{\rho_s} \right) f \right\|_{L^p(\Omega)} \right)$$

$$\times \left(\prod_{i=1}^{k} \sum_{\substack{\zeta_i \in \mathbb{N}^{\alpha_i} \\ |\zeta_i| = \alpha_i - \gamma_i}} \sum_{\substack{\delta_i', \kappa_i' \in \mathbb{N}^k \\ |\delta_i'| = \delta_i \\ |\kappa_i'| = \kappa_i}} \left\| \left(\prod_{\ell_i=1}^{k} D^{\delta_{i,\ell_i}'} B^{\kappa_{i,\ell_i}'} \right) \left(\prod_{r_i=1}^{\alpha_i} (D^{\zeta_{i,r_i}} v) \right) \right\|_{L^\infty(\Omega)} \right)$$

$$\lesssim \sum_{\substack{\gamma \in \mathbb{N}^k \\ 1^k \leq \gamma \leq \alpha}} \sum_{0 \leq n' \leq n+|\gamma|} \sum_{\substack{\delta, \kappa \in \mathbb{N}^k \\ 0 \leq m' \leq |\beta| \\ |\kappa| = |\beta| - m'}} \left(\mathcal{C}_f \mathcal{M}\left(n', N_x, \lambda, \widetilde{\lambda} \right) \mathcal{M}\left(m', N_t, \mu, \widetilde{\mu} \right) \right)$$

$$\times \left(\prod_{i=1}^{k} \mathcal{C}_v^{\alpha_i} \mathcal{M}\left(\alpha_i - \gamma_i + \delta_i, N_x, \lambda, \widetilde{\lambda} \right) \mathcal{M}\left(\kappa_i, N_t, \mu, \widetilde{\mu} \right) \right)$$

$$\lesssim \mathcal{C}_f \sum_{\substack{0 \leq n' \leq n+|\alpha| \\ 0 \leq m' \leq |\beta|}} \left(\mathcal{C}_f \mathcal{M}\left(n', N_x, \lambda, \widetilde{\lambda} \right) \mathcal{M}\left(m', N_t, \mu, \widetilde{\mu} \right) \right)$$

$$\times \left(\mathcal{C}_v^{|\alpha|} \mathcal{M}\left(|\alpha| + n - n', N_x, \lambda, \widetilde{\lambda} \right) \mathcal{M}\left(|\beta| - m', N_t, \mu, \widetilde{\mu} \right) \right)$$

$$\lesssim \mathcal{C}_f \mathcal{C}_v^{|\alpha|} \mathcal{M}\left(|\alpha| + n, N_x, \lambda, \widetilde{\lambda} \right) \mathcal{M}\left(|\beta|, N_t, \mu, \widetilde{\mu} \right),$$

which is precisely the bound claimed in $(A.42)$. Estimate $(A.43)$ follows immediately, while the bound (A.44) is a consequence of the above and (A.39).   □

A.7   COMMUTATORS WITH MATERIAL DERIVATIVES

Let D represent a pure spatial derivative and let

$$D_t = \partial_t + v \cdot \nabla$$

denote a material derivative along the smooth (incompressible) vector field v. This vector field v is fixed throughout this section. The question we would like to address in this section is the following: *Assume that for the vector field v we have $D^a D_t^b Dv$ estimates available. Can we then bound the operator norm of $D_t^b D^a$ in terms of the operator norm of $D^a D_t^b$?*

Following Komatsu [47, Lemma 5.2], a useful ingredient for bounding commutators of Eulerian and material derivatives is the following lemma. We use the following commutator notation:

$$(\operatorname{ad} D_t)^0(D) = D$$

$$(\operatorname{ad} D_t)^1(D) = [D_t, D] = -Dv \cdot \nabla$$

$$(\operatorname{ad} D_t)^a(D) = (\operatorname{ad} D_t)((\operatorname{ad} D_t)^{a-1}(D)) = [D_t, (\operatorname{ad} D_t)^{a-1}(D)]$$

for all $a \geq 2$. Note that for any $a \geq 0$, $(\operatorname{ad} D_t)^a(D)$ is a differential operator of order 1.

Lemma A.12. *Let $m, n \geq 0$. Then we have that the commutator of D_t^m and D^n is given by*

$$[D_t^m, D^n] = \sum_{\{\alpha \in \mathbb{N}^n \,:\, 1 \leq |\alpha| \leq m\}} \frac{m!}{\alpha!(m - |\alpha|)!} \left(\prod_{\ell=1}^{n} (\operatorname{ad} D_t)^{\alpha_\ell}(D) \right) D_t^{m-|\alpha|}. \quad \text{(A.48)}$$

By the product in (A.48) we mean the product/composition of operators

$$\prod_{\ell=1}^{n} (\operatorname{ad} D_t)^{\alpha_\ell}(D) = (\operatorname{ad} D_t)^{\alpha_n}(D)(\operatorname{ad} D_t)^{\alpha_{n-1}}(D) \ldots (\operatorname{ad} D_t)^{\alpha_1}(D),$$

so that on the right side of (A.48) we have a sum of differential operators of order at most n.

For the above lemma to be useful, we need to be able to characterize the operator $(\operatorname{ad} D_t)^a(D)$.

Lemma A.13. *Let $a \in \mathbb{N}$. Then the order 1 differential operator $(\operatorname{ad} D_t)^a(D)$*

may be expressed as

$$(\operatorname{ad} D_t)^a(D) = \sum_{k=1}^{a} \sum_{\{\beta \in \mathbb{N}^k \,:\, |\beta| = a-k\}} c_{a,k,\beta} \prod_{j=1}^{k} (D_t^{\beta_j} Dv) \cdot \nabla, \qquad (\text{A.49})$$

where the $\prod$ in (A.49) *denotes the product of matrices and $c_{a,k,\beta}$ are coefficients which depend only on a, k, β.*

Proof of Lemma A.13. When $a = 1$ we know that $(\operatorname{ad} D_t)(D) = -Dv \cdot \nabla$, so that the lemma trivially holds. We proceed by induction on a. Using the fact that $[D_t, \nabla] = -Dv \cdot \nabla$, we obtain

$$(\operatorname{ad} D_t)^{a+1}(D) = D_t \left(\sum_{k=1}^{a} \sum_{\beta \in \pi(k,a)} c_{a,k,\beta} \prod_{j=1}^{k} (D_t^{\beta_j} Dv) \right) \cdot \nabla$$

$$+ \sum_{k=1}^{a} \sum_{\beta \in \pi(k,a)} c_{a,k,\beta} \prod_{j=1}^{k} (D_t^{\beta_j} Dv) \cdot [D_t, \nabla]$$

$$= D_t \left(\sum_{k=1}^{a} \sum_{\beta \in \pi(k,a)} c_{a,k,\beta} \prod_{j=1}^{k} (D_t^{\beta_j} Dv) \right) \cdot \nabla$$

$$- \sum_{k=1}^{a} \sum_{\beta \in \pi(k,a)} c_{a,k,\beta} \prod_{j=1}^{k} (D_t^{\beta_j} Dv) Dv \cdot \nabla,$$

where we have denoted by

$$\pi(k,a) = \left\{ \beta \in \mathbb{N}^k : |\beta| = a - k \right\}$$

the set of all partitions of a set of $a - k$ elements into k sets. For the first term we use the Leibniz rule for D_t, so that for any $\beta \in \pi(k,a)$, we obtain an element $\beta + e_j \in \pi(k, a+1)$, with $e_j = (0, \ldots, 0, 1, 0, \ldots, 0) \in \mathbb{N}^k$, and the 1 lies in the jth coordinate. For $1 \leq k \leq a$, this in fact lists all the elements in $\pi(k, a+1)$. For the second sum, we identify $\beta \in \pi(k,a)$ with $\beta \in \pi(k+1, a+1)$, upon padding it with a 0 in the $k+1^{\text{st}}$ entry. Changing variables $k+1 \to k$ then recovers an element $\beta \in \pi(k, a+1)$, including the case $k = a+1$, which was missing from the first sum. $\qquad \square$

From Lemma A.12 and Lemma A.13 we deduce the following:

Lemma A.14. *Let $p \in [1, \infty]$. Fix $N_x, N_t, N_*, M_* \in \mathbb{N}$, let v be a vector field, let $D_t = \partial_t + v \cdot \nabla$ be the associated material derivative, and let Ω be a space-time domain. Assume that the vector field v obeys*

$$\left\| D^N D_t^M Dv \right\|_{L^\infty(\Omega)} \lesssim C_v \mathcal{M} \left(N+1, N_x, \lambda_v, \tilde{\lambda}_v \right) \mathcal{M} \left(M, N_t, \mu_v, \tilde{\mu}_v \right) \qquad (\text{A.50})$$

for $N \leq N_*$ and $M \leq M_*$. Moreover, let f be a function which obeys

$$\left\|D^N D_t^M f\right\|_{L^p(\Omega)} \lesssim \mathcal{C}_f \mathcal{M}\left(N, N_x, \lambda_f, \widetilde{\lambda}_f\right) \mathcal{M}\left(M, N_t, \mu_f, \widetilde{\mu}_f\right) \qquad (A.51)$$

for all $N \leq N_*$ and $M \leq M_*$. Denote

$$\lambda = \max\{\lambda_f, \lambda_v\}, \quad \widetilde{\lambda} = \max\{\widetilde{\lambda}_f, \widetilde{\lambda}_v\}, \quad \mu = \max\{\mu_f, \mu_v\}, \quad \widetilde{\mu} = \max\{\widetilde{\mu}_f, \widetilde{\mu}_v\}.$$

Let $m, n, \ell \geq 0$ be such that $n + \ell \leq N_*$ and $m \leq M_*$. Then, we have that the commutator $[D_t^m, D^n]$ is bounded as

$$\left\|D^\ell [D_t^m, D^n] f\right\|_{L^p(\Omega)} \lesssim \mathcal{C}_f \mathcal{C}_v \widetilde{\lambda}_v \mathcal{M}\left(\ell + n, N_x, \lambda, \widetilde{\lambda}\right)$$
$$\times \mathcal{M}(m - 1, N_t, \max\{\mu, \mathcal{C}_v \widetilde{\lambda}_v\}, \max\{\widetilde{\mu}, \mathcal{C}_v \widetilde{\lambda}_v\}) \quad (A.52)$$
$$\lesssim \mathcal{C}_f \mathcal{M}\left(\ell + n, N_x, \lambda, \widetilde{\lambda}\right)$$
$$\times \mathcal{M}\left(m, N_t, \max\{\mu, \mathcal{C}_v \widetilde{\lambda}_v\}, \max\{\widetilde{\mu}, \mathcal{C}_v \widetilde{\lambda}_v\}\right). \quad (A.53)$$

Moreover, we have that for $k \geq 2$, and any $\alpha, \beta \in \mathbb{N}^k$ with $|\alpha| \leq N_*$ and $|\beta| \leq M_*$, the estimate

$$\left\|\left(\prod_{i=1}^k D^{\alpha_i} D_t^{\beta_i}\right) f\right\|_{L^p(\Omega)}$$
$$\lesssim \mathcal{C}_f \mathcal{M}\left(|\alpha|, N_x, \lambda, \widetilde{\lambda}\right) \mathcal{M}\left(|\beta|, N_t, \max\{\mu, \mathcal{C}_v \widetilde{\lambda}_v\}, \max\{\widetilde{\mu}, \mathcal{C}_v \widetilde{\lambda}_v\}\right) \quad (A.54)$$

holds.

Remark A.15. If instead of (A.50) and (A.51) holding for $N \leq N_*$ and $M \leq M_*$, we know that both of these inequalities hold for all $N + M \leq N_\circ$ for some $N_\circ \geq 1$, then the conclusions of the lemma hold as follows: the bounds (A.52) and (A.53) hold for $\ell + n + m \leq N_\circ$, while (A.54) holds for $|\alpha| + |\beta| \leq N_\circ$. This fact follows immediately from the proof of the lemma, but may alternatively also be derived from its statement (see also Lemma A.3).

Remark A.16. In Lemma A.14, if the assumption (A.51) is replaced by

$$\left\|D^N D_t^M f\right\|_{L^p(\Omega)} \lesssim \mathcal{C}_f \mathcal{M}\left(N - 1, N_x, \lambda_f, \widetilde{\lambda}_f\right) \mathcal{M}\left(M, N_t, \mu_f, \widetilde{\mu}_f\right), \quad (A.55)$$

whenever $1 \leq N \leq N_*$, then the conclusion (A.54) changes, and it instead becomes

$$\left\|\left(\prod_{i=1}^k D^{\alpha_i} D_t^{\beta_i}\right) f\right\|_{L^p(\Omega)}$$

$$\lesssim \mathcal{C}_f \mathcal{M}\left(|\alpha| - 1, N_x, \lambda, \widetilde{\lambda}\right) \mathcal{M}\left(|\beta|, N_t, \max\{\mu, \mathcal{C}_v\widetilde{\lambda}_v\}, \max\{\widetilde{\mu}, \mathcal{C}_v\widetilde{\lambda}_v\}\right) \quad (A.56)$$

whenever $|\alpha| \geq 1$. This follows for instance by noting that the sum on the second line of (A.61) only contains terms with $j \geq 1$, so that (A.55) is not required when $N = 0$.

Proof of Lemma A.14. First, we deduce from (A.49) that for any $\alpha_i \geq 1$ and $1 \leq i \leq n$, we have

$$(\mathrm{ad}\, D_t)^{\alpha_i}(D) = \sum_{\kappa_i=1}^{\alpha_i} f_{\kappa_i,\alpha_i} \cdot \nabla\,, \qquad (A.57)$$

where the functions f_{κ_i,α_i} are computed as

$$f_{\kappa_i,\alpha_i} = \sum_{\{\beta \in \mathbb{N}^{\kappa_i}\,:\,|\beta|=\alpha_i-\kappa_i\}} c_{(\dots)} \prod_{j=1}^{\kappa_i} (D_t^{\beta_j} Dv)$$

for suitable combinatorial coefficients (tensors) $c_{(\dots)}$ which depend on κ_i, α_i, and β. In particular, in view of assumption (A.50), and the Leibniz rule, we have that

$$\left\|D^\ell f_{\kappa_i,\alpha_i}\right\|_{L^\infty(\Omega)} \lesssim \mathcal{C}_v^{\kappa_i} \mathcal{M}\left(\kappa_i + \ell, N_x, \lambda_v, \widetilde{\lambda}_v\right) \mathcal{M}\left(\alpha_i - \kappa_i, N_t, \mu_v, \widetilde{\mu}_v\right). \quad (A.58)$$

Next, from (A.57) we deduce that for any $\alpha \in \mathbb{N}^n$ with $|\alpha| \geq 1$, one may write

$$\prod_{i=1}^{n}(\mathrm{ad}\, D_t)^{\alpha_i}(D) = \sum_{j=1}^{n} g_{j,\alpha} D^j\,, \qquad (A.59)$$

where

$$g_{j,\alpha} = \sum_{\{\kappa \in \mathbb{N}^n\,:\,1^n \leq \kappa \leq \alpha\}} \sum_{\{\gamma \in \mathbb{N}^n\,:\,|\gamma|=n-j\}} \widetilde{c}_{(\dots)} \prod_{i=1}^{n} D^{\gamma_i} f_{\kappa_i,\alpha_i}\,.$$

As a consequence of (A.58) we see that

$$\left\|D^\ell g_{j,\alpha}\right\|_{L^\infty(\Omega)} \lesssim \sum_{|\kappa|=1}^{|\alpha|} \mathcal{C}_v^{|\kappa|} \mathcal{M}\left(\ell + n - j + |\kappa|, N_x, \lambda_v, \widetilde{\lambda}_v\right)$$

$$\times \mathcal{M}\left(|\alpha| - |\kappa|, N_t, \mu_v, \widetilde{\mu}_v\right). \qquad (A.60)$$

From (A.48), assumption (A.51), identity (A.59), and bound (A.60), we see that

$$\left\|D^\ell [D_t^m, D^n] f\right\|_{L^p(\Omega)}$$

$$\lesssim \sum_{|\alpha|=1}^{m} \sum_{j=1}^{n} \left\| D^\ell \left(g_{j,\alpha} D^j D_t^{m-|\alpha|} \right) f \right\|_{L^p(\Omega)}$$

$$\lesssim \sum_{|\alpha|=1}^{m} \sum_{j=1}^{n} \left\| D^\ell g_{j,\alpha} \right\|_{L^\infty(\Omega)} \left\| D^j D_t^{m-|\alpha|} f \right\|_{L^p(\Omega)}$$

$$+ \left\| g_{j,\alpha} \right\|_{L^\infty(\Omega)} \left\| D^{\ell+j} D_t^{m-|\alpha|} f \right\|_{L^p(\Omega)}$$

$$\lesssim \sum_{k=1}^{m} \sum_{j=1}^{n} \mathcal{C}_f \mathcal{C}_v^k \mathcal{M} \left(\ell + n - j + k, N_x, \lambda_v, \widetilde{\lambda}_v \right) \mathcal{M} \left(j, N_x \lambda, \widetilde{\lambda} \right)$$

$$\times \mathcal{M} \left(m - k, N_t, \mu, \widetilde{\mu} \right)$$

$$+ \mathcal{C}_f \mathcal{C}_v^k \mathcal{M} \left(n - j + k, N_x, \lambda_v, \widetilde{\lambda}_v \right) \mathcal{M} \left(j + \ell, N_x \lambda, \widetilde{\lambda} \right)$$

$$\times \mathcal{M} \left(m - k, N_t, \mu, \widetilde{\mu} \right)$$

$$\lesssim \mathcal{C}_f \mathcal{M} \left(\ell + n, N_x, \lambda, \widetilde{\lambda} \right) \sum_{k=1}^{m} (\mathcal{C}_v \widetilde{\lambda}_v)^k \mathcal{M} \left(m - k, N_t, \mu, \widetilde{\mu} \right), \qquad (A.61)$$

from which (A.53) follows directly.

In order to prove (A.54) we proceed by induction on k. For $k = 1$ the statement holds in view of (A.51). We assume that (A.54) holds for $k' \leq k-1$, and denote

$$P_{k'} = \left(\prod_{i=1}^{k'} D^{\alpha_i} D_t^{\beta_i} \right) f.$$

With this notation we have

$$P_k = D^{\alpha_k} D_t^{\beta_k} D^{\alpha_{k-1}} D_t^{\beta_{k-1}} P_{k-2}$$
$$= D^{\alpha_k + \alpha_{k-1}} D_t^{\beta_k + \beta_{k-1}} P_{k-2} + D^{\alpha_k} \left[D_t^{\beta_k}, D^{\alpha_{k-1}} \right] D_t^{\beta_{k-1}} P_{k-2}.$$

Using (A.54) with $k-1$ gives the desired estimate for the first term above. For the second term, we appeal to the commutator bound (A.53), applied to $D_t^{\beta_{k-1}} P_{k-2}$, which obeys condition (A.51) in view of (A.54) at level $k-1$. This concludes the proof of (A.54) at level k. $\qquad\square$

A.8 INTERMITTENCY-FRIENDLY INVERSION OF THE DIVERGENCE

Given a vector field G^i, a zero mean periodic function ϱ, and an incompressible flow Φ, our goal in this section is to write $G^i(x)\varrho(\Phi(x))$ as the divergence of a symmetric tensor.

Proposition A.17 (Inverse divergence iteration step). *Fix two zero-mean*
$\mathbb{T}^3$*-periodic functions* ϱ *and* ϑ*, which are related by* $\varrho = \Delta\vartheta$*. Let* Φ *be a volume*
preserving transformation of $\mathbb{T}^3$*, such that* $\|\nabla\Phi - \mathrm{Id}\|_{L^\infty(\mathbb{T}^3)} \leq 1/2$*. Define the*
matrix $A = (\nabla\Phi)^{-1}$*. Given a vector field* G^i*, we have*

$$G^i \varrho \circ \Phi = \partial_n \mathring{R}^{in} + \partial_i P + E^i \,, \tag{A.62}$$

where the traceless symmetric stress R^{in} *is given by*

$$\mathring{R}^{in} = \left(G^i A_\ell^n + G^n A_\ell^i - A_k^i A_k^n G^p \partial_p \Phi^\ell\right)(\partial_\ell \vartheta) \circ \Phi - P\delta_{in} \,, \tag{A.63}$$

where the pressure term is given by

$$P = \left(2G^n A_\ell^n - A_k^n A_k^n G^p \partial_p \Phi^\ell\right)(\partial_\ell \vartheta) \circ \Phi \tag{A.64}$$

and the error term E^i *is given by*

$$E^i = \left(\partial_n \left(G^p A_k^i A_k^n - G^n A_k^i A_k^p\right) \partial_p \Phi^\ell - \partial_n G^i A_\ell^n\right)(\partial_\ell \vartheta) \circ \Phi \,. \tag{A.65}$$

Proof of Proposition A.17. Note that by definition we have $A_\ell^k \partial_j \Phi^\ell = \delta_{kj}$. Since
Φ is volume preserving, $\det(\nabla\Phi) = 1$, and so each entry of the matrix A equals
the corresponding cofactor of $\nabla\Phi$, which in three dimensions is a quadratic
function of entries of $\nabla\Phi$ given explicitly by $A_j^i = \frac{1}{2}\varepsilon_{ipq}\varepsilon_{jk\ell}\partial_k \Phi^p \partial_\ell \Phi^q$. In two
dimensions, A is a linear map in $\nabla\Phi$. Moreover, since Φ is volume preserving,
the Piola identity $\partial_j A_i^j = 0$ holds for every $i \in \{1, 2, 3\}$. The main identity
that we use in the proof is that for any scalar function φ we have $(\partial_i \varphi) \circ \Phi = A_i^m \partial_m(\varphi \circ \Phi) = \partial_m(A_i^m \varphi \circ \Phi)$.

Starting from $\varrho = \Delta\vartheta$, we have

$$\begin{aligned}
G^i \varrho \circ \Phi &= G^i(\partial_{kk}\vartheta) \circ \Phi \\
&= G^i A_k^n \partial_n(\partial_k \vartheta) \circ \Phi \\
&= \partial_n \left(G^i A_k^n(\partial_k \vartheta) \circ \Phi\right) - \partial_n G^i A_k^n(\partial_k \vartheta) \circ \Phi \\
&= \partial_n \left(G^i A_k^n(\partial_k \vartheta) \circ \Phi + G^n A_k^i(\partial_k \vartheta) \circ \Phi\right) \\
&\quad - \partial_n \left(G^n A_k^i(\partial_k \vartheta) \circ \Phi\right) - \partial_n G^i A_k^n(\partial_k \vartheta) \circ \Phi \,.
\end{aligned}$$

Next, we have

$$\begin{aligned}
\partial_n \left(G^n A_k^i(\partial_k \vartheta) \circ \Phi\right) &= \partial_n \left(G^n A_k^i A_k^p \partial_p(\vartheta \circ \Phi)\right) \\
&= \partial_p \partial_n \left(G^n A_k^i A_k^p \vartheta \circ \Phi\right) - \partial_n \left(\partial_p(G^n A_k^i A_k^p)\vartheta \circ \Phi\right) \\
&= \partial_p \left(G^n A_k^i A_k^p \partial_n(\vartheta \circ \Phi)\right) + \partial_p \left(\partial_n(G^n A_k^i A_k^p)\vartheta \circ \Phi\right) \\
&\quad - \partial_n \left(\partial_p(G^n A_k^i A_k^p)\vartheta \circ \Phi\right) \\
&= \partial_n \left(G^p A_k^i A_k^n \partial_p(\vartheta \circ \Phi)\right) + \partial_n \left(\partial_p(G^p A_k^i A_k^n)\vartheta \circ \Phi\right)
\end{aligned}$$

$$- \partial_n \left(\partial_p (G^n A_k^i A_k^p) \vartheta \circ \Phi \right) ,$$

where in the last equality we have just switched the letters of summation n and p. We further massage the last term in the above equality:

$$\partial_n \left(\partial_p (G^n A_k^i A_k^p) \vartheta \circ \Phi \right) = \partial_p \left(G^n A_k^i A_k^p \right) \partial_n (\vartheta \circ \Phi) + \partial_{np} \left(G^n A_k^i A_k^p \right) \vartheta \circ \Phi$$
$$= \partial_p \left(G^n A_k^i A_k^p \right) \partial_n (\vartheta \circ \Phi) + \partial_p \left(\partial_n \left(G^n A_k^i A_k^p \right) \vartheta \circ \Phi \right)$$
$$- \partial_n \left(G^n A_k^i A_k^p \right) \partial_p (\vartheta \circ \Phi).$$

Combining the above three equalities, we arrive at

$$G^i \varrho \circ \Phi = \partial_n \left((G^i A_k^n + G^n A_k^i)(\partial_k \vartheta) \circ \Phi - A_k^i A_k^n G^p \partial_p (\vartheta \circ \Phi) \right)$$
$$+ \partial_n \left(G^p A_k^i A_k^n - G^n A_k^i A_k^p \right) \partial_p (\vartheta \circ \Phi) - \partial_n G^i A_k^n (\partial_k \vartheta) \circ \Phi$$
$$= \partial_n \left((G^i A_k^n + G^n A_k^i)(\partial_k \vartheta) \circ \Phi - A_k^i A_k^n G^p \partial_p \Phi^\ell (\partial_\ell \vartheta) \circ \Phi \right)$$
$$+ \partial_n \left(G^p A_k^i A_k^n - G^n A_k^i A_k^p \right) \partial_p \Phi^\ell (\partial_\ell \vartheta) \circ \Phi - \partial_n G^i A_\ell^n (\partial_\ell \vartheta) \circ \Phi .$$

In the last equality, we have exchanged the order of summation. Identities (A.62)–(A.65) follow upon declaring that the trace part of the symmetric stress is the pressure.   □

Proposition A.17 allows us to obtain the following result, which is the main conclusion of this section.

Proposition A.18 (Inverse divergence with error term). *Fix an incompressible vector field v and denote its material derivative by $D_t = \partial_t + v \cdot \nabla$. Fix integers $N_* \geq M_* \geq 1$. Also fix $\mathsf{N}_{\mathrm{dec}}, \mathsf{d} \geq 1$ such that $N_* - \mathsf{d} \geq 2\mathsf{N}_{\mathrm{dec}} + 4$.*

Let G be a vector field and assume there exists a constant $\mathcal{C}_G > 0$ and parameters $\lambda, \nu \geq 1$ such that

$$\left\| D^N D_t^M G \right\|_{L^1} \lesssim \mathcal{C}_G \lambda^N \mathcal{M} (M, M_t, \nu, \tilde{\nu}) \qquad (A.66)$$

for all $N \leq N_*$ and $M \leq M_*$.

Let Φ be a volume preserving transformation of $\mathbb{T}^3$, such that

$$D_t \Phi = 0 \qquad \text{and} \qquad \| \nabla \Phi - \mathrm{Id} \|_{L^\infty (\mathrm{supp}\, G)} \leq 1/2 .$$

Denote by Φ^{-1} the inverse of the flow Φ, which is the identity at a time slice which intersects the support of G. Assume that the velocity field v and the flow functions Φ and Φ^{-1} satisfy the bounds

$$\left\| D^{N+1} \Phi \right\|_{L^\infty (\mathrm{supp}\, G)} + \left\| D^{N+1} \Phi^{-1} \right\|_{L^\infty (\mathrm{supp}\, G)} \lesssim \lambda'^N \qquad (A.67)$$
$$\left\| D^N D_t^M Dv \right\|_{L^\infty (\mathrm{supp}\, G)} \lesssim \nu \lambda'^N \mathcal{M} (M, M_t, \nu, \tilde{\nu}) , \qquad (A.68)$$

for all $N \leq N_*$, $M \leq M_*$, and some $\lambda' > 0$.

Lastly, let $\varrho, \vartheta \colon \mathbb{T}^3 \to \mathbb{R}$ be two zero mean functions with the following properties:

1. *there exists* $\mathsf{d} \geq 1$ *and a parameter* $\zeta \geq 1$ *such that* $\varrho(x) = \zeta^{-2\mathsf{d}} \Delta^{\mathsf{d}} \vartheta(x)$;
2. *there exists a parameter* $\mu \geq 1$ *such that* ϱ *and* ϑ *are* $(\mathbb{T}/\mu)^3$*-periodic;*
3. *there exist parameters* $\Lambda \geq \zeta$ *and* $\mathcal{C}_* \geq 1$ *such that*

$$\left\| D^N \varrho \right\|_{L^1} \lesssim \mathcal{C}_* \Lambda^N \qquad \text{and} \qquad \left\| D^N \vartheta \right\|_{L^1} \lesssim \mathcal{C}_* \mathcal{M}\left(N, 2\mathsf{d}, \zeta, \Lambda\right) \quad \text{(A.69)}$$

for all $0 \leq N \leq \mathsf{N}_{\mathrm{fin}}$, *except for the case* $N = 2\mathsf{d}$, *when the Calderón-Zygmund inequality fails. In this exceptional case, the second inequality in* (A.69) *is allowed to be weaker by a factor of* Λ^α, *for an arbitrary* $\alpha \in (0, 1]$; *that is, we only require that* $\left\| D^{2\mathsf{d}} \vartheta \right\|_{L^1} \lesssim \mathcal{C}_* \Lambda^\alpha \zeta^{2\mathsf{d}}$.

If the above parameters satisfy

$$\lambda' \leq \lambda \ll \mu \leq \zeta \leq \Lambda, \qquad \text{(A.70)}$$

where by the second inequality in (A.70) *we mean that*

$$\Lambda^4 \left(\frac{\mu}{2\pi\sqrt{3}\lambda} \right)^{-\mathsf{N}_{\mathrm{dec}}} \leq 1, \qquad \text{(A.71)}$$

then we have that

$$G \, \varrho \circ \Phi = \mathrm{div}\, \mathring{R} + \nabla P + E =: \mathrm{div}\left(\mathcal{H}\left(G\varrho \circ \Phi \right) \right) + \nabla P + E, \qquad \text{(A.72)}$$

where the traceless symmetric stress $\mathring{R} = \mathcal{H}(G\varrho \circ \Phi)$ *and the scalar pressure* P *are supported in* $\mathrm{supp}\, G$, *and for any fixed* $\alpha \in (0, 1)$ *they satisfy*

$$\left\| D^N D_t^M \mathring{R} \right\|_{L^1} + \left\| D^N D_t^M P \right\|_{L^1}$$
$$\lesssim \Lambda^\alpha \mathcal{C}_G \mathcal{C}_* \zeta^{-1} \mathcal{M}\left(N, 1, \zeta, \Lambda\right) \mathcal{M}\left(M, M_t, \nu, \tilde{\nu}\right) \qquad \text{(A.73)}$$

for all $N \leq N_* - \mathsf{d}$ *and* $M \leq M_*$. *The implicit constants depend on* N, M, *and* α *but not on* G, ϱ, *or* Φ. *Lastly, for* $N \leq N_* - \mathsf{d}$ *and* $M \leq M_*$ *the error term* E *in* (A.72) *satisfies*

$$\left\| D^N D_t^M E \right\|_{L^1} \lesssim \mathcal{C}_G \mathcal{C}_* \Lambda^\alpha \lambda^\mathsf{d} \zeta^{-\mathsf{d}} \Lambda^N \mathcal{M}\left(M, M_t, \nu, \tilde{\nu}\right). \qquad \text{(A.74)}$$

We emphasize that the range of M *in* (A.73) *and* (A.74) *is exactly the same as the one in* (A.66), *while the range of permissible values for* N *shrinks from* N_* *to* $N_* - \mathsf{d}$.

Lastly, let $N_\circ, M_\circ$ *be integers such that* $1 \leq M_\circ \leq N_\circ \leq M_*/2$. *Assume that in addition to the bound* (A.68) *we have the following global lossy estimates:*

$$\left\| D^N \partial_t^M v \right\|_{L^\infty(\mathbb{T}^3)} \lesssim \mathcal{C}_v \tilde{\lambda}_q^N \tilde{\tau}_q^{-M} \qquad \text{(A.75)}$$

for all $M \leq M_\circ$ and $N + M \leq N_\circ + M_\circ$, where

$$\mathcal{C}_v \widetilde{\lambda}_q \lesssim \widetilde{\tau}_q^{-1}, \qquad \text{and} \qquad \lambda' \leq \widetilde{\lambda}_q \leq \Lambda \leq \lambda_{q+1}. \tag{A.76}$$

If d *is chosen* large enough *so that*

$$\mathcal{C}_G \mathcal{C}_* \Lambda \left(\frac{\lambda}{\zeta}\right)^{d-1} \left(1 + \frac{\max\{\widetilde{\tau}_q^{-1}, \widetilde{\nu}, \mathcal{C}_v \Lambda\}}{\tau_q^{-1}}\right)^{M_\circ} \leq \frac{\delta_{q+2}}{\lambda_{q+1}^5}, \tag{A.77}$$

then we may write

$$E = \operatorname{div} \mathring{R}_{\text{nonlocal}} + \fint_{\mathbb{T}^3} G\varrho \circ \Phi dx$$

$$=: \operatorname{div} \left(\mathcal{R}^*(G\varrho \circ \Phi)\right) + \fint_{\mathbb{T}^3} G\varrho \circ \Phi dx, \tag{A.78}$$

where $\mathring{R}_{\text{nonlocal}} = \mathcal{R}^*(G\varrho \circ \Phi)$ *is a traceless symmetric stress which satisfies*

$$\left\|D^N D_t^M \mathring{R}_{\text{nonlocal}}\right\|_{L^1} \leq \frac{\delta_{q+2}}{\lambda_{q+1}^5} \lambda_{q+1}^N \tau_q^{-M} \tag{A.79}$$

for $N \leq N_\circ$ and $M \leq M_\circ$.

Before turning to the proof of Lemma A.18, let us make three remarks. First, we highlight certain parameter values which will occur commonly in applications of the proposition. Second, we comment on a technical aspect of the application of the proposition in Section 8.3. Finally, we comment on the assumptions (1)–(3) and (A.71) and (A.77) for the functions ϱ and ϑ, which in applications are related to the transversal densities of the pipe flows.

Remark A.19. Frequently, G will come with derivative bounds which are satisfied for $N + M \leq N^\sharp$. In this case, we set $N_* = M_* = N^\sharp/2$, so that (A.66) is satisfied. The bounds in (A.67) and (A.68) will be true (due to Corollary 6.27 and estimate (6.60)) for much higher order derivatives than $N^\sharp/2$, and so we ignore them. The bounds in (A.69) are given by construction in Proposition 4.4. Then the bounds (A.73) and (A.74) are satisfied for $N \leq N^\sharp/2 - $d and $M \leq N^\sharp/2$, and in particular for $N + M \leq N^\sharp/2 - $d. In (A.75) we will then set $N_\circ = M_\circ \leq N^\sharp/4$, which in practice will give $N_\circ = M_\circ = 3N_{\text{ind},v}$. Arguing in the same way used to produce the bound (5.18) shows that for $N + M \leq N_{\text{fin}}$,

$$\left\|D^N \partial_t^M v_{\ell_q}\right\|_{L^\infty} \lesssim \left(\lambda_q^4 \delta_q^{1/2}\right) \widetilde{\lambda}_q^N \widetilde{\tau}_q^{-M} \tag{A.80}$$

and so (A.75) is satisfied with $\mathcal{C}_v = \lambda_q^4 \delta_q^{1/2}$ up to $N + M \leq 2N_{\text{fin}}$ (which will in fact be far beyond anything required for the inverse divergence). The inequalities in (A.76) follow from (9.43), (9.39), and the definitions of $\lambda' = \widetilde{\lambda}_q$ and $\Lambda = \lambda_{q+1}$.

In applications, $\widetilde{\nu} = \widetilde{\tau}_q^{-1}\Gamma_{q+1}^{-1}$, so that from (9.39) and (9.43) we have that

$$\max\{\widetilde{\tau}_q^{-1}, \widetilde{\nu}, \mathcal{C}_v\Lambda\} \leq \tau_q^{-1}\widetilde{\lambda}_q^3\widetilde{\lambda}_{q+1} \leq \tau_q^{-1}\lambda_{q+1}^4\,,$$

which holds as soon as ε_Γ is taken to be sufficiently small. Then, (A.77) will follow from (9.55). Finally, (A.79) will hold for all $N, M \leq N^\sharp/4$, which will be taken larger than $3N_{\mathrm{ind},v}$. In summary, if (A.66) is known to hold for $N + M \leq N^\sharp$, then (A.73) holds for $N \leq N^\sharp/2 - \mathsf{d}$ and $M \leq N^\sharp/2$, while (A.79) is valid for $N, M \leq N^\sharp/4$.

Remark A.20. In the identification of the error terms in Section 8.3, we apply Proposition A.18 to write

$$G\varrho \circ \Phi = \mathrm{div}\left(\mathcal{H}(G\varrho \circ \Phi)\right) + \nabla P + \mathrm{div}\left(\mathcal{R}^*\left(G\varrho \circ \Phi\right)\right) + \fint_{\mathbb{T}^3} G\varrho \circ \Phi dx.$$

The estimates on G, ϱ, and Φ, and then the right-hand side of the above equality, will be checked in later sections. We emphasize that $\mathcal{H}$ is a *local* operator and is thus well suited to working with estimates on the support of a cutoff function. Conversely, $\mathcal{R}^*$ is nonlocal but will always produce extremely small errors which can be absorbed into $\mathring{R}_{q+1}$ and for which the cutoff functions are not relevant.

Remark A.21. We consider examples of functions ϑ and ϱ with which Proposition A.18 is used.

1. *This is the case corresponding to the density of a pipe flow.* Recalling the construction of pipe flows from Proposition 4.4, we let $\varrho = \varrho_{\xi,\lambda,r}^k$ and $\vartheta = \vartheta_{\xi,\lambda,r}^k$. Set $\zeta = \Lambda = \lambda$ (where the λ refers to Proposition 4.4, not the λ from Proposition A.18) and $\mu = \lambda r$. To verify (1), we appeal to item (1) from Proposition 4.4 and our choice of Λ and μ. The periodicity requirement in (2) follows from item (2) from Proposition 4.4 and, referring back, from item (1) from Proposition 4.3. Next, (A.69) is satisfied with $\mathcal{C}_* = r$ using (4.11). Finally, (A.71) and (A.77) will follow from large choice of N_{dec} and d and the fact that our choice of λ can always be related to ζ and μ by a power strictly less than 1 (see (9.48) and (9.55)).

2. *This is the case corresponding to the Littlewood-Paley projection for the square of the density of a pipe flow.* Fix $1 \leq \mu \leq \zeta < \Lambda$, and a constant $\mathcal{C}_* > 0$. Let $\eta(x)$ be any $(\mathbb{T}/\mu)^3$-periodic function (which need not have zero mean), with $\|\eta\|_{L^p(\mathbb{T}^3)} \leq \mathcal{C}_*$. In applications, we shall refer to (4.15) from Proposition 4.4 and set $\eta = \left(\varrho_{\xi,\lambda,r}^k\right)^2$ and $\mu = \lambda r$. This means that we may write $\eta(x) = \eta_\mu(\mu x)$ where η_μ is $\mathbb{T}^3$-periodic, with $\|\eta_\mu\|_{L^1(\mathbb{T}^3)} \leq \mathcal{C}_*$. Following (4.15) from Proposition 4.4 with $\lambda_1 = \zeta$, $\lambda_2 = \Lambda$, we may define

$$\varrho(x) = \left(\mathbb{P}_{[\zeta,\Lambda]}\eta\right)(x) = \left(\mathbb{P}_{[\frac{\zeta}{\mu},\frac{\Lambda}{\mu}]}\eta_\mu\right)(\mu x),$$

a function which is $(\mathbb{T}/\mu)^3$-periodic and has zero mean (since $\zeta \geq \mu > 0$), and clearly

$$\left\| D^N \varrho \right\|_{L^1(\mathbb{T}^3)} \leq \mathcal{C}_* \Lambda^N.$$

We now define the associated function ϑ by first defining the zero mean $\mathbb{T}^3$-periodic function

$$\vartheta_\mu = \left(\frac{\zeta}{\mu}\right)^{2\mathsf{d}} \Delta^{-\mathsf{d}} \mathbb{P}_{\left[\frac{\zeta}{\mu}, \frac{\Lambda}{\mu}\right]} \eta_\mu \,,$$

where the negative powers of the Laplacian are defined simply as a Fourier multiplier (since the periodic function we apply it to has zero mean). Then we let

$$\vartheta(x) = \vartheta_\mu(\mu x) \,,$$

which has zero mean, is $(\mathbb{T}/\mu)^3$-periodic, and clearly satisfies $\Delta^{\mathsf{d}}\vartheta = \zeta^{2\mathsf{d}}\varrho$, as required. It only remains to estimate the $\dot{W}^{N,1}$ norms of ϑ, which up to paying a factor of μ^N is equivalent to estimating the $\dot{W}^{N,1}$ norms of ϑ_μ. When $0 \leq N < 2\mathsf{d}$, the operator

$$D^N \Delta^{-\mathsf{d}} \mathbb{P}_{\left[\frac{\zeta}{\mu}, \frac{\Lambda}{\mu}\right]}$$

is a bounded operator on L^1, whose operator norm is $\lesssim (\zeta/\mu)^{N-2\mathsf{d}}$. This may be verified via a standard Littlewood-Paley argument. The exceptional case $N = 2\mathsf{d}$ leads to a logarithmic loss since there are roughly $\log(\Lambda/\mu)$-many Littlewood-Paley shells to estimate; we absorb this loss into a factor of Λ^α, with $\alpha > 0$ arbitrarily small. Since $\|\eta_\mu\|_{L^1} \leq \mathcal{C}_*$, the second estimate in (3) above clearly follows, at least when $N \leq 2\mathsf{d}$. The case $N > 2\mathsf{d}$ follows similarly, except that now $D^N \Delta^{-\mathsf{d}}$ is a positive order operator, and thus the L^1 operator norm of $D^N \Delta^{-\mathsf{d}} \mathbb{P}_{\left[\frac{\zeta}{\mu}, \frac{\Lambda}{\mu}\right]}$ is bounded by $\approx (\Lambda/\mu)^{N-2\mathsf{d}}$. We remark that as in the previous case, (A.71) and (A.77) will follow from large choices of $\mathsf{N}_{\mathrm{dec}}$ and d and the fact that our choice of λ can always be related to ζ and μ by a power strictly less than 1.

Proof of Proposition A.18. Since $D_t\Phi \equiv 0$, we have $D^N D_t^m \nabla \Phi = D^N [D_t^M, \nabla]\Phi$. We may now appeal to Lemma A.14, more precisely to Remark A.16. Let $\Omega = \operatorname{supp} G$ and $f = \Phi$, so that (A.67) implies that (A.55) holds with $\mathcal{C}_f = 1$, $\lambda_f = \widetilde{\lambda}_f = \lambda'$, and $\mu_f = \widetilde{\mu}_f = 1$ (in fact, whenever $M \geq 1$ we may replace the right side of (A.55) by 0). Moreover, (A.68) implies that (A.50) holds with $\mathcal{C}_v = \nu/\lambda'$, $\lambda_v = \widetilde{\lambda}_v = \lambda'$, $N_t = M_t$, $\mu_v = \nu$, and $\widetilde{\mu}_v = \widetilde{\nu}$. We deduce from (A.56) that

$$\left\| D^{N''} D_t^M D^{N'} D\Phi \right\|_{L^\infty(\operatorname{supp} G)} \lesssim \lambda'^{N'+N''} \mathcal{M}(M, M_t, \nu, \widetilde{\nu}) \qquad \text{(A.81)}$$

whenever $N' + N'' \leq N_*$ and $M \leq M_*$. Similarly, we use Lemma A.14 with $f =$

G, so that due to (A.66) we know that (A.51) holds with $\mathcal{C}_f = \mathcal{C}_G$, $\lambda_f = \widetilde{\lambda}_f = \lambda$, $\mu_f = \nu$, $\widetilde{\mu}_f = \widetilde{\nu}$, and $N_t = M_t$. With $\Omega = \operatorname{supp} G$, since $\lambda' \leq \lambda$, as before we have that (A.68) implies that (A.50) holds with $\mathcal{C}_v = \nu/\lambda$, $\lambda_v = \widetilde{\lambda}_v = \lambda$, $N_t = M_t$, $\mu_v = \nu$, and $\widetilde{\mu}_v = \widetilde{\nu}$. Therefore, from (A.54) we obtain that

$$\left\| D^{N''} D_t^M D^{N'} G \right\|_{L^1} \lesssim \mathcal{C}_G \lambda^{N'+N''} \mathcal{M}\left(M, M_t, \nu, \widetilde{\nu}\right) \tag{A.82}$$

whenever $N' + N'' \leq N_*$ and $M \leq M_*$. With (A.81) and (A.82), we turn to the proof of (A.73).

Instead of defining $\mathring{R}$ and P separately, we shall simply construct a symmetric stress R with a prescribed divergence, and use the convention that $P = \operatorname{tr}(R)$ and $\mathring{R} = R - \operatorname{tr}(R)\mathrm{Id}$. The construction is based on iterating Proposition A.17, d times. For notational purposes, let $\varrho_{(0)} = \varrho$, and for $1 \leq k \leq d$ we let $\varrho_{(k)} = (\zeta^{-2}\Delta)^{d-k}\vartheta$. Then $\varrho_{(k-1)} = \zeta^{-2}\Delta\varrho_{(k)}$ and $\varrho_{(d)} = \vartheta$. We also define $G_{(0)} = G$.

Since $\rho_{(0)} = \zeta^{-2}\Delta\rho_{(1)}$, we deduce from Proposition A.17, identities (A.62)–(A.65), that

$$G_{(0)}^i \varrho_{(0)} \circ \Phi = \partial_n R_{(0)}^{in} + G_{(1)}^{i\ell}(\zeta^{-1}\partial_\ell \varrho_{(1)}) \circ \Phi\,, \tag{A.83}$$

where the symmetric stress $R_{(0)}$ is given by

$$R_{(0)}^{in} = \zeta^{-1} \underbrace{\left(G_{(0)}^i A_\ell^n + G_{(0)}^n A_\ell^i - A_k^i A_k^n G_{(0)}^p \partial_p \Phi^\ell \right)}_{=: S_{(0)}^{in\ell}} (\zeta^{-1}\partial_\ell \varrho_{(1)}) \circ \Phi \tag{A.84}$$

and the error terms are computed as

$$G_{(1)}^{i\ell} = \zeta^{-1} \left(\partial_n \left(G_{(0)}^p A_k^i A_k^n - G_{(0)}^n A_k^i A_k^p \right) \partial_p \Phi^\ell \right) - \partial_n G_{(0)}^i A_\ell^n\,, \tag{A.85}$$

where, as before, we denote $(\nabla\Phi)^{-1} = A$. We first show that the symmetric stress $R_{(0)}$ defined in (A.84) satisfies the estimate (A.73). First, we note that the ζ^{-1} factor has already been accounted for explicitly in (A.84). Second, we note that since $D_t\Phi = 0$, material derivatives may only land on the components of the 3-tensor $S_{(0)}$. Third, the function $\zeta^{-1}D\varrho_{(1)}$ has zero mean, is $(\mathbb{T}/\mu)^3$ periodic, and satisfies

$$\left\| D^N(\zeta^{-1}D\varrho_{(1)}) \right\|_{L^1} \lesssim C_* \mathcal{M}\left(N, 1, \zeta, \Lambda\right) \tag{A.86}$$

for $1 \neq N \leq N_{\mathrm{fin}}$, in view of (A.69). For $N = 1$, the above estimate incurs a logarithmic loss of Λ, which we can absorb with Λ^α for any $\alpha > 0$ to produce the estimate

$$\left\| D(\zeta^{-1}D\varrho_{(1)}) \right\|_{L^1} \lesssim \Lambda^\alpha C_* \mathcal{M}\left(N, 1, \zeta, \Lambda\right). \tag{A.87}$$

The implicit constants depend on α and degenerate as $\alpha \to 0$. Fourth, the components of the 3-tensor $S_{(0)}$ are sums of terms of two kinds: $G_{(0)} \otimes A$ is a linear function of $G_{(0)}$ multiplied by a homogeneous quadratic polynomial in $D\Phi$, while $G \otimes A \otimes A \otimes D\Phi$ is a linear function of G multiplied by a homogeneous polynomial of degree 5 in the entries of $D\Phi$. In particular, due to our assumption (A.66) and the previously established bound (A.81), upon applying the Leibniz rule and using that $\lambda' \leq \lambda$, we obtain that

$$\left\| D^N D_t^M S_{(0)} \right\|_{L^1} \lesssim \mathcal{C}_G \lambda^N \mathcal{M}(M, M_t, \nu, \tilde{\nu}) \tag{A.88}$$

for $N \leq N_*$ and $M \leq M_*$. Having collected these estimates, the L^1 norm of the space and material derivatives of $R_{(0)}$ is obtained from Lemma A.7. As dictated by (A.84) we apply this lemma with $f = \zeta^{-1} S_{(0)}$ and $\varphi = \zeta^{-1} \nabla \varrho_{(1)}$. Due to (A.88), the bound (A.30) holds with $\mathcal{C}_f = \mathcal{C}_G \zeta^{-1}$. Due to (A.67) and $\lambda' \leq \lambda$, the assumptions (A.31) and (A.32) are verified. Next, due to (A.86) and (A.87), the assumption (A.33) is verified, with $N_x = 1$, $\tilde{\zeta} = \Lambda$, and $\mathcal{C}_\varphi = \mathcal{C}_* \Lambda^\alpha$. Lastly, assumption (A.71) verifies the condition (A.34) of Lemma A.7. Thus, applying estimate (A.36) we deduce that

$$\left\| D^N D_t^M R_{(0)} \right\|_{L^1} \lesssim \mathcal{C}_G \mathcal{C}_* \Lambda^\alpha \zeta^{-1} \mathcal{M}(N, 1, \zeta, \Lambda) \mathcal{M}(M, M_t, \nu, \tilde{\nu}) \tag{A.89}$$

for all $N \leq N_*$ and $M \leq M_*$, which is precisely the bound stated in (A.73). Here we have used that $N_* \geq 2N_{\mathrm{dec}} + 4$, which was required due to (A.35).

Next we analyze the second term in (A.83). The point is that this term has the *same structure* as what we started with; for every fixed $\ell \in \{1, 2, 3\}$, we may replace $G_{(0)}^i$ by $G_{(1)}^{i\ell}$, and we replace $\varrho_{(0)}$ with $\zeta^{-1} \partial_\ell \varrho_{(1)}$; the only difference is that the bounds for this term are better. Indeed, from (A.85) we see that the 2-tensor $G_{(1)}$ is the sum of entries in $\zeta^{-1} DG_{(0)} \otimes A$, $\zeta^{-1} DG_{(0)} \otimes A \otimes A \otimes D\Phi$, and $\zeta^{-1} G_{(0)} \otimes DA \otimes A \otimes D\Phi$. Recalling that the entries of A are homogeneous quadratic polynomials in the entries of $D\Phi$, from (A.81), (A.82), $\lambda' \leq \lambda$, and the Leibniz rule we deduce that

$$\left\| D^{N''} D_t^M D^{N'} G_{(1)}^{i\ell} \right\|_{L^1} \lesssim \mathcal{C}_G (\lambda \zeta^{-1}) \lambda^{N'+N''} \mathcal{M}(M, M_t, \nu, \tilde{\nu}) \tag{A.90}$$

for $N' + N'' \leq N_* - 1$ and $M \leq M_*$. Compare the above estimate with (A.82), and notice that since $\lambda \zeta^{-1} \ll 1$, the bounds for $G_{(1)}$ are indeed better than those for $G_{(0)}$; the only caveat is the bounds hold for one less spatial derivative. In order to iterate Proposition A.17, for simplicity we ignore the ℓ index; since the argument works in exactly the same way for all values of ℓ, we write $G_{(1)}^{i\ell}$ simply as $G_{(1)}^i$, and $\partial_\ell \varrho_{(1)}$ as $D\varrho_{(1)}$. We start by noting that $\zeta^{-1} D\varrho_{(1)} = \zeta^{-2} \Delta(\zeta^{-1} D\varrho_{(2)})$. Thus, using identities (A.62)–(A.65) we obtain that the second term in (A.83) may be written as

$$G_{(1)}^i (\zeta^{-1} D\varrho_{(1)}) \circ \Phi = \partial_n R_{(1)}^{in} + G_{(2)}^{i\ell} (\zeta^{-2} \partial_\ell D\varrho_{(2)}) \circ \Phi, \tag{A.91}$$

where the symmetric stress $R_{(1)}$ is given by

$$R_{(1)}^{in} = \zeta^{-1} \underbrace{\left(G_{(1)}^i A_\ell^n + G_{(1)}^n A_\ell^i - A_k^i A_k^n G_{(1)}^p \partial_p \Phi^\ell \right)}_{=:S_{(1)}^{in\ell}} (\zeta^{-2} \partial_\ell D\varrho_{(2)}) \circ \Phi \qquad \text{(A.92)}$$

and the error terms are computed as

$$G_{(2)}^{i\ell} = \zeta^{-1} \left(\partial_n \left(G_{(1)}^p A_k^i A_k^n - G_{(1)}^n A_k^i A_k^p \right) \partial_p \Phi^\ell \right) - \partial_n G_{(1)}^i A_\ell^n . \qquad \text{(A.93)}$$

We emphasize that by combining (A.85) with (A.92) and (A.93), we may compute the 3-tensor $S_{(1)}$ and the 2-tensor $G_{(2)}$ *explicitly in terms of just space derivatives* of G and $D\Phi$. Using a similar argument to the one which was used to prove (A.88), but by appealing to (A.90) instead of (A.82), we deduce that for $N \le N_* - 1$ and $M \le M_*$,

$$\left\| D^N D_t^M S_{(1)} \right\|_{L^1} \lesssim \mathcal{C}_G(\lambda \zeta^{-1}) \lambda^N \mathcal{M}\left(M, M_t, \nu, \widetilde{\nu}\right) . \qquad \text{(A.94)}$$

Using the bound (A.94) and the estimate

$$\left\| D^N (\zeta^{-2} \partial_\ell D\varrho_{(2)}) \right\|_{L^1} \lesssim \mathcal{C}_* \mathcal{M}\left(N, 2, \zeta, \Lambda\right) ,$$

which is a consequence of (A.69)—in the case $N = 2$, as before, we may weaken the bound by a factor of Λ^α—we may deduce from Lemma A.7 that

$$\left\| D^N D_t^M R_{(1)} \right\|_{L^1} \lesssim \mathcal{C}_G \mathcal{C}_* \Lambda^\alpha (\lambda \zeta^{-2}) \mathcal{M}\left(N, 2, \zeta, \Lambda\right) \mathcal{M}\left(M, M_t, \nu, \widetilde{\nu}\right) \qquad \text{(A.95)}$$

for $N \le N_* - 1$ and $M \le M_*$, which is an estimate that is even better than (A.89), since $\lambda \ll \zeta \le \Lambda$. This shows that the first term in (A.91) satisfies the expected bound. The second term in (A.91) may in turn be shown to satisfy

$$\left\| D^{N''} D_t^M D^{N'} G_{(2)}^{i\ell} \right\|_{L^1} \lesssim \mathcal{C}_G(\lambda^2 \zeta^{-2}) \lambda^{N'+N''} \mathcal{M}\left(M, M_t, \nu, \widetilde{\nu}\right) . \qquad \text{(A.96)}$$

for $N' + N'' \le N_* - 2$ and $M \le M_*$, and it is clear that this procedure may be iterated d times.

Without spelling out the details, the iteration procedure described above produces

$$G_{(0)} \varrho_{(0)} \circ \Phi = \sum_{k=0}^{\mathsf{d}-1} \operatorname{div} R_{(k)} + \underbrace{G_{(\mathsf{d})} \otimes (\zeta^{-\mathsf{d}} D^{\mathsf{d}} \vartheta) \circ \Phi}_{=:E} , \qquad \text{(A.97)}$$

where each of the d symmetric stresses satisfies

$$\left\| D^N D_t^M R_{(k)} \right\|_{L^1} \lesssim \mathcal{C}_G \mathcal{C}_* \Lambda^\alpha (\lambda^k \zeta^{-k+1}) \mathcal{M}\left(N, 1, \zeta, \Lambda\right) \mathcal{M}\left(M, M_t, \nu, \widetilde{\nu}\right) , \qquad \text{(A.98)}$$

for $N \leq N_* - k$ and $M \leq M_*$. Each component of the the error tensor $G_{(\mathsf{d})}$ in (A.97) is recursively computable solely in terms of G and $D\Phi$ and their spatial derivatives, and satisfies

$$\left\| D^{N''} D_t^M D^{N'} G_{(\mathsf{d})} \right\|_{L^1} \lesssim \mathcal{C}_G (\lambda^{\mathsf{d}} \zeta^{-\mathsf{d}}) \lambda^{N'+N''} \mathcal{M}\left(M, M_t, \nu, \tilde\nu\right) \qquad \text{(A.99)}$$

for $N' + N'' \leq N_* - \mathsf{d}$ and $M \leq M_*$. Lastly, since $\left\| D^N (\zeta^{-\mathsf{d}} D^{\mathsf{d}} \vartheta) \right\|_{L^1} \lesssim \mathcal{C}_* \Lambda^\alpha \mathcal{M}\left(N, \mathsf{d}, \zeta, \Lambda\right)$ and $D^{\mathsf{d}} \vartheta$ is $(\mathbb{T}/\mu)^3$-periodic, a final application of Lemma A.7 combined with (A.99) and the assumption that $N_* - \mathsf{d} \geq 2N_{\mathrm{dec}} + 4$ show that estimate (A.74) holds.

Next, we turn to the proof of (A.78) and (A.79). Recall that E is defined by the second term in (A.97), and thus $\fint_{\mathbb{T}^3} G\varrho \circ \Phi dx = \fint_{\mathbb{T}^3} E dx$. Using the standard nonlocal inverse divergence operator

$$\mathcal{R}v = \Delta^{-1}\left(\nabla v + (\nabla v)^T\right) - \frac{1}{2}\left(\mathrm{Id} + \nabla\nabla\Delta^{-1}\right)\Delta^{-1}\mathrm{div}\, v\,, \qquad \text{(A.100)}$$

we may define

$$\mathring{R}_{\mathrm{nonlocal}} = \mathcal{R}E\,.$$

By the definition of $\mathcal{R}$ we have that $\mathring{R}_{\mathrm{nonlocal}}$ is traceless, symmetric, and satisfies $\mathrm{div}\, \mathring{R}_{\mathrm{nonlocal}} = E - \fint_{\mathbb{T}^3} E dx$, i.e., (A.78) holds. In the last equality we have used the fact that, by assumption, $G\varrho \circ \Phi$ has zero mean, and thus so does E. The idea here is very simple: because d is very large, the gain of $(\lambda\zeta^{-1})^{\mathsf{d}}$ present in the E estimate (A.74) is so strong that we may simply convert D and D_t bounds on E to (terrible) ∂_t bounds, which commute with $\mathcal{R}$, and get away with it.

Using the formulas (5.17a) and (5.17b) and the assumption (A.75), since D and ∂_t commute with $\mathcal{R}$, we deduce that for every $N \leq N_\circ$ and $M \leq M_\circ$ we have

$$\left\| D^N D_t^M \mathring{R}_{\mathrm{nonlocal}} \right\|_{L^1}$$

$$\lesssim \sum_{\substack{M' \leq M \\ N'+M' \leq N+M}} \sum_{K=0}^{M-M'} \mathcal{C}_v^K \widetilde\lambda_q^{N-N'+K} \widetilde\tau_q^{-(M-M'-K)} \left\| D^{N'} \partial_t^{M'} \mathcal{R}E \right\|_{L^1}$$

$$\lesssim \sum_{\substack{M' \leq M \\ N'+M' \leq N+M}} \widetilde\lambda_q^{N-N'} \widetilde\tau_q^{-(M-M')} \left\| D^{N'} \partial_t^{M'} E \right\|_{L^p} \qquad \text{(A.101)}$$

for any $p \in (1, {}^3\!/\!_2)$, where in the last inequality we have used the facts that, by assumption, $\mathcal{C}_v \widetilde\lambda_q \lesssim \widetilde\tau_q^{-1}$ and that $\mathcal{R} \colon L^p(\mathbb{T}^3) \to L^1(\mathbb{T}^3)$ is a bounded operator.

Our goal is to appeal to estimate (A.44) in Lemma A.10, with $A = -v \cdot \nabla$, $B = D_t$, and $f = E$ in order to estimate the L^p norm of $D^{N'} \partial_t^{M'} E = D^{N'}(A + B)^{M'} E$.

First, we claim that v satisfies the lossy estimate

$$\left\| D^N D_t^M v \right\|_{L^\infty} \lesssim C_v \widetilde{\lambda}_q^N \widetilde{\tau}_q^{-M} \tag{A.102}$$

for $M \leq M_\circ$ and $N + M \leq N_\circ + M_\circ$. This estimate does not follow from (A.68), which provides bounds for only Dv, instead of v. For this purpose, we apply Lemma A.10 with $f = v$, $B = \partial_t$, $A = v \cdot \nabla$, and $p = \infty$. Using (A.75), and the fact that $B = \partial_t$ and D commute, we obtain that bounds (A.40) and (A.41) hold with $C_f = C_v$, $\lambda_v = \widetilde{\lambda}_v = \lambda_f = \widetilde{\lambda}_f = \widetilde{\lambda}_q$, and $\mu_v = \widetilde{\mu}_v = \mu_f = \widetilde{\mu}_f = \widetilde{\tau}_q^{-1}$. Since $A + B = D_t$, we obtain from the bound (A.44) and our assumption $C_v \widetilde{\lambda}_q \lesssim \widetilde{\tau}_q^{-1}$ that (A.102) holds.

Second, we claim that for any $k \geq 1$ we have

$$\left\| \left(\prod_{i=1}^k D^{\alpha_i} D_t^{\beta_i} \right) v \right\|_{L^\infty(\operatorname{supp} G)} \lesssim C_v \widetilde{\lambda}_q^{|\alpha|} (\max\{\widetilde{\nu}, \widetilde{\tau}_q^{-1}\})^{|\beta|} \tag{A.103}$$

whenever $|\beta| \leq M_\circ$ and $|\alpha| + |\beta| \leq N_\circ + M_\circ$. To see this, we use Lemma A.14 with $f = v$, $p = \infty$, and $\Omega = \operatorname{supp} G$. From (A.68) we have that (A.50) holds with $C_v = \nu/\lambda'$, $\lambda_v = \widetilde{\lambda}_v = \lambda'$, $\mu_v = \nu$, and $\widetilde{\mu}_v = \widetilde{\nu}$. On the other hand, from (A.102) we have that (A.51) holds with $C_f = C_v$, $\lambda_f = \widetilde{\lambda}_f = \widetilde{\lambda}_q$, and $\mu_f = \widetilde{\mu}_f = \widetilde{\tau}_q^{-1}$. Since $\widetilde{\lambda}_q \geq \lambda'$, we deduce from (A.54) that (A.103) holds.

Third, we claim that

$$\left\| \left(\prod_{i=1}^k D^{\alpha_i} D_t^{\beta_i} \right) E \right\|_{L^p(\operatorname{supp} G)} \lesssim C_G C_*(\lambda \zeta^{-1})^{\mathsf{d}} \Lambda^{|\alpha|+1} \mathcal{M}(|\beta|, M_t, \nu, \widetilde{\nu}) \tag{A.104}$$

holds whenever $|\alpha| \leq N_* - d$ and $|\beta| \leq M_*$. This estimate again follows from Lemma A.14, this time with $f = E$, by appealing to the previously established bound (A.74) and the Sobolev embedding $W^{1,1}(\mathbb{T}^3) \subset L^p(\mathbb{T}^3)$ for $p \in (1, 3/2)$.

At last, we are in the position to apply Lemma A.10. The bound (A.103) implies that assumption (A.40) holds with $B = D_t$, $\lambda_v = \widetilde{\lambda}_v = \widetilde{\lambda}_q$, and $\mu_v = \widetilde{\mu}_v = \max\{\widetilde{\tau}_q^{-1}, \widetilde{\nu}\}$. The bound (A.104) implies that assumption (A.41) of Lemma A.10 holds with $C_f = C_G C_*(\lambda \zeta^{-1})^{\mathsf{d}} \Lambda$, $\lambda_f = \widetilde{\lambda}_f = \Lambda$, $\mu_f = \nu$, and $\widetilde{\mu}_f = \widetilde{\nu}$. We may now use estimate (A.44), and the assumption that $\Lambda \geq \widetilde{\lambda}_q$, to deduce that

$$\left\| D^{N'} \partial_t^{M'} E \right\|_{L^p} \lesssim C_G C_*(\lambda \zeta^{-1})^{\mathsf{d}} \Lambda^{N'+1} (\max\{C_v \Lambda, \widetilde{\nu}, \widetilde{\tau}_q^{-1}\})^{M'} \tag{A.105}$$

holds whenever $M' \leq M_\circ$ and $N' + M' \leq N_\circ + M_\circ$. Combining (A.101) and (A.105) we deduce that

$$\left\| D^N D_t^M \mathring{R}_{\text{nonlocal}} \right\|_{L^1}$$

$$\lesssim \mathcal{C}_G \mathcal{C}_*(\lambda\zeta^{-1})^{\mathsf{d}} \sum_{\substack{M' \le M \\ N'+M' \le N+M}} \widetilde{\lambda}_q^{N-N'} \widetilde{\tau}_q^{-(M-M')} \Lambda^{N'+1}(\max\{\mathcal{C}_v\Lambda, \widetilde{\nu}, \widetilde{\tau}_q^{-1}\})^{M'}$$

$$\lesssim \mathcal{C}_G \mathcal{C}_*(\lambda\zeta^{-1})^{\mathsf{d}} \Lambda^{N+1}(\max\{\mathcal{C}_v\Lambda, \widetilde{\nu}, \widetilde{\tau}_q^{-1}\})^{M} \tag{A.106}$$

whenever $N \le N_\circ$ and $M \le M_\circ$. Estimate (A.79) follows by appealing to the assumption (A.77), which ensures that the gain from $(\lambda\zeta^{-1})^{\mathsf{d}-1}$ is already a sufficiently strong amplitude gain, and we use the leftover factor of $\lambda\zeta^{-1}$ to absorb implicit constants. $\qquad\square$

Bibliography

[1] F. Anselmet, Y. Gagne, E. J. Hopfinger, and R. A. Antonia. High-order velocity structure functions in turbulent shear flows. *J. Fluid Mech.*, 140:63–89, 1984.

[2] R. Beekie, T. Buckmaster, and V. Vicol. Weak solutions of ideal MHD which do not conserve magnetic helicity. *Annals of PDE*, 6(1): Paper No. 1, 40 pp., 2020.

[3] R. Beekie and M. Novack. Non-conservative solutions of the Euler-α equations. *arXiv:2111.01027*, 2021.

[4] E. Bruè and M. Colombo. Nonuniqueness of solutions to the Euler equations with vorticity in a Lorentz space. *arXiv:2108.09469*, 2021.

[5] E. Bruè, M. Colombo, and C. De Lellis. Positive solutions of transport equations and classical nonuniqueness of characteristic curves. *arXiv:2003.00539*, 2020.

[6] T. Buckmaster. Onsager's conjecture almost everywhere in time. *Comm. Math. Phys.*, 333(3):1175–1198, 2015.

[7] T. Buckmaster, M. Colombo, and V. Vicol. Wild solutions of the Navier-Stokes equations whose singular sets in time have Hausdorff dimension strictly less than 1. *arXiv:1809.00600. Journal of the EMS*, to appear, 2022.

[8] T. Buckmaster, C. De Lellis, P. Isett, and L. Székelyhidi, Jr. Anomalous dissipation for 1/5-Hölder Euler flows. *Ann. of Math.*, 182(1):127–172, 2015.

[9] T. Buckmaster, C. De Lellis, and L. Székelyhidi, Jr. Dissipative Euler flows with Onsager-critical spatial regularity. *Comm. Pure Appl. Math.*, 69(9):1613–1670, 2016.

[10] T. Buckmaster, C. De Lellis, and L. Székelyhidi Jr. Transporting microstructure and dissipative Euler flows. *arXiv:1302.2815*, 2013.

[11] T. Buckmaster, C. De Lellis, L. Székelyhidi Jr., and V. Vicol. Onsager's conjecture for admissible weak solutions. *Comm. Pure Appl. Math.*,

72(2):229–274, July 2018.

[12] T. Buckmaster and V. Vicol. Convex integration and phenomenologies in turbulence. *EMS Surv. Math. Sci.*, 6(1):173–263, 2019.

[13] T. Buckmaster and V. Vicol. Nonuniqueness of weak solutions to the Navier-Stokes equation. *Ann. of Math.*, 189(1):101–144, 2019.

[14] T. Buckmaster and V. Vicol. Convex integration constructions in hydrodynamics. *Bull. Amer. Math. Soc.*, 58(1):1–44, 2020.

[15] S. Chen, B. Dhruva, S. Kurien, K. Sreenivasan, and M. Taylor. Anomalous scaling of low-order structure functions of turbulent velocity. *J. Fluid Mech.*, 533:183–192, 2005.

[16] A. Cheskidov, P. Constantin, S. Friedlander, and R. Shvydkoy. Energy conservation and Onsager's conjecture for the Euler equations. *Nonlinearity*, 21(6):1233–1252, 2008.

[17] A. Cheskidov and X. Luo. Nonuniqueness of weak solutions for the transport equation at critical space regularity. *arXiv:2004.09538*, 2020.

[18] A. Cheskidov and X. Luo. Sharp nonuniqueness for the Navier-Stokes equations. *arXiv:2009.06596*, 2020.

[19] A. Cheskidov and X. Luo. L^2-critical nonuniqueness for the 2D Navier-Stokes equations. *arXiv:2105.12117*, 2021.

[20] A. Cheskidov and R. Shvydkoy. Euler equations and turbulence: Analytical approach to intermittency. *SIAM J. Math. Anal.*, 46(1):353–374, 2014.

[21] P. Constantin. On the Euler equations of incompressible fluids. *Bull. Amer. Math. Soc. (N.S.)*, 44(4):603–621, 2007.

[22] P. Constantin, W. E, and E. Titi. Onsager's conjecture on the energy conservation for solutions of Euler's equation. *Comm. Math. Phys.*, 165(1):207–209, 1994.

[23] P. Constantin and C. Fefferman. Scaling exponents in fluid turbulence: Some analytic results. *Nonlinearity*, 7(1):41, 1994.

[24] P. Constantin, Q. Nie, and S. Tanveer. Bounds for second order structure functions and energy spectrum in turbulence. *Phys. Fluids*, 11(8):2251–2256, 1999. The International Conference on Turbulence (Los Alamos, NM, 1998).

[25] G. Constantine and T. Savits. A multivariate Faà di Bruno formula with applications. *Trans. Amer. Math. Soc.*, 348(2):503–520, 1996.

[26] M. Dai. Non-uniqueness of Leray-Hopf weak solutions of the 3D Hall-MHD

system. *arXiv:1812.11311*, 2018.

[27] S. Daneri and L. Székelyhidi, Jr. Non-uniqueness and h-principle for Hölder-continuous weak solutions of the Euler equations. *Arch. Rational Mech. Anal.*, 224(2):471–514, 2017.

[28] C. De Lellis and H. Kwon. On non-uniqueness of Hölder continuous globally dissipative Euler flows. *arXiv:2006.06482*, 2020.

[29] C. De Lellis and L. Székelyhidi, Jr. The Euler equations as a differential inclusion. *Ann. of Math. (2)*, 170(3):1417–1436, 2009.

[30] C. De Lellis and L. Székelyhidi, Jr. The h-principle and the equations of fluid dynamics. *Bull. Amer. Math. Soc. (N.S.)*, 49(3):347–375, 2012.

[31] C. De Lellis and L. Székelyhidi, Jr. Dissipative continuous Euler flows. *Invent. Math.*, 193(2):377–407, 2013.

[32] C. De Lellis and L. Székelyhidi Jr. High dimensionality and h-principle in PDE. *Bull. Amer. Math. Soc.*, 54(2):247–282, 2017.

[33] C. De Lellis and L. Székelyhidi Jr. On turbulence and geometry: From Nash to Onsager. *arXiv:1901.02318*, 01 2019.

[34] T. Drivas and G. Eyink. An Onsager singularity theorem for Leray solutions of incompressible Navier-Stokes. *Nonlinearity*, 32(11):4465, 2019.

[35] J. Duchon and R. Robert. Inertial energy dissipation for weak solutions of incompressible Euler and Navier-Stokes equations. *Nonlinearity*, 13(1):249, 2000.

[36] G. Eyink. Energy dissipation without viscosity in ideal hydrodynamics I. Fourier analysis and local energy transfer. *Physica D: Nonlinear Phenomena*, 78(3–4):222–240, 94.

[37] G. Eyink and K. Sreenivasan. Onsager and the theory of hydrodynamic turbulence. *Rev. Modern Phys.*, 78(1):87–135, 2006.

[38] C. Foias, U. Frisch, and R. Temam. Existence de solutions C^∞ des équations d'euler. *C. R. Acad. Sci. Paris, Ser. A*, 280:505–508, 1975.

[39] U. Frisch. *Turbulence*. Cambridge University Press, Cambridge, 1995. The legacy of A. N. Kolmogorov.

[40] M. Gromov. *Partial Differential Relations*, volume 9. Springer Science & Business Media, 1986.

[41] P. Isett. Hölder continuous Euler flows in three dimensions with compact support in time. *arXiv:1211.4065*, 11 2012.

[42] P. Isett. On the endpoint regularity in Onsager's conjecture. *arXiv preprint arXiv:1706.01549*, 2017.

[43] P. Isett. A proof of Onsager's conjecture. *Annals of Mathematics*, 188(3):871, 2018.

[44] T. Ishihara, T. Gotoh, and Y. Kaneda. Study of high–Reynolds number isotropic turbulence by direct numerical simulation. *Annual Review of Fluid Mechanics*, 41:165–180, 2009.

[45] Y. Kaneda, T. Ishihara, M. Yokokawa, K. Itakura, and A. Uno. Energy dissipation rate and energy spectrum in high resolution direct numerical simulations of turbulence in a periodic box. *Physics of Fluids*, 15(2):L21–L24, 2003.

[46] A. Kolmogorov. Local structure of turbulence in an incompressible fluid at very high reynolds number. *Dokl. Acad. Nauk SSSR*, 30(4):299–303, 1941. Translated from the Russian by V. Levin, Turbulence and stochastic processes: Kolmogorov's ideas 50 years on.

[47] G. Komatsu. Analyticity up to the boundary of solutions of nonlinear parabolic equations. *Comm. Pure Appl. Math.*, 32(5):669–720, 1979.

[48] L. Lichtenstein. Über einige Existenzprobleme der Hydrodynamik homogener, unzusammendrückbarer, reibungsloser Flüssigkeiten und die Helmholtzschen Wirbelsätze. *Mathematische Zeitschrift*, 23(1):89–154, 1925.

[49] X. Luo. Stationary solutions and nonuniqueness of weak solutions for the Navier-Stokes equations in high dimensions. *Arch. Ration. Mech. Anal.*, 233(2):701–747, 2019.

[50] C. Meneveau and K. Sreenivasan. The multifractal nature of turbulent energy dissipation. *J. Fluid Mech.*, 224:429–484, 1991.

[51] S. Modena and L. Székelyhidi, Jr. Non-renormalized solutions to the continuity equation. *arXiv:1806.09145*, 2018.

[52] S. Modena and L. Székelyhidi, Jr. Non-uniqueness for the transport equation with Sobolev vector fields. *Ann. PDE*, 4(2):Paper No. 18, 38, 2018.

[53] S. Müller and V. Šverák. Convex integration for Lipschitz mappings and counterexamples to regularity. *Annals of Mathematics*, 157(3):pp. 715–742, 2003.

[54] J. Nash. C^1 isometric imbeddings. *Ann. of Math.*, 60(3):383–396, 1954.

[55] F. Nguyen, J.-P. Laval, P. Kestener, A. Cheskidov, R. Shvydkoy, and B. Dubrulle. Local estimates of Hölder exponents in turbulent vector fields.

Physical Review E, 99(5):053114, 2019.

[56] M. Novack. Nonuniqueness of weak solutions to the 3-dimensional quasi-geostrophic equations *SIAM Journal on Mathematical Analysis*, 52(4), 2020.

[57] M. Novack and V. Vicol. An intermittent Onsager theorem. *arXiv:2203.13115*, 2022.

[58] L. Onsager. Statistical hydrodynamics. *Nuovo Cimento (9)*, 6 (Supplemento, 2 (Convegno Internazionale di Meccanica Statistica)):279–287, 1949.

[59] V. Scheffer. An inviscid flow with compact support in space-time. *J. Geom. Anal.*, 3(4):343–401, 1993.

[60] A. Shnirelman. Weak solutions with decreasing energy of incompressible Euler equations. *Comm. Math. Phys.*, 210(3):541–603, 2000.

[61] R. Shvydkoy. Lectures on the Onsager conjecture. *Discrete Contin. Dyn. Syst. Ser. S*, 3(3):473–496, 2010.

[62] K. Sreenivasan, S. Vainshtein, R. Bhiladvala, I. San Gil, S. Chen, and N. Cao. Asymmetry of velocity increments in fully developed turbulence and the scaling of low-order moments. *Phys. Rev. Lett.*, 77(8):1488–1491, 1996.

[63] P.-L. Sulem and U. Frisch. Bounds on energy flux for finite energy turbulence. *J. Fluid Mech.*, 72(3):417–423, 1975.

[64] L. Székelyhidi Jr. From isometric embeddings to turbulence. *HCDTE Lecture Notes. Part II. Nonlinear Hyperbolic PDEs, Dispersive and Transport Equations*, 7:63, 2012.

[65] T. Tao. 255b, notes 2: Onsager's conjecture, 2019.

[66] E. Wiedemann. Weak-strong uniqueness in fluid dynamics. *arXiv:1705.04220*, 2017.

Index